Advanced Testing of Soft Polymer Materials

Advanced Testing of Soft Polymer Materials

Editors

Radek Stoček
Gert Heinrich
Reinhold Kipscholl

MDPI • Basel • Beijing • Wuhan • Barcelona • Belgrade • Manchester • Tokyo • Cluj • Tianjin

Editors
Radek Stoček
Centre of Polymer Systems
Tomas Bata University in Zlín
Zlín
Czech Republic

Gert Heinrich
Institut für Textilmaschinen
und Textile
Hochleistungswerkstofftechnik
Technische Universität Dresden
Dresden
Germany

Reinhold Kipscholl
Coesfeld GmbH & Co.KG
Dortmund
Germany

Editorial Office
MDPI
St. Alban-Anlage 66
4052 Basel, Switzerland

This is a reprint of articles from the Special Issue published online in the open access journal *Polymers* (ISSN 2073-4360) (available at: www.mdpi.com/journal/polymers/special_issues/Test_Soft_Polym_Mater).

For citation purposes, cite each article independently as indicated on the article page online and as indicated below:

LastName, A.A.; LastName, B.B.; LastName, C.C. Article Title. *Journal Name* **Year**, *Volume Number*, Page Range.

ISBN 978-3-0365-6224-7 (Hbk)
ISBN 978-3-0365-6223-0 (PDF)

© 2022 by the authors. Articles in this book are Open Access and distributed under the Creative Commons Attribution (CC BY) license, which allows users to download, copy and build upon published articles, as long as the author and publisher are properly credited, which ensures maximum dissemination and a wider impact of our publications.

The book as a whole is distributed by MDPI under the terms and conditions of the Creative Commons license CC BY-NC-ND.

Contents

About the Editors vii

Preface to "Advanced Testing of Soft Polymer Materials" ix

Niclas Lindemann, Jürgen E. K. Schawe and Jorge Lacayo-Pineda
Kinetics of the Glass Transition of Silica-Filled Styrene–Butadiene Rubber: The Effect of Resins
Reprinted from: *Polymers* 2022, 14, 2626, doi:10.3390/polym14132626 1

Sahbi Aloui, Horst Deckmann, Jürgen Trimbach and Jorge Lacayo-Pineda
Influence of the Polarity of the Plasticizer on the Mechanical Stability of the Filler Network by Simultaneous Mechanical and Dielectric Analysis
Reprinted from: *Polymers* 2022, 14, 2126, doi:10.3390/polym14102126 21

E. Harea, S. Datta, M. Stěnička, J. Maloch and R. Stoček
The Influence of Local Strain Distribution on the Effective Electrical Resistance of Carbon Black Filled Natural Rubber
Reprinted from: *Polymers* 2021, 13, 2411, doi:10.3390/polym13152411 35

Jonathan A. Sotomayor-del-Moral, Juan B. Pascual-Francisco, Orlando Susarrey-Huerta, Cesar D. Resendiz-Calderon, Ezequiel A. Gallardo-Hernández and Leonardo I. Farfan-Cabrera
Characterization of Viscoelastic Poisson's Ratio of Engineering Elastomers via DIC-Based Creep Testing
Reprinted from: *Polymers* 2022, 14, 1837, doi:10.3390/polym14091837 53

Jean-Benoît Le Cam
Fast Evaluation and Comparison of the Energy Performances of Elastomers from Relative Energy Stored Identification under Mechanical Loadings
Reprinted from: *Polymers* 2022, 14, 412, doi:10.3390/polym14030412 67

William Amoako Kyei-Manu, Charles R. Herd, Mahatab Chowdhury, James J. C. Busfield and Lewis B. Tunnicliffe
The Influence of Colloidal Properties of Carbon Black on Static and Dynamic Mechanical Properties of Natural Rubber
Reprinted from: *Polymers* 2022, 14, 1194, doi:10.3390/polym14061194 81

Dana Bakošová and Alžbeta Bakošová
Testing of Rubber Composites Reinforced with Carbon Nanotubes
Reprinted from: *Polymers* 2022, 14, 3039, doi:10.3390/polym14153039 101

Marek Pöschl, Shibulal Gopi Sathi, Radek Stoček and Ondřej Kratina
Rheometer Evidences for the Co-Curing Effect of a Bismaleimide in Conjunction with the Accelerated Sulfur on Natural Rubber/Chloroprene Rubber Blends
Reprinted from: *Polymers* 2021, 13, 1510, doi:10.3390/polym13091510 119

Marek Pöschl, Shibulal Gopi Sathi and Radek Stoček
Identifying the Co-Curing Effect of an Accelerated-Sulfur/ Bismaleimide Combination on Natural Rubber/Halogenated Rubber Blends Using a Rubber Process Analyzer
Reprinted from: *Polymers* 2021, 13, 4329, doi:10.3390/polym13244329 131

Sanjoy Datta, Radek Stocek and Kinsuk Naskar
Influence of Ultraviolet Radiation on Mechanical Properties of a Photoinitiator Compounded High Vinyl Styrene–Butadiene–Styrene Block Copolymer
Reprinted from: *Polymers* 2021, 13, 1287, doi:10.3390/polym13081287 145

Klara Loos, Vivianne Marie Bruère, Benedikt Demmel, Yvonne Ilmberger, Alexander Lion and Michael Johlitz
Future-Oriented Experimental Characterization of 3D Printed and Conventional Elastomers Based on Their Swelling Behavior
Reprinted from: *Polymers* **2021**, *13*, 4402, doi:10.3390/polym13244402 . . . . **163**

Tomáš Gejguš, Jonas Schröder, Klara Loos, Alexander Lion and Michael Johlitz
Advanced Characterisation of Soft Polymers under Cyclic Loading in Context of Engine Mounts
Reprinted from: *Polymers* **2022**, *14*, 429, doi:10.3390/polym14030429 . . . . **183**

About the Editors

Radek Stoček

Radek Stoček obtained his diploma degree as engineer in 2005 from the Czech Technical University in Prague and received his Ph.D. in engineering science in 2012 from the Technical University Chemnitz (Germany), working with M. Gehde and parallel with G. Heinrich at IPF Dresden (G). Then, he started an industrial career at Polymer Research Lab (PRL), Zlin, Czech Republic, and parallel an independent academic career at the Tomas Bata University (TBU) in Zlin. He finished his Habilitation in 2019. Currently, he is holding the two positions as General manager at PRL and Head of the Rubber Department at TBU. His research and scientific interests are focused on characterization of rubber material properties with respect to fatigue and fracture mechanics and on the development of new and advanced testing methodologies, hardware and equipment. One main goal is to optimize industrial rubber products in terms of performance and durability as well as to hasten development cycles and minimizing extensive real rubber product tests before production. His work has been recognized by awards from The Tire Society (USA). R. Stocek is author of 59 publications (according to Scopus) and holds four Utility Models.

Gert Heinrich

Gert Heinrich graduated at the University in Jena (G) in quantum physics in 1973. At the University of Technology (TH) Leuna-Merseburg, he finished his doctorate in 1978 in polymer network physics and his Habilitation in 1986 about theory of polymer networks and topological constraints. In 1990, he received a position at the tire manufacturer Continental in Hanover (G) as senior research scientist and head of Materials Research. Heinrich continued his academic activities as lecturer at the Universities of Hanover (G) and Halle/Wittenberg (G). In 2002, he was appointed as a full professor for "Polymer Materials and Rubber Technology" at the University of Technology Dresden and as director of the Institute of Polymer Materials at the Leibniz Institute of Polymer Research Dresden e. V. (IPF). Since 2017, he is a Senior Professor. His work has been recognized by several grants and awards, e.g., the George Stafford Whitby Award for distinguished teaching and research from the Rubber Division of the ACS, the Colwyn Medal in UK for outstanding services to the rubber industry; the Carl Dietrich Harries Medal from the German Rubber Society, and the Lifetime Achievement Award from Tire Technology International Magazine.

Reinhold Kipscholl

Reinhold Kipscholl graduated as Dipl.-Ing. in engineering of data processing and electronics. He has been active for more than 20 years in leading industrial positions, especially in the field of testing and characterization of materials with respect of their physical behavior. For 20 years he has been General Manager of Coesfeld GmbH & Co. KG (Dortmund), a German Company developing and producing material testing equipment for plastics and elastomers. Since 2012, R. Kipscholl is founder and General Manager of PRL Polymer Research Lab., a Czech Company researching and developing new testing methods for characterization of fracture and wear behavior of rubbers. He has been awarded with the 2018 Fernley H. Bunbury Award (Rubber Division, American Chemical Society).

Preface to "Advanced Testing of Soft Polymer Materials"

Manufacturers of soft polymer products, as well as suppliers and processors of polymers, raw materials, and compounds or blends are compelled to use predictive and advanced laboratory testing in their search for high-performance soft polymer materials for future applications. Ideally, predictive laboratory testing balances accuracy, relevance, instrument productivity and cost-effectiveness, while providing new mechanistic insights and opportunities for modelling the overall properties of materials and products. In this context, new concepts for soft polymer materials are of great importance, taking into account new trends in many modern technological fields. New advanced test methods and techniques will link to fundamental scientific principles, even showing how test results from individual pieces of uncured/cured elastomers or other soft polymers relate to real geometry and loading conditions of polymer parts, creating new opportunities to link laboratory test data from soft polymer materials to the real product performance. Furthermore, the rapid development of simulation tools offers great prospects for predicting the behaviour of soft polymer materials and their durability based on unique data sets obtained through new advanced testing methods, including the upcoming possibilities of artificial intelligence.

The collection of publications contained in this edition therefore presents different methods used to solve problems in the characterization of various phenomena in soft polymer materials.

This reprint presents recent research results (Lindemann et al.) on the evaluation of the effect of resin content on the glass transition of rubber compounds. Broadband dielectric spectroscopy (BDS) and fast differential scanning calorimetry (FDSC) are applied for the characterization of the dielectric and thermal relaxations as well as for the corresponding vitrification kinetics. The dielectric behaviour of rubber is currently receiving considerable scientific attention and effort. Therefore, the following two publications are devoted to this phenomenon. Aloui et al. experimentally investigated the effect of plasticizer polarity on the mechanical stability of the filler network using simultaneous mechanical and dielectric analysis, while Harea et al. studied the effect of local strain distribution on the effective electrical resistivity of carbon black-filled natural rubber, combining for the first time the digital image correlation (DIC) method with measurements of dielectric behavior under mechanical stress. The DIC method is a very effective tool for describing the strain fields or deformation of stressed bodies and Sotomayor-del-Moral et al. used this method very effectively to analyze the viscoelastic Poisson's ratio of different types of elastomers and also the thermal effect under creep loading. J-B. Le Cam presents a simple and fast approach to characterize the mechanical and energetic behavior of elastomers, i.e., how they consume the applied mechanical energy. The methodology consists of performing a single uniaxial cyclic tensile test with simultaneous temperature measurements, whereby the temperature measurements at the sample surface are processed using the thermal diffusion equation to reconstruct the heat source fields, which in effect amounts to surface calorimetry. The mechanical properties of elastomers are also the subject of research by Kyei-Manu et al. to study the effect of colloidal properties of carbon black on statically and dynamically loaded natural rubber. While standard test methods were used to characterize the properties under static loading, a torsion rheometer was used to describe the properties under dynamic cyclic loading. The effect of fillers on the various rubber properties is evident from the studies presented, where the basic filler is mainly carbon black. Their characteristics and properties have been studied in detail in many scientific papers. Carbon nanotubes are one of the very promising materials that have a significant positive effect on rubber properties. Therefore, Bakošová and

Bakošová studied the effect of reinforcement of rubber composites by carbon nanotubes, investigating this phenomenon by atomic force microscopy (AFM), tensile tests, hardness tests and dynamic mechanical analysis (DMA). The mechanical properties of the rubber are achieved through the curing process, and therefore the process must be optimized to obtain a suitable rubber network. This topic has long been extensively addressed in the studies of Poschl et al., who presented two articles in the issue. The first publication deals with the rheometric evidence of the co-curing effect of bismaleimide in conjunction with accelerated sulphur on natural rubber and chloroprene rubber blends, both from a methodological and material point of view. The second publication deals with the identical topic, but in combination with different halogenated rubbers. Within both topics, rotational rheometers were used as a device to characterize the cure kinetics. Another type of curing is the effect of ultraviolet radiation. This issue is addressed in the study by Datta et al. and here the effect of ultraviolet radiation on rubber is characterized by determining basic mechanical properties, which are complemented by infrared spectroscopy, contact angle analysis and scanning electron microscopy analyses. Loos et al. investigated various elastomers in terms of their behaviour towards liquids such as moisture, fuels or fuel components. For this purpose, an analytical procedure using sorption experiments in combination with gas chromatography and mass spectrometry was presented, which is thus able to accurately analyse the swelling behaviour of the elastomers. Gejguš et al. introduced high-frequency dynamic stiffness measurements up to 3000 Hz on a newly developed test bench to characterize the rubber material in the context of the engine mounts.

Thus, the present volume provides a comprehensive overview of the recent developments in the field and will be of interest to both academic researchers and industrial professionals.

This work was supported by the Ministry of Education, Youth and Sports of the Czech Republic—DKRVO (RP/CPS/2022/006).

Radek Stoček, Gert Heinrich, and Reinhold Kipscholl
Editors

Article

Kinetics of the Glass Transition of Silica-Filled Styrene–Butadiene Rubber: The Effect of Resins

Niclas Lindemann [1,2,*], Jürgen E. K. Schawe [3] and Jorge Lacayo-Pineda [2,4]

1 Institut für Physikalische Chemie und Elektrochemie, Leibniz Universität Hannover, Callinstraße 3A, 30167 Hanover, Germany
2 Continental Reifen Deutschland GmbH, Jädekamp 30, 30419 Hanover, Germany; jorge.lacayo-pineda@conti.de
3 Mettler-Toledo GmbH, Heuwinkelstrasse 3, 8606 Nänikon, Switzerland; juergen.schawe@mt.com
4 Institut für Anorganische Chemie, Leibniz Universität Hannover, Callinstraße 9, 30167 Hanover, Germany
* Correspondence: niclas.lindemann@pci.uni-hannover.de

Abstract: Resins are important for enhancing both the processability and performance of rubber. Their efficient utilization requires knowledge about their influence on the dynamic glass transition and their miscibility behavior in the specific rubber compound. The resins investigated, poly-(α-methylstyrene) (AMS) and indene-coumarone (IC), differ in molecular rigidity but have a similar aromaticity degree and glass transition temperature. Transmission electron microscopy (TEM) investigations show an accumulation of IC around the silanized silica in styrene–butadiene rubber (SBR) at high contents, while AMS does not show this effect. This higher affinity between IC and the silica surface leads to an increased compactness of the filler network, as determined by dynamic mechanical analysis (DMA). The influence of the resin content on the glass transition of the rubber compounds is evaluated in the sense of the Gordon–Taylor equation and suggests a rigid amorphous fraction for the accumulated IC. Broadband dielectric spectroscopy (BDS) and fast differential scanning calorimetry (FDSC) are applied for the characterization of the dielectric and thermal relaxations as well as for the corresponding vitrification kinetics. The cooling rate dependence of the vitrification process is combined with the thermal and dielectric relaxation time by one single Vogel–Fulcher–Tammann–Hesse equation, showing an increased fragility of the rubber containing AMS.

Keywords: glass transition; kinetics; rubber; resin; BDS; FDSC

1. Introduction

The properties of elastomer-based materials can be modified by blending different polymers [1–3] and mixing them with various additives, such as fillers [4,5], plasticizers [6–8] and different vulcanization systems [9–12] for a wide variety of technical applications. A frequently used form of modification is the coupling of the rubber matrix with reinforcing fillers in order to tailor the mechanical properties to the application [4]. Apart from carbon black as a conventional filler, precipitated silica with a silane coupling agent is state-of-the-art in tire compounds [4,13]. The advantage of silica arises with an adaption of the polymer to solution styrene–butadiene rubber (SBR) [14,15]. The silica-filled rubber provides a lower rolling resistance and higher wet traction without decreasing the abrasion resistance [13].

High amounts of fillers can disturb the processability of rubber compounds due to their higher viscosity. Oils and resins are used to counteract this rheological behavior. Additionally, the tackiness of the rubber compounds can be increased by some types of resins [16–18]. Hydrocarbon resins, with a high glass transition temperature, T_g, and a melting point, T_m, at the processing temperature are beneficial in preserving the rubber compound hardness at the service temperature [16]. This is where the possibility to decrease the rolling resistance of a tire or to lower fuel consumption occurs and therefore, contributes to a reduction in CO_2 emissions. On the other hand, the hardness of the rubber compound does not necessarily decrease the braking performance to the same degree. The rolling resistance mainly correlates with a dynamic excitation at low frequencies of around 100 Hz, while higher frequencies of around 10^5 Hz are characteristic of traction [19]. Hence, the material properties at different frequencies are important parameters, which are strongly linked to molecular dynamics and the local structure in the elastomer system. The glass transition is a phenomenon sensitive to molecular dynamics. Its modification, due to local structural changes, is, therefore, the focus of many investigations [19,20]. Two manifestations are characteristic of the glass' transition: (i) the relaxation process, measured by frequency-dependent dynamical experiments in the rubbery state, which is also called "dynamic glass transition"; and (ii) the vitrification process, occurring during cooling as the transformation from a soft rubbery state into a solid glassy state [19].

The addition of plasticizers in the rubber matrix increases the flexibility of the polymer chains and usually decreases the T_g of the rubber compound [8,21]. The influence on the dynamic properties depends on the specific combination of plasticizer and rubber. For a flexible plasticizer having a small molecular size, the strength of the attractive interactions between the polymer and the plasticizer is of great importance for the dynamical glass transition [22]. In contrast to plasticizers, resins usually increase the T_g of the rubber compound [23,24]. Furthermore, the miscibility between resin and the host polymer is more often crucial [25].

With the addition of nanosized filler particles, the rubber compound becomes a polymeric nanocomposite showing additional interfacial phenomena. The surface of silica fillers mostly leads to a reduced mobility with a slower relaxation process of the host polymer [26–29]. The enhanced properties of the rubber compound are related to these interfacial interactions [30,31]. The interfacial effects result from both the interactions between the host polymer and the silica fillers (polymer–filler interaction) and interactions between the silica fillers among each other (filler–filler interaction). To increase the compatibility between silica and the host polymer, surface modifications of the silica are necessary [4,32]. Increasingly, the host polymer is functionalized as well [33,34]. Filler–filler interactions are necessary to build a network structure which provides reinforcing properties. Besides the surface modification, the surface area of the particles is critical for the mechanical properties of the rubber compound [35,36].

In this study, we characterize the variations in the molecular dynamics of a silica-filled styrene–butadiene rubber (SBR) system, which is mixed with two different resins: poly-(α-methylstyrene) (AMS) and indene-coumarone (IC). These resins differ in rigidity [37] but have a similar aromaticity degree and glass transition temperature ($T_g \approx 45$ °C). The efficient use of the resins depends on the miscibility between the resin and the polymer.

The morphology of the resulting rubbers is investigated by transmission electron microscopy (TEM). The influence of the composition on the relaxation behavior and glass transition is evaluated by conventional differential scanning calorimetry (DSC), fast differential scanning calorimetry (FDSC), temperature-modulated FDSC and broadband dielectric spectroscopy (BDS).

Dynamic glass transition takes place in the structurally equilibrated super-cooled melt as a thermal relaxation process, characterized by the relaxation time, τ, and the dynamic glass transition temperature $T_{g,\omega}$ [38]. During vitrification, the structurally equilibrated super-cooled melt transforms into a non-equilibrated glassy state. This transformation depends on the cooling rate β_c [19,39] and correlates with the relaxation time [40]. The correlation between β_c and τ has been described for thermoplastics [41–43] and unfilled solution styrene–butadiene rubber (SBR) [44] elsewhere, and is valid for the silica-filled SBR used in this study.

In this article, we investigate the influence of AMS and IC on the glass transition and the kinetics of relaxation and vitrification in vulcanized-SBR filled with silica. Furthermore, the affinity of the resin to accumulate at the silanized silica surface and the consequences for the filler network are studied.

2. Materials and Methods

2.1. Materials

The materials for this investigation are the solution styrene–butadiene rubbers (SBR) vulcanized with sulfur and filled with silica. They consist of a systematic variation in resin content. The resins are poly-(α-methylstyrene) (AMS) and indene-coumarone (IC). The chemical structures of SBR, AMS and IC are shown in Figure 1.

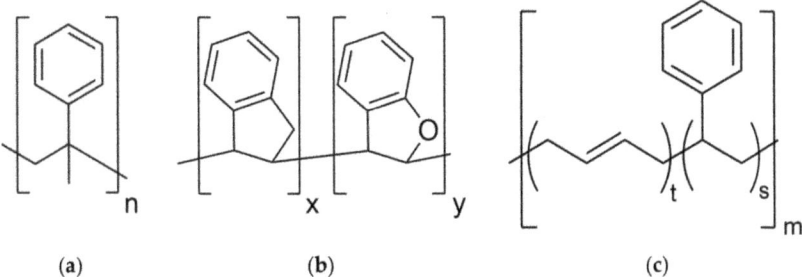

Figure 1. (a) Chemical structure of poly-(α-methylstyrene) (AMS), $n \approx 10$; (b) chemical structure of indene-coumarone (IC) resin $x + y \approx 10$ with a proportion of 95% indene and 5% coumarone; (c) chemical structure of styrene–butadiene rubber (SBR). Styrene groups (s), chain part in trans-orientation (t), m > 6000.

The formulations are given in Table 1. It is common practice in the rubber industry to develop compound formulations using the non-SI unit "parts per hundred rubber" (phr) for the weight of a component per 100 units of rubber. The relation between phr and the weight percentage for a component i is given by

$$\text{wt\%}_i = \text{phr}_i / \sum\nolimits_j \text{phr}_j \tag{1}$$

and shown for the resins in Table 2.

Table 1. Formulation of the rubber compounds used in this study.

Ingredients	Quantity [phr [1]]
SBR [2]	100
Silica	60
TESPD [3]	4.3
6PPD [4]	2.0
Wax [5]	2.0
Zinc oxide	2.5
Stearic acid	2.5
DPG [6]	1.0
CBS [7]	2.0
Sulfur	2.0
AMS [8] or IC [9]	0/20/40/60/80

[1] Non-SI unit, parts per hundred rubber (phr); [2] microstructure: 30% cis, 28–32% vinyl, 15% styrene, 42% trans; [3] bis-[3-(Triethoxysilyl)-propyl]-disulfid; [4] N-(1,3-Dimethylbutyl)-N′-phenyl-p-phenylenediamine; [5] mixture of refined hydrocarbons and plastics; [6] 1,3-Diphenylguanidine; [7] N-Cyclohexylbenzothiazol-2-sulfenamid; [8] poly-(α-methylstyrene), M_w = 1296 g/mol, PDI = 1.78; [9] indene-coumarone (IC) resin with a proportion of 95% indene, M_w = 1092 g/mol, PDI = 3.07.

Table 2. Amount of resin in phr and wt% as well as the amount of the total rubber compound in phr.

Amount Resin [phr]	Amount Total Mixture [phr]	Amount Resin [wt%]
0	178.3	0
20	198.3	10.1
40	218.3	18.3
60	238.3	25.2
80	258.3	31.0

2.2. Mixing and Vulcanization

The ingredients were mixed in a two-step mixing process with a 300 mL miniature internal mixer Haake Rheomix (Thermo Fisher Scientific, Waltham, MA, USA). In the first step, all ingredients, except the vulcanization system (DPG, CBS and sulfur), were mixed at around 140 °C for 3 min. After adding the vulcanization system in the second step, the rubber compound was mixed at 80 °C for 3 min to avoid premature crosslinking. Afterwards, the samples were vulcanized at 160 °C, according to t_{90}, the time for the 90% crosslinking, as listed in Table 3. The t_{90} time was determined according to ASTM D5289 [45].

Table 3. Vulcanization times t_{90} for the SBR compounds with variating resin content.

| Amount Resin [phr] | t_{90} [min] | |
	AMS	IC
0		13
20	18	14
40	19	17
60	21	19
80	22	20

2.3. Methods

2.3.1. Broadband Dielectric Spectroscopy (BDS)

The dielectric measurements were performed with an Alpha-A High-Performance Frequency Analyzer with a Novocool cryo-system (Novocontrol Technologies, Montabaur, Germany). The isothermal frequency sweeps, between 0.1 Hz and 2×10^6 Hz, were performed in a temperature range from -100 °C to 70 °C with an increment of 5 K. Specimens with a thickness from 150 µm to 250 µm were mounted between two round gold-plated electrodes in a plate-capacitor arrangement with a diameter of 30 mm.

2.3.2. Conventional Differential Scanning Calorimetry (DSC)

Conventional DSC measurements were performed with a DSC 1 (Mettler-Toledo, Greifensee, Switzerland) equipped with the liquid nitrogen cooling option and the HSS-8 sensor. The device was adjusted with n-octan, water, indium and zinc. The scanning rate was 10 K/min in a temperature range between -140 °C and 40 °C. The specimen was cooled and subsequently heated. In between these scanning segments, the instrument was equilibrated for 3 min. The specimens were prepared as cylindric sheets with a thickness of about 0.3 mm and a diameter of 4 mm. They were measured in a hermetically sealed standard Al-crucible.

2.3.3. Fast Differential Scanning Calorimetry (FDSC)

The FDSC experiments were performed using a Flash DSC 1 (Mettler-Toledo, Greifensee, Switzerland) equipped with an Intracooler TC100 (Huber, Offenburg, Germany) to reach the low temperature needed for the analysis of the glass transition in elastomers. The UFS 1 sensor was purged with a 20 mL/min nitrogen gas. The sensor's support temperature during the measurement was set at -95 °C.

Samples of the rubber compounds with a resin content of up to 40 phr were prepared as slices of 6 µm thickness using a cryo-microtome MT-990 (RMC Boeckeler, Tucson, AZ, USA) equipped with a glass knife operated at −60 °C and a cutting speed of 1 mm/s. The microtomic slices were cut with a scalpel to attain a final specimen shape smaller than (150 µm)^2, which is comparable to the area of the center of the active zone of the sensor. The stickier specimens, prepared from the rubber compounds with higher resin contents, were first shaped in the cryo-microtome using an angulated diamond knife. A slice of 6 µm thickness was cut and carefully placed on the chip sensor, which was stored inside the cryo-chamber of the microtome. In this way, flat and thin specimens were produced that exhibited a good thermal contact when placed within the active zone of the chip sensor [46].

The prepared specimens were cooled from 40 °C to −95 °C at rates between 1500 K/s and 0.1 K/s, and were subsequently heated at a rate of 1000 K/s to determine the cooling rate dependence of the glass transition. The glass transition temperature is defined as the limiting fictive temperature [47–49]. To evaluate the thermal contact between the specimen and sensor, measurements with a cooling and heating rate of 1000 K/s were performed for each specimen. As expected for a sufficient thermal contact, the fictive temperatures that were measured during the cooling and subsequent heating were identical within the limits of experimental uncertainty. Thus, the preparation was considered to be successful [47].

Temperature-modulated fast differential scanning calorimetry (TM-FDSC) was performed for the selected specimens using a sawtooth-modulation function (Figure 2). The temperature amplitude was 2 K, and the period was 0.1 s. The underlying cooling rate was −2 K/s between 0 °C and −60 °C. The TM-FDSC measurements were evaluated using the first harmonic of Fourier analysis.

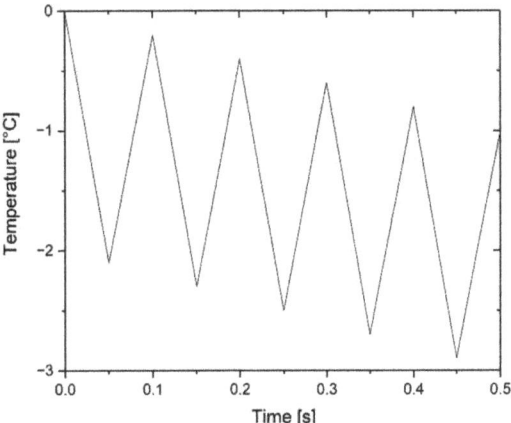

Figure 2. Sequence of the temperature program for temperature-modulated fast differential scanning calorimetry (TM-FDSC).

The resulting scanning rates were fast enough to obtain a suitable signal and slow enough to achieve a high resolution without any smearing effects (see ref. [41]). The temperature program was devised as a sequence of heating and cooling steps and were calculated as follows: (1) Cooling step of 2.1 K with a cooling rate of −42 K/s; (2) heating step of 1.9 K with a heating rate of 38 K/s; (3) repetition of steps 1 and 2 until the lowest temperature of −60 °C is reached.

The calibration of the sensor was performed with a post-measurement calibration using adamantane as a reference substance. Further details on the sample preparation and calibration are given in ref. [44].

2.3.4. Transmission Electron Microscope (TEM)

The TEM investigation was performed on a JEM-1400 (Jeol, Tokyo, Japan) using an acceleration voltage of 100 kV. Specimens of 60 nm thickness were cut with a cryo-ultramicrotome Leica EM UC6/EM FC6 (Leica Microsystems, Wetzlar, Germany) equipped with a diamond knife. The cutting temperature was −55 °C.

2.3.5. Dynamic Mechanical Analysis (DMA)

DMA investigations of the vulcanized specimens were performed in compression mode on a DMA Gabo Eplexor® 150N (Netzsch, Ahlden, Germany). Strain sweeps between 0.1% and 12% and at a frequency of 10 Hz were performed at 55 °C with a static strain of 20%. The samples were prepared as cylindrical specimens with a diameter and height of 10 mm, respectively.

3. Results and Discussion

3.1. Structural Investigation

The structure of the rubber compounds at high concentrations of resin was visualized using TEM imaging. Figure 3a,b show the TEM images of the rubber compounds containing 80 phr AMS and IC, respectively. The image of the rubber compound containing AMS (Figure 3a) shows a homogenous matrix with silica-filler particles forming aggregates in the matrix. The AMS is indistinguishable from the polymer. In the case of the IC compound (Figure 3b), the silica-filler particles are surrounded by a substance of 5 to 10 nm thickness.

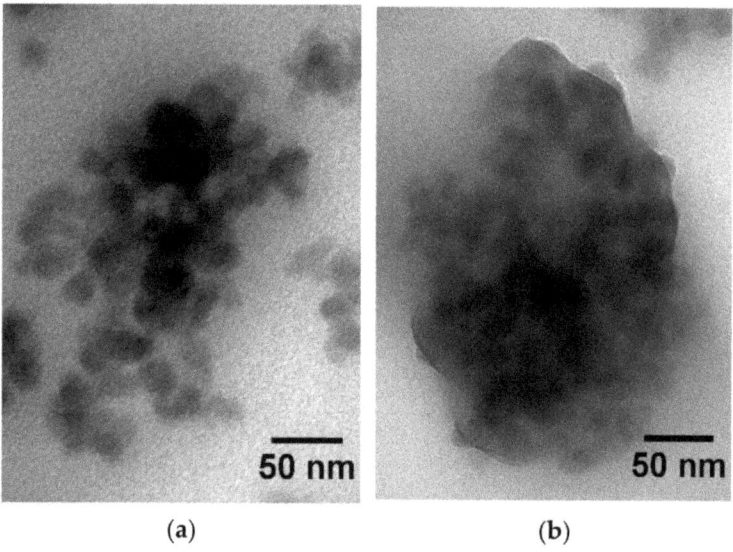

Figure 3. (a) TEM image of the rubber compound containing 80 phr AMS as resin; (b) TEM image of the rubber compound containing 80 phr IC as resin.

To identify this substance, the filler particles were irradiated with the focused electron beam of the TEM. The substance around the filler particles was easily damaged (Figure 4), as is known for organic matter. While the primary damage mechanism is caused by inelastic scattering, the damage of the organic substance is due to heat and bond scission [50]. This organic substance in the rubber compound containing IC tends to accumulate at the silica–polymer interface. It has an affinity for the silica particles.

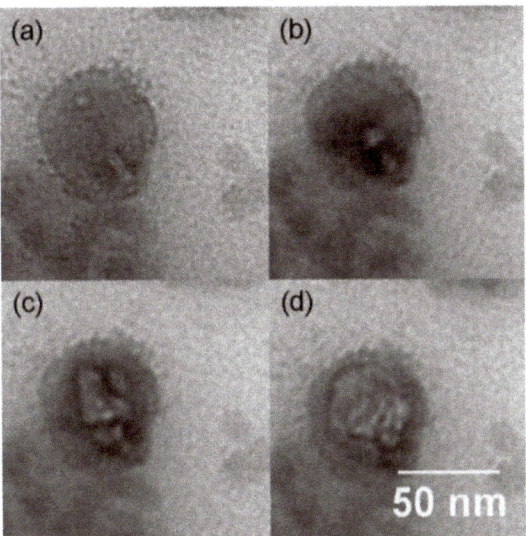

Figure 4. TEM image of the rubber compound containing 80 phr IC as resin showing the organic matter being sensitive to beam damages. The images were taken after different times of radiation treatment. (**a**) shows the untreated sample, and between (**b**–**d**), the treatment time was extended by 5 s each.

3.2. Linearity of the Mechanical Response

Rheological linearity occurs when the modulus is invariant with respect to the strain amplitude. Elastomers containing reinforcing fillers show a decrease in the dynamic storage modulus, E', with an increasing strain amplitude, ε_a (Payne-effect) [4,51,52]. The E'-ε_a diagram for both the AMS (a) and the IC rubber compounds (b) is displayed in Figure 5. As expected, the modulus decreases with the increasing resin content. The linearity limit, indicated on the curves in Figure 5, is defined as the strain amplitude at which E' is reduced by 2%. This limit is always lower for SBR-IC (Figure 6).

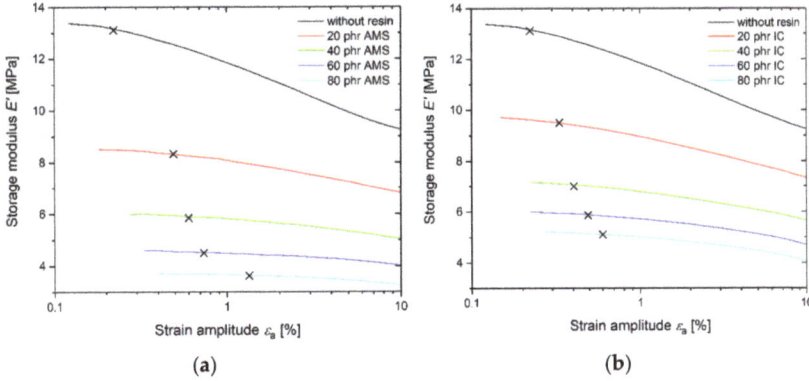

Figure 5. (**a**) Strain sweeps of AMS compounds; (**b**) strain sweeps of IC compounds. The linearity limit is indicated on the curves.

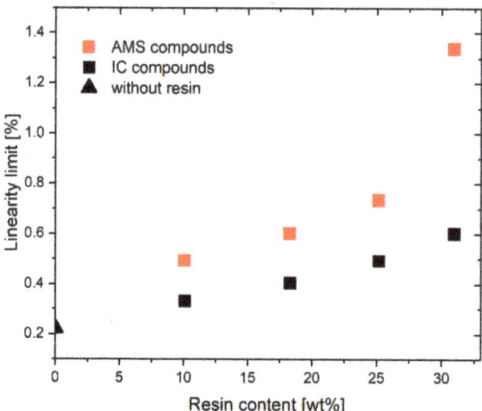

Figure 6. Linearity limit of the mechanical response as a function of the resin content.

The nonlinear behavior is due to the disruption of the filler–filler network and, therefore, is related to the percolation threshold [53]. Syed et al. showed a reduced filler percolation threshold for carbon black filled rubber with an increasing resin content [54]. The resin interacts with the surface of the filler, acts as an activator, and builds a more compact filler network [54].

For the rubber compound in this study, the IC that accumulated at the silica surface likely acts in a similar way and led to a more compact filler network. This higher compactness of the filler can lead to a stronger nonlinearity of the SBR-IC, as shown in Figures 5 and 6.

3.3. Composition Dependence of the Glass Transition

The glass transition temperatures, T_g, are measured by DSC at a cooling rate of 10 K/min. As shown in Figure 7, T_g increases with the increasing resin content. For the determination of the weight fraction, only the amorphous components (polymer and resin) are considered. The initial slope in the diagram in Figure 7 is larger for the SBR-AMS compared with the SBR-IC. Similar behavior was found for the AMS and IC in polybutadiene rubber [37,55]. The glass transition dependence of a mixture of amorphous components is usually described by the Gordon–Taylor (GT) equation [56,57]:

$$T_{g,\text{mix}} = \frac{w_c T_{g,c} + k w_r T_{g,r}}{w_c + k w_r} \quad (2)$$

where w stands for the weight fractions and T_g for the glass transition temperatures, the indices c and r refer to the polymer components and the pure resin, respectively. The GT parameter k is a fitting parameter. The fitting curves using Equation (2) are shown in Figure 7. The values of the GT-parameters are calculated as $k_{\text{fit,IC}} = 0.30$ for the SBR-IC and $k_{\text{fit,AMS}} = 0.44$ for the SBR-AMS.

For the athermic mixtures, the GT parameter is [58]:

$$k = \frac{\Delta c_{p,r}}{\Delta c_{p,c}}. \quad (3)$$

With the intensity of the glass transition for SBR, $\Delta c_{p,c} = 0.51$ J/gK, the calculated k_{calc} values are obtained and listed in Table 4.

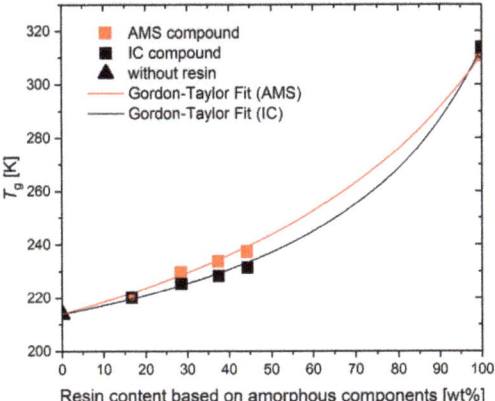

Figure 7. Glass transition temperatures of the different rubber compounds as a function of the resin content in relation to the amorphous components. The lines represent the Gordon–Taylor fits.

Table 4. Intensity of the glass transition of the pure resins and the calculated and fitted k values.

Resin	$\Delta c_{p,r}$ [J/gK]	k_0	k_{fit}
AMS	0.35	0.65	0.44
IC	0.33	0.69	0.30

Both resins show significant differences between k_{fit} and k_0. Hence, the specific molecular interactions between the resin and the polymer are expected [59], resulting in the rubber compounds being thermic mixtures. The difference between k_{fit} and k_0 increases for the SBR-IC compared to the SBR-AMS. This could be a consequence of the stronger molecular interactions between the SBR and IC, or a decreased effective resin content in the polymer-resin mixture caused by the increased amount of IC at the silanized silica surface (Figures 3 and 4). However, the reduced IC content is most likely not sufficient for the large difference in k_{fit}.

The increase in the width of the calorimetric glass transition, ΔT_w, with an increasing resin content (Figure 8) is stronger for the SBR-AMS compared with SBR-IC. Besides the effect of the reduced effective IC content, the IC is expected to have stronger specific molecular interactions with the SBR compared to AMS. The width of the calorimetric glass transition can be understood as a more reliable value for the determination of the miscibility behavior in the polymer blends compared to the shift in the glass transition temperature [60].

The width of the calorimetric glass transition is related to the average temperature fluctuation in the cooperative rearrangement regions (CRR) [40]. The size of those regions decreases with an increasing temperature fluctuation, and consequently, the size of the CRR is expected to be bigger for the IC compound compared to the AMS compound at the same resin level [61]. The interactions of IC with the polymer might yield a decrease in the volume of the independently movable regions, the CRRs. This effect is less pronounced for AMS. Thus, the less flexible IC in the SBR matrix may reduce the mobility of the polymer chain segments responsible for the glass transition more than AMS at the same content. Since the aromaticity degree and the glass transition temperature of both resins, AMS and IC, are very similar, it can be assumed that the reduced interactions are due to the differences in their molecular rigidity.

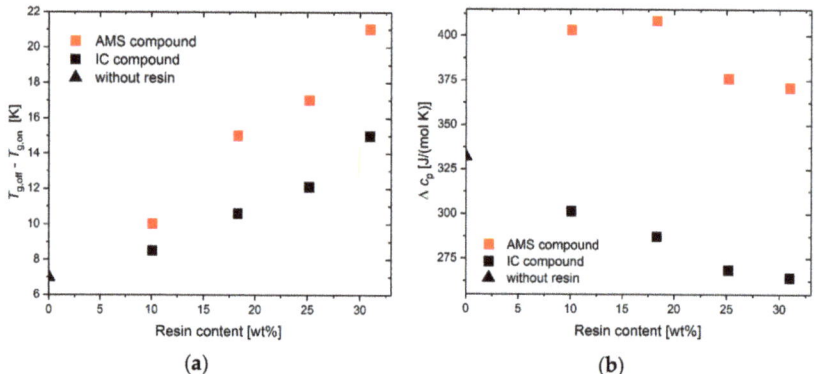

Figure 8. (a) Width of the glass transition determined as the difference between the onset and the offset as a function of the resin content. (b) Intensity of the glass transition as a function of the resin content.

The intensity of the glass transition, Δc_p, decreases in the case of the SBR-IC, while AMS increases the intensity of the glass transition (Figure 8b). The decrease in Δc_p of the SBR-IC indicates a reduced contribution of amorphous material for this glass transition. In the case of partial-phase separation, a second glass transition at higher temperatures, or at least a significant broadening of the glass transition, is expected. Such behavior was not found. The accumulation of the IC-based material, together with the decrease in Δc_p, indicates the formation of a rigid amorphous fraction on the silica surface [27,62].

3.4. Dielectric Relaxation

To characterize the relaxation behavior in a wide frequency range, dielectric measurements were performed. The dielectric loss ε'' is normalized to the peak maximum and plotted in Figure 9 as a function of the angular frequency ω at $-10\,^\circ$C for all rubber compounds under investigation. The peak is caused by the α-relaxation. The peak frequency decreases with the increasing resin content. The peak shift is stronger for the SBR-AMS compared with SBR-IC.

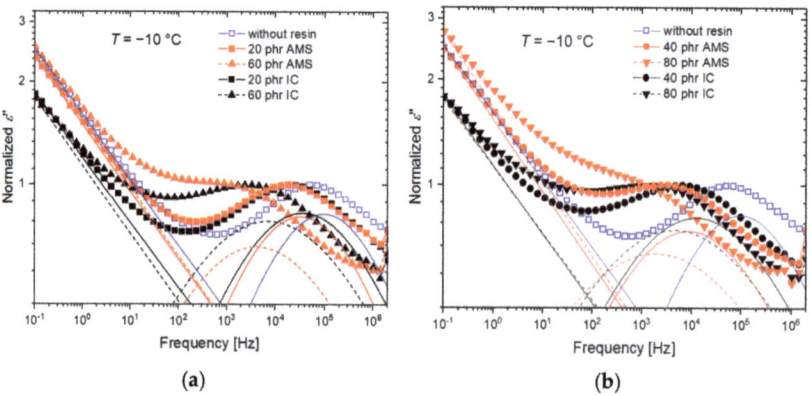

Figure 9. Dielectric losses as a function of the frequency normalized to the peak maximum of the α-relaxation for all rubber compounds measured at $-10\,^\circ$C. (a) and (b) show curves of samples with different resin contents in two groups to improve visibility. The fits, according to Equation (6) of the conductivity contributions and the relaxation processes, are indicated separately.

The decay of the curves at low frequencies is caused by both the contribution of conductivity

$$\sigma(\omega) = \frac{\sigma_0}{i\omega\varepsilon_0} \qquad (4)$$

and the Maxwell–Wagner–Sillars relaxations, which are triggered by the tapping of the charge carriers at the silica/polymer interface [26,63–67]. The latter effect can be taken into account in the dielectric loss equation by adding the exponent N to the frequency dependence of the conductivity contribution resulting in [28,68]

$$\sigma(\omega) = \left(\frac{\sigma_0}{i\omega\varepsilon_0}\right)^N. \qquad (5)$$

The accumulation of charge carriers at the interface can lead to a formation of a high dipole moment [65,69]. This leads to strong signals in the BDS measurement compared to the rubber compounds having a low polarity.

Symmetric relaxation processes, such as the α-relaxation in SBR [10,70], can be described by the Cole–Cole equation with a shape constant α. The complex permittivity function can be described by

$$\varepsilon^*(\omega) = \varepsilon_\infty + \frac{\Delta\varepsilon}{1+(i\omega\tau)^\alpha} + \left(\frac{\sigma_0}{i\omega\varepsilon_0}\right)^N \qquad (6)$$

where i is the imaginary unit, ε_∞ is the high-frequency limit of the permittivity, $\Delta\varepsilon$ is the relaxation strength, and τ is the characteristic relaxation time. The characteristic relaxation time τ can be determined from the peak maximum of the dielectric loss peak by $\omega_{max}\tau \approx 1$, where ω_{max} is the angular frequency at the maximum of the fitted relaxation function.

The vulcanization accelerator, DPG, is known to show a dielectric response that is slightly slower compared to the α-relaxation of SBR, which is possibly coupled to the segmental dynamics of the polymer [10,71]. For the silica-filled rubber compounds, DPG is assumed to be adsorbed by silica, which decreases the relaxation strength of this slow process [10]. Together, with the increasing strength of MWS and conductivity contribution, the slow process becomes indistinguishable within the curves.

3.5. Thermal Relaxation

3.5.1. Temperature Modulation

Thermal relaxation was measured by temperature-modulated DSC (TM-DSC) using the approach of the frequency-dependent complex heat capacity [72,73]

$$c_p^*(\omega, T) = c_p'(\omega, T) - i\, c_p''(\omega, T). \qquad (7)$$

The FDSC measurements were performed by means of sawtooth modulation. The evaluation was carried out by Fourier analysis of the first harmonic at a frequency of $f = 10$ Hz and an underlying cooling rate of 2 K/s. As an example, the complex heat capacity component c_p^* of the rubber compound containing 80 phr AMS is shown in Figure 10. The characteristic relaxation time is $\tau = 1/(2\pi f) = 16$ ms. The respective temperature is taken from the inflection point of the $c_p^*(T)$ curve.

3.5.2. Vitrification

The cooling rate dependence of the glass transition characterizes the thermal relaxation behavior [40]. The characteristic glass temperature, T_g, of the vitrification is indicated by the limiting fictive temperature, T_f:

$$T_g = T_{rl} - \int_{T_{rg}}^{T_{rl}} \frac{\phi(T) - \phi_g(T)}{\phi_l(T) - \phi_g(T)} dT, \qquad (8)$$

where $\phi(T)$ is the measured heat flow curve, $\phi_l(T)$ is the extrapolation of the liquid state, and $\phi_g(T)$ is the extrapolation of the glassy state. T_{rl} and T_{rg} are the reference temperatures in the super-cooled liquid and glassy state, respectively [74,75].

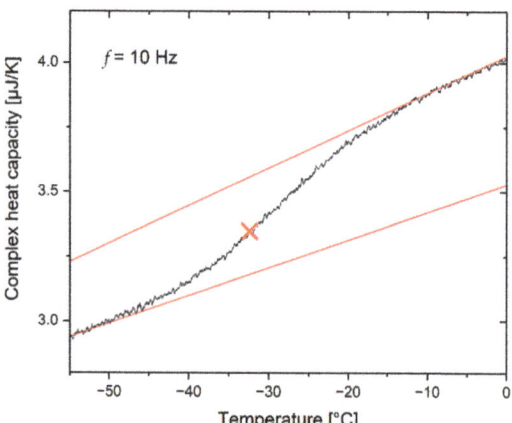

Figure 10. Complex heat capacity curve of the rubber compound containing 80 phr AMS. The intersection point at $T_g = -32.3\ °C$ is indicated.

The cooling rate dependence of T_g is measured in a range between 0.1 and 1500 K/s. To determine T_g, the specimens were subsequently heated at 1000 K/s. This method can be applied because the limiting fictive temperature of the heating curve is identical to that of the previous cooling if no aging in the glassy state occurs. This is a consequence of the conservation of energy [76].

Figure 11a shows the selected heating curves that were measured after cooling at different rates. As expected, T_g increases with an increasing cooling rate. Due to the hysteresis of the glass process, an overheating peak appears at the high-temperature side of the glass transition interval if the cooling rate β_c is lower than the heating rate β_h ($|\beta_c| < \beta_h$). The intensity of this peak increases with growing differences between the cooling and heating rates. The glass transition temperature, defined as the limiting fictive temperature, is a measure of the configurational entropy of the glass. Both properties decrease with the decreasing cooling rate.

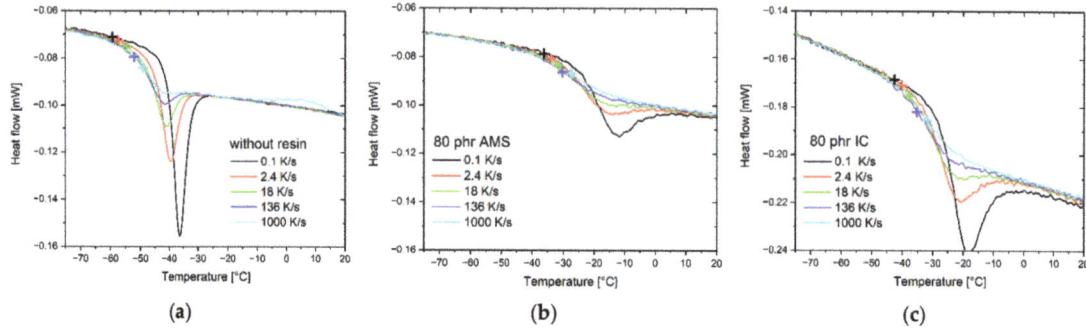

Figure 11. Selected FDSC curves at 1000 K/s measured after cooling at the indicated rates (**a**) for the rubber compound without resins; (**b**) for the rubber compound containing 80 phr AMS; (**c**) for the rubber compound containing 80 phr IC. The glass transition temperatures are indicated.

Figure 11b,c show the selected heating curves that were measured after cooling at different rates for the rubber compounds containing 80 phr AMS and IC, respectively. In agreement with the conventional DSC measurements, both the shift and the broadening of the glass transition step increase stronger in the rubber with AMS compared with the IC. The enthalpic overshoot appears to be less pronounced for the rubber compounds containing 80 phr AMS compared with the sample containing 80 phr IC. This indicates a variation in the relaxation spectrum in the composites.

The cooling rate dependence of the glass transition temperatures shows differences between the two resins. The sample containing 80 phr IC exhibits a shift between the T_g measured after cooling at 1000 K/s and 0.1 K/s of 10.9 K. This is significantly larger than the same shift of the composite containing 80 phr AMS of 8.8 K.

3.6. Influence of the Composition on the Relaxation Kinetics

In the structurally equilibrated super-cooled liquid, the temperature dependence of the relaxation frequency $1/\tau$ follows the Vogel–Fulcher–Tammann–Hesse (VFTH) equation [77–80]:

$$\log\left(\tau^{-1} \cdot 1\,\mathrm{s}\right) = A - \frac{B}{T - T_V} \tag{9}$$

where A is the logarithm of the pre-exponent factor, B is the curvature parameter and T_V is the Vogel temperature. The curvature parameter is related to the dynamic fragility m as [40,81–84]:

$$m = \frac{BT}{(T - T_V)^2}, \tag{10}$$

which describes the deviation from Arrhenius behavior.

The activation diagram of the dielectric relaxation process is plotted in Figure 12. The DC conductivity and the Maxwell–Wagner–Sillars effect limit the measurement at low frequencies. The frequency range is, therefore, expanded using the data of the thermal relaxation. It has been shown for many materials that the activation curves of the dielectric permeability and the frequency-dependent dynamic heat capacity $c_p{}^*$ are comparable [42,73,85,86].

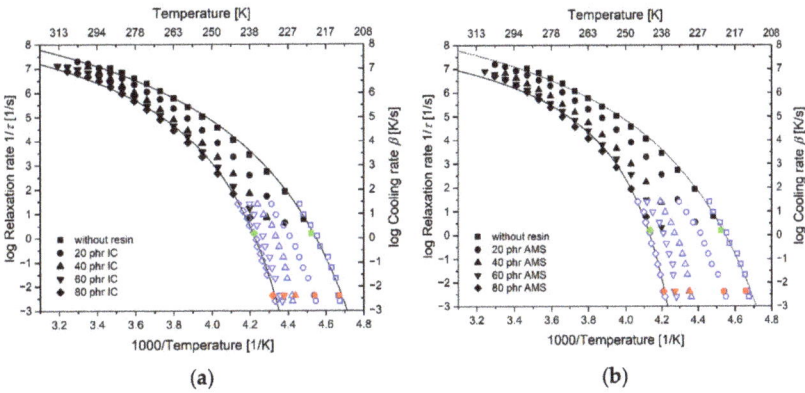

Figure 12. Activation diagrams of the different rubber compounds. The left ordinate is the logarithm of the reciprocal dielectric relaxation time of both BDS and TM-FDSC. The right ordinate is the logarithm of the cooling rate for both DSC and FDSC. The abscissa characterizes the measurement temperature of the dielectric measurements and the fictive temperature determined by the DSC and FDSC measurements, respectively. Data determined by: BDS (black) FDSC (blue), TM-FDSC (green), DSC (red). (**a**) Compounds containing IC; (**b**) compounds containing AMS. VFTH-fits are shown for the samples without resin and the samples containing 80 phr IC and AMS, respectively.

The cooling rate dependence of the vitrification process is related to the thermal relaxation time [40,42,44,87,88]. For the thermo-rheologically simple materials, the relation between the relaxation time, τ, and the cooling rate of the vitrification process follows the Frenkel–Kobeko–Reiner (FKR) equation [40,44]:

$$\log(\beta_c \tau / 1\text{ K}) = C \qquad (11)$$

The logarithmic shift of $C = 1.6$ is determined by the best overlap between the cooling rate-dependent vitrification data and the dielectric and thermal relaxation frequencies, respectively. This fact agrees with our previous findings for the unfilled SBR without resin [44] and indicates the thermo-rheological simplicity of the investigated materials. Hence, the confinement effects do not play a role in the systems in this investigation [89].

The combined dataset describes the temperature dependence of the relaxation time in a wide range of about ten orders of magnitude and can be described by a single VFTH equation (Equation (9)). The fit parameters are listed in Table 5. Additionally, the fragility parameter m is determined at $T = T_g$ using Equation (10).

Table 5. VFTH parameters of fitting the combined data in Figure 12 with Equation (9). The fragility index m is calculated using Equation (10).

Sample	A	B [K]	T_V [K]	m	T_g (100 mHz) [°C]
Without resin	10.4	355	186	92	−59
20 AMS	10.8	397	189	91	−52
20 IC	10.7	379	191	94	−52
40 AMS	9.3	242	208	121	−44
40 IC	10.2	316	200	106	−47
60 AMS	9.1	222	214	129	−39
60 IC	9.7	277	206	114	−43
80 AMS	8.9	197	220	142	−35
80 IC	9.6	269	208	115	−41

The high-frequency limit for all rubber compounds is approximately identical (Figure 12), while the low-temperature limit (the Vogel temperature) differs with the changes in the composition. This leads to the assumption that the information of the variation in the Vogel temperature, T_V, in the system of investigation, is comparable with that of curvature parameter B and the dynamic fragility m. The linear correlation between these parameters is shown in Figure 13.

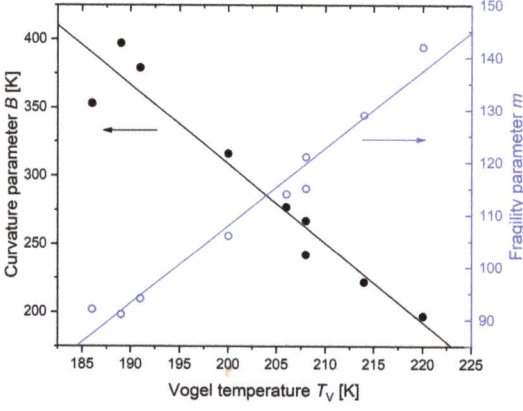

Figure 13. Indication of the linear correlation of both the curvature parameter and the fragility parameter with the Vogel temperature for the system of investigation.

The fragility index *m* is plotted versus the amount of resin in Figure 14. A higher amount of resin leads to a higher dynamic fragility of the rubber compound. This effect appears to be less pronounced for the SBR-IC compared to the SBR-AMS and vanishes at high IC concentrations.

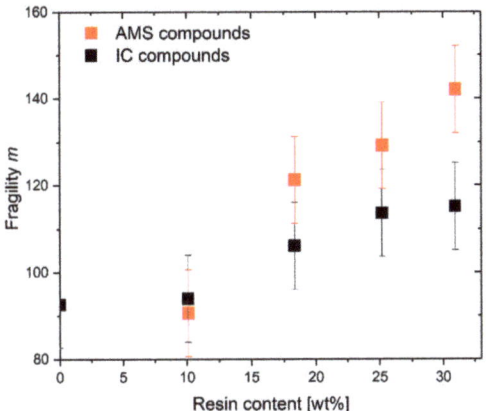

Figure 14. Fragility as a function of the resin content.

4. Conclusions

Resins are important additives in rubber compounds for enhancing both processability and the material properties. For efficient use, knowledge is needed about the effect of resins on the dynamic glass transition and the miscibility behavior in the rubber compound.

The resins AMS and IC, having a similar aromaticity degree and different molecular rigidity, are used as additives in vulcanized, silica-filled SBR. The structural investigations by TEM show an accumulation of IC around the filler particles at high contents, whereas no additional substance could be detected around the filler particles for the rubber compounds with AMS. The accumulation of IC on the silica particles generates a more compact filler network, which leads to a reduced filler percolation threshold, determined by the DMA measurements of the Payne-effect in a compression mode.

The phase diagram of the SBR-resin mixtures results in an increased difference between the theoretical GT parameter of an athermal mixture and the corresponding fit value. This indicates an increased specific interaction between the SBR and IC and, consequently, a higher affinity of IC to accumulate at the silica surface. The reduced intensity of the glass transition indicates the formation of an IC-enriched rigid amorphous fraction on the surface of the filler particles.

For both systems, the dielectric and thermal relaxation measurements result in the same activation curves, which differ depending on the type of resin and its content. The kinetics of vitrification were studied by the measurement of the cooling rate dependence of the glass transition by FDSC. According to the FKR equation, all activation curves of relaxation and vitrification overlap after shifting by the constant factor $C = 1.6$. This value agrees with the findings for unfilled SBR [44]. The validity of the FKR equation indicates thermo-rheological simplicity and enables the description of the glass process by a single VFTH equation in a frequency range of over ten orders of magnitude.

The effect of resin on the frequency dependence of T_g is strong at low frequencies, while the high-frequency limit is almost unaffected by the composition. This finding might open possibilities of efficiently tuning the material properties of rubber regarding the frequency response characteristics.

Author Contributions: Conceptualization, N.L., J.E.K.S. and J.L.-P.; methodology, N.L., J.E.K.S. and J.L.-P.; investigation, N.L. and J.L.-P.; data curation, N.L.; writing—original draft preparation, N.L. writing—review and editing, J.E.K.S. and J.L.-P.; supervision, J.L.-P.; funding acquisition, N.L. All authors have read and agreed to the published version of the manuscript.

Funding: This research received funding from the European Union's Horizon 2020 research and innovation program, grant number 760907. The APC was funded by the Open Access Publishing Fund of Leibniz Universität, Hannover.

Institutional Review Board Statement: Not applicable.

Informed Consent Statement: Not applicable.

Data Availability Statement: Restrictions apply to the availability of these data. Data are available from the corresponding author with the permission of Continental Reifen Deutschland GmbH.

Acknowledgments: The authors gratefully acknowledge the helpful discussions with Sebastian Finger and Ali Karimi, as well as the permission for publication granted by Continental Tires.

Conflicts of Interest: The authors declare no conflict of interest.

References

1. Datta, S. Elastomer Blends. In *The Science and Technology of Rubber*, 4th ed.; Erman, B., Mark, J.E., Roland, C.M., Eds.; Elsevier Acad. Press: Amsterdam, The Netherlands, 2013; pp. 547–589. [CrossRef]
2. Klat, D.; Kępas-Suwara, A.; Lacayo-Pineda, J.; Cook, S. Morphology and nanomechanical characteristics of nr/sbr blends. *Rubber Chem. Technol.* **2018**, *91*, 151–166. [CrossRef]
3. Utracki, L.A.; Mukhopadhyay, P.; Gupta, R.K. Polymer Blends: Introduction. In *Polymer Blends Handbook*, 2nd ed.; Utracki, L.A., Wilkie, C.A., Eds.; Springer: Dordrecht, The Netherlands, 2014; pp. 3–136. [CrossRef]
4. Donnet, J.-B.; Custodero, E. Reinforcement of Elastomers by Particulate Fillers. In *The Science and Technology of Rubber*, 4th ed.; Erman, B., Mark, J.E., Roland, C.M., Eds.; Elsevier Acad. Press: Amsterdam, The Netherlands, 2013; pp. 383–415. [CrossRef]
5. Bokobza, L. The Reinforcement of Elastomeric Networks by Fillers. *Macromol. Mater. Eng.* **2004**, *289*, 607–621. [CrossRef]
6. Sisanth, K.S.; Thomas, M.G.; Abraham, J.; Thomas, S. General introduction to rubber compounding. In *Progress in Rubber Nanocomposites*; Thomas, S., Maria, H.J., Eds.; Woodhead Publishing: Amsterdam, The Netherlands, 2017; pp. 1–39.
7. Rathi, A.; Hernández, M.; Garcia, S.J.; Dierkes, W.K.; Noordermeer, J.W.M.; Bergmann, C.; Trimbach, J.; Blume, A. Identifying the effect of aromatic oil on the individual component dynamics of S-SBR/BR blends by broadband dielectric spectroscopy. *J. Polym. Sci. Part B Polym. Phys.* **2018**, *56*, 842–854. [CrossRef]
8. Sharma, P.; Roy, S.; Karimi-Varzaneh, H.A. Impact of Plasticizer Addition on Molecular Properties of Polybutadiene Rubber and its Manifestations to Glass Transition Temperature. *Macromol. Theory Simul.* **2019**, *28*, 1900003. [CrossRef]
9. Coran, A.Y. Vulcanization. In *The Science and Technology of Rubber*, 4th ed.; Erman, B., Mark, J.E., Roland, C.M., Eds.; Elsevier Acad. Press: Amsterdam, The Netherlands, 2013; pp. 337–381. [CrossRef]
10. Ortega, L.; Cerveny, S.; Sill, C.; Isitman, N.A.; Rodriguez-Garraza, A.L.; Meyer, M.; Westermann, S.; Schwartz, G.A. The effect of vulcanization additives on the dielectric response of styrene-butadiene rubber compounds. *Polymer* **2019**, *172*, 205–212. [CrossRef]
11. Mostoni, S.; Milana, P.; Credico, B.; D'Arienzo, M.; Scotti, R. Zinc-Based Curing Activators: New Trends for Reducing Zinc Content in Rubber Vulcanization Process. *Catalysts* **2019**, *9*, 664. [CrossRef]
12. Sotta, P.; Albouy, P.-A.; Abou Taha, M.; Moreaux, B.; Fayolle, C. Crosslinked Elastomers: Structure-Property Relationships and Stress-Optical Law. *Polymers* **2021**, *14*, 9. [CrossRef]
13. Blume, A. Analytical properties of silica-a key for understanding silica reinforcement. *Kautsch. Gummi Kunstst.* **2000**, *53*, 338–344.
14. Heinrich, G.; Vilgis, T.A. Why Silica Technology Needs S-SBR in High Performance Tires? The Physics of Confined Polymers in Filled Rubbers. *Kautsch. Gummi Kunstst.* **2008**, *61*, 370–376.
15. Dhanorkar, R.J.; Mohanty, S.; Gupta, V.K. Synthesis of Functionalized Styrene Butadiene Rubber and Its Applications in SBR–Silica Composites for High Performance Tire Applications. *Ind. Eng. Chem. Res.* **2021**, *60*, 4517–4535. [CrossRef]
16. Rodgers, B.; Waddell, W. The Science of Rubber Compounding. In *The Science and Technology of Rubber*, 4th ed.; Erman, B., Mark, J.E., Roland, C.M., Eds.; Elsevier Acad. Press: Amsterdam, The Netherlands, 2013; p. 417. [CrossRef]
17. Khan, I.; Poh, B.T. Effect of molecular weight and testing rate on adhesion property of pressure-sensitive adhesives prepared from epoxidized natural rubber. *Mater. Des.* **2011**, *32*, 2513–2519. [CrossRef]
18. Wypych, G. (Ed.) Tackifiers. In *Handbook of Surface Improvement and Modification*; ChemTec Publishing: Toronto, ON, Canada, 2018; pp. 73–95. [CrossRef]
19. Donth, E. *The Glass Transition. Relaxation Dynamics in Liquids and Disordered Materials*; Springer: Berlin/Heidelberg, Germany, 2001. [CrossRef]
20. Zheng, Q.; Zhang, Y.; Montazerian, M.; Gulbiten, O.; Mauro, J.C.; Zanotto, E.D.; Yue, Y. Understanding Glass through Differential Scanning Calorimetry. *Chem. Rev.* **2019**, *119*, 7848–7939. [CrossRef]

21. Wypych, G. (Ed.) Effect of Plasticizers on Properties of Plasticized Materials. In *Handbook of Plasticizers*, 3rd ed.; ChemTec Publishing: Toronto, ON, Cananda, 2017; pp. 209–332. [CrossRef]
22. Stukalin, E.B.; Douglas, J.F.; Freed, K.F. Plasticization and antiplasticization of polymer melts diluted by low molar mass species. *J. Chem. Phys.* **2010**, *132*, 84504. [CrossRef]
23. Shee, B.; Chanda, J.; Dasgupta, M.; Sen, A.K.; Bhattacharyya, S.K.; Das Gupta, S.; Mukhopadhyay, R. A study on hydrocarbon resins as an advanced material for performance enhancement of radial passenger tyre tread compound. *J. Appl. Polym. Sci.* **2022**, *139*, 51950. [CrossRef]
24. Vleugels, N.; Pille-Wolf, W.; Dierkes, W.K.; Noordermeer, J.W.M. Understanding the influence of oligomeric resins on traction and rolling resistance of silica-reinforced tire treads. *Rubber Chem. Technol.* **2015**, *88*, 65–79. [CrossRef]
25. L'Heveder, S.; Sportelli, F.; Isitman, N.A. Investigation of solubility in plasticised rubber systems for tire applications. *Plast. Rubber Compos.* **2016**, *45*, 319–325. [CrossRef]
26. Genix, A.-C.; Baeza, G.P.; Oberdisse, J. Recent advances in structural and dynamical properties of simplified industrial nanocomposites. *Eur. Polym. J.* **2016**, *85*, 605–619. [CrossRef]
27. Koutsoumpis, S.; Raftopoulos, K.N.; Oguz, O.; Papadakis, C.M.; Menceloglu, Y.Z.; Pissis, P. Dynamic glass transition of the rigid amorphous fraction in polyurethane-urea/SiO2 nanocomposites. *Soft Matter* **2017**, *13*, 4580–4590. [CrossRef]
28. Füllbrandt, M.; Purohit, P.J.; Schönhals, A. Combined FTIR and Dielectric Investigation of Poly(vinyl acetate) Adsorbed on Silica Particles. *Macromolecules* **2013**, *46*, 4626–4632. [CrossRef]
29. Zou, H.; Wu, S.; Shen, J. Polymer/silica nanocomposites: Preparation, characterization, properties, and applications. *Chem. Rev.* **2008**, *108*, 3893–3957. [CrossRef]
30. Stöckelhuber, K.W.; Wießner, S.; Das, A.; Heinrich, G. Filler flocculation in polymers—A simplified model derived from thermodynamics and game theory. *Soft Matter* **2017**, *13*, 3701–3709. [CrossRef]
31. Torbati-Fard, N.; Hosseini, S.M.; Razzaghi-Kashani, M. Effect of the silica-rubber interface on the mechanical, viscoelastic, and tribological behaviors of filled styrene-butadiene rubber vulcanizates. *Polym. J.* **2020**, *52*, 1223–1234. [CrossRef]
32. Presto, D.; Meyerhofer, J.; Kippenbrock, G.; Narayanan, S.; Ilavsky, J.; Moctezuma, S.; Sutton, M.; Foster, M.D. Influence of Silane Coupling Agents on Filler Network Structure and Stress-Induced Particle Rearrangement in Elastomer Nanocomposites. *ACS Appl. Mater. Interfaces* **2020**, *12*, 47891–47901. [CrossRef]
33. Zhang, S.; Leng, X.; Han, L.; Li, C.; Lei, L.; Bai, H.; Ma, H.; Li, Y. The effect of functionalization in elastomers: Construction of networks. *Polymer* **2021**, *213*, 123331. [CrossRef]
34. Mazumder, A.; Chanda, J.; Bhattacharyya, S.; Dasgupta, S.; Mukhopadhyay, R.; Bhowmick, A.K. Improved tire tread compounds using functionalized styrene butadiene rubber-silica filler/hybrid filler systems. *J. Appl. Polym. Sci.* **2021**, *138*, 51236. [CrossRef]
35. Pourhossaini, M.-R.; Razzaghi-Kashani, M. Effect of silica particle size on chain dynamics and frictional properties of styrene butadiene rubber nano and micro composites. *Polymer* **2014**, *55*, 2279–2284. [CrossRef]
36. Padmanathan, H.R.; Federico, C.E.; Addiego, F.; Rommel, R.; Kotecký, O.; Westermann, S.; Fleming, Y. Influence of Silica Specific Surface Area on the Viscoelastic and Fatigue Behaviors of Silica-Filled SBR Composites. *Polymers* **2021**, *13*, 3094. [CrossRef]
37. Giunta, G. Multiscale Modelling of Polymers at Interfaces. Ph.D. Thesis, The University of Manchester, Manchester, UK, 2020.
38. Schawe, J.E.K. Investigations of the glass transitions of organic and inorganic substances. *J. Therm. Anal.* **1996**, *47*, 475–484. [CrossRef]
39. Gutzow, I.S.; Schmelzer, J.W.P. *The Vitreous State. Thermodynamics, Structure, Rheology, and Crystallization*, 2nd ed.; Springer: Dordrecht, The Netherlands, 2013. [CrossRef]
40. Schawe, J.E.K. Vitrification in a wide cooling rate range: The relations between cooling rate, relaxation time, transition width, and fragility. *J. Chem. Phys.* **2014**, *141*, 184905. [CrossRef]
41. Schawe, J.E.K. Measurement of the thermal glass transition of polystyrene in a cooling rate range of more than six decades. *Thermochim. Acta* **2015**, *603*, 128–134. [CrossRef]
42. Chua, Y.Z.; Schulz, G.; Shoifet, E.; Huth, H.; Zorn, R.; Schmelzer, J.W.P.; Schick, C. Glass transition cooperativity from broad band heat capacity spectroscopy. *Colloid Polym. Sci.* **2014**, *292*, 1893–1904. [CrossRef]
43. Shamim, N.; Koh, Y.P.; Simon, S.L.; McKenna, G.B. Glass transition temperature of thin polycarbonate films measured by flash differential scanning calorimetry. *J. Polym. Sci. Part B Polym. Phys.* **2014**, *52*, 1462–1468. [CrossRef]
44. Lindemann, N.; Schawe, J.E.K.; Lacayo-Pineda, J. Kinetics of the glass transition of styrene-butadiene-rubber: Dielectric spectroscopy and fast differential scanning calorimetry. *J. Appl. Polym. Sci.* **2021**, *138*, 49769. [CrossRef]
45. ASTM D5289; Standard Test Method for Rubber Property—Vulcanization Using Rotorless Cure Meters; Annual Book of Standards, Vol. 09.01; ASTM International: West Conshohocken, PA, USA, 2021.
46. Poel, G.V.; Istrate, D.; Magon, A.; Mathot, V. Performance and calibration of the Flash DSC 1, a new, MEMS-based fast scanning calorimeter. *J. Anal. Calorim.* **2012**, *110*, 1533–1546. [CrossRef]
47. Schawe, J.E.K.; Hess, K.-U. The kinetics of the glass transition of silicate glass measured by fast scanning calorimetry. *Thermochim. Acta* **2019**, *677*, 85–90. [CrossRef]
48. Perez-de-Eulate, N.G.; Di Lisio, V.; Cangialosi, D. Glass Transition and Molecular Dynamics in Polystyrene Nanospheres by Fast Scanning Calorimetry. *ACS Macro Lett.* **2017**, *6*, 859–863. [CrossRef]
49. Gao, S.; Simon, S.L. Measurement of the limiting fictive temperature over five decades of cooling and heating rates. *Thermochim. Acta* **2015**, *603*, 123–127. [CrossRef]

50. Reimer, L. *Transmission Electron Microscopy. Physics of Image Formation and Microanalysis*, 2nd ed.; Springer: Berlin/Heidelberg, Germany, 1989. [CrossRef]
51. Payne, A.R. Effect of dispersion on the dynamic properties of filler-loaded rubbers. *J. Appl. Polym. Sci.* **1965**, *9*, 2273–2284. [CrossRef]
52. Harwood, J.A.C.; Mullins, L.; Payne, A.R. Stress softening in natural rubber vulcanizates. Part II. Stress softening effects in pure gum and filler loaded rubbers. *J. Appl. Polym. Sci.* **1965**, *9*, 3011–3021. [CrossRef]
53. Klüppel, M. The Role of Disorder in Filler Reinforcement of Elastomers on Various Length Scales. In *Filler-Reinforced Elastomers Scanning Force Microscopy*; Capella, B., Geuss, M., Klüppel, M., Munz, M., Schulz, E., Sturm, H., Eds.; Springer: Berlin/Heidelberg, Germany, 2003; pp. 1–86. [CrossRef]
54. Syed, I.H.; Klat, D.; Braer, A.; Fleck, F.; Lacayo-Pineda, J. Characterizing the influence of reinforcing resin on the structure and the mechanical response of filled isoprene rubber. *Soft Mater.* **2018**, *16*, 275–288. [CrossRef]
55. Lindemann, N.; Finger, S.; Karimi-Varzaneh, H.A.; Lacayo-Pineda, J. Rigidity of plasticizers and their miscibility in silica-filled polybutadiene rubber by broadband dielectric spectroscopy. *J. Appl. Polym. Sci.* **2022**, *10*, 52215. [CrossRef]
56. Gordon, M.; Taylor, J.S. Ideal copolymers and the second-order transitions of synthetic rubbers. i. non-crystalline copolymers. *J. Appl. Chem.* **1952**, *2*, 493–500. [CrossRef]
57. Schneider, H.A. The Gordon-Taylor equation. Additivity and interaction in compatible polymer blends. *Makromol. Chem.* **1988**, *189*, 1941–1955. [CrossRef]
58. Couchman, P.R.; Karasz, F.E. A Classical Thermodynamic Discussion of the Effect of Composition on Glass-Transition Temperatures. *Macromolecules* **1978**, *11*, 117–119. [CrossRef]
59. Lu, X.; Weiss, R.A. Relationship between the glass transition temperature and the interaction parameter of miscible binary polymer blends. *Macromolecules* **1992**, *25*, 3242–3246. [CrossRef]
60. Utracki, L.A. Thermodynamics of Polymer Blends. In *Polymer Blends Handbook*, 2nd ed.; Utracki, L.A., Wilkie, C.A., Eds.; Springer Netherlands: Dordrecht, The Netherlands, 2014; pp. 171–290. [CrossRef]
61. Yin, H.; Schönhals, A. Broadband Dielectric Spectroscopy on Polymer Blends. In *Polymer Blends Handbook*, 2nd ed.; Utracki, L.A., Wilkie, C.A., Eds.; Springer Netherlands: Dordrecht, The Netherlands, 2014; pp. 1299–1356. [CrossRef]
62. Omar, H.; Smales, G.J.; Henning, S.; Li, Z.; Wang, D.-Y.; Schönhals, A.; Szymoniak, P. Calorimetric and Dielectric Investigations of Epoxy-Based Nanocomposites with Halloysite Nanotubes as Nanofillers. *Polymers* **2021**, *13*, 1634. [CrossRef]
63. Otegui, J.; Schwartz, G.A.; Cerveny, S.; Colmenero, J.; Loichen, J.; Westermann, S. Influence of Water and Filler Content on the Dielectric Response of Silica-Filled Rubber Compounds. *Macromolecules* **2013**, *46*, 2407–2416. [CrossRef]
64. Baeza, G.P.; Oberdisse, J.; Alegria, A.; Saalwächter, K.; Couty, M.; Genix, A.-C. Depercolation of aggregates upon polymer grafting in simplified industrial nanocomposites studied with dielectric spectroscopy. *Polymer* **2015**, *73*, 131–138. [CrossRef]
65. Steeman, P.A.M.; van Turnhout, J. Dielectric Properties of Inhomogeneous Media. In *Broadband Dielectric Spectroscopy*; Kremer, F., Schönhals, A., Eds.; Springer: Berlin/Heidelberg, Germany, 2003; pp. 495–520. [CrossRef]
66. Carretero-González, J.; Ezquerra, T.A.; Amnuaypornsri, S.; Toki, S.; Verdejo, R.; Sanz, A.; Sakdapipanich, J.; Hsiao, B.S.; López-Manchado, M.A. Molecular dynamics of natural rubber as revealed by dielectric spectroscopy: The role of natural cross–linking. *Soft Matter* **2010**, *6*, 3636. [CrossRef]
67. Cheng, S.; Mirigian, S.; Carrillo, J.-M.Y.; Bocharova, V.; Sumpter, B.G.; Schweizer, K.S.; Sokolov, A.P. Revealing spatially heterogeneous relaxation in a model nanocomposite. *J. Chem. Phys.* **2015**, *143*, 194704. [CrossRef]
68. Schönhals, A.; Kremer, F. Analysis of Dielectric Spectra. In *Broadband Dielectric Spectroscopy*; Kremer, F., Schönhals, A., Eds.; Springer: Berlin/Heidelberg, Germany, 2003; pp. 59–98. [CrossRef]
69. Psarras, G.C.; Gatos, K.G. Relaxation Phenomena in Elastomeric Nanocomposites. In *Recent Advances in Elastomeric Nanocomposites*; Mittal, V., Kim, J.K., Pal, K., Eds.; Springer: Berlin/Heidelberg, Germany, 2011; pp. 89–118. [CrossRef]
70. Schwartz, G.A.; Ortega, L.; Meyer, M.; Isitman, N.A.; Sill, C.; Westermann, S.; Cerveny, S. Extended Adam–Gibbs Approach To Describe the Segmental Dynamics of Cross-Linked Miscible Rubber Blends. *Macromolecules* **2018**, *51*, 1741–1747. [CrossRef]
71. Schönhals, A.; Szymoniak, P. *Dynamics of Composite Materials*; Springer International Publishing: Cham, Switzerland, 2022. [CrossRef]
72. Schawe, J.E.K. Principles for the interpretation of modulated temperature DSC measurements. Part 1. Glass transition. *Thermochim. Acta* **1995**, *261*, 183–194. [CrossRef]
73. Hensel, A.; Dobbertin, J.; Schawe, J.E.K.; Boller, A.; Schick, C. Temperature modulated calorimetry and dielectric spectroscopy in the glass transition region of polymers. *J. Therm. Anal.* **1996**, *46*, 935–954. [CrossRef]
74. Moynihan, C.T.; Easteal, A.J.; De Bolt, M.A.; Tucker, J. Dependence of the Fictive Temperature of Glass on Cooling Rate. *J Am. Ceram. Soc.* **1976**, *59*, 12–16. [CrossRef]
75. Richardson, M.J.; Savill, N.G. Derivation of accurate glass transition temperatures by differential scanning calorimetry. *Polymer* **1975**, *16*, 753–757. [CrossRef]
76. Schawe, J.E.K. An analysis of the meta stable structure of poly(ethylene terephthalate) by conventional DSC. *Thermochim. Acta* **2007**, *461*, 145–152. [CrossRef]
77. Vogel, H. Das Temperaturabhängigkeitsgesetz der Viskosität von Flüssigkeiten. *Phys. Z.* **1921**, *22*, 645–646.
78. Fulcher, G.S. Analysis of recent measurements of the viscosity of glasses. *J. Am. Ceram. Soc.* **1925**, *8*, 339–355. [CrossRef]

79. Tammann, G.; Hesse, W. Die Abhängigkeit der Viscosität von der Temperatur bei unterkühlten Flüssigkeiten. *Z. Anorg. Allg. Chem.* **1926**, *156*, 245–257. [CrossRef]
80. Alberdi, J.M.; Alegríaa, A.; Macho, E.; Colmenero, J. Relationship between relaxation time and viscosity above the glass-transition in two glassy polymers (polyarylate and polysulfone). *J. Polym. Sci. C Polym. Lett.* **1986**, *24*, 399–402. [CrossRef]
81. Böhmer, R.; Angell, C.A. Correlations of the nonexponentiality and state dependence of mechanical relaxations with bond connectivity in Ge-As-Se supercooled liquids. *Phys. Rev. B Condens. Matter* **1992**, *45*, 10091–10094. [CrossRef]
82. Spieckermann, F.; Steffny, I.; Bian, X.; Ketov, S.; Stoica, M.; Eckert, J. Fast and direct determination of fragility in metallic glasses using chip calorimetry. *Heliyon* **2019**, *5*, e01334. [CrossRef] [PubMed]
83. Qin, Q.; McKenna, G.B. Correlation between dynamic fragility and glass transition temperature for different classes of glass forming liquids. *J. Non-Cryst. Solids* **2006**, *352*, 2977–2985. [CrossRef]
84. Angell, C.A. Spectroscopy simulation and scattering, and the medium range order problem in glass. *J. Non-Cryst. Solids* **1985**, *73*, 1–17. [CrossRef]
85. Robles-Hernández, B.; Monnier, X.; Pomposo, J.A.; Gonzalez-Burgos, M.; Cangialosi, D.; Alegría, A. Glassy Dynamics of an All-Polymer Nanocomposite Based on Polystyrene Single-Chain Nanoparticles. *Macromolecules* **2019**, *52*, 6868–6877. [CrossRef]
86. Huth, H.; Beiner, M.; Donth, E. Temperature dependence of glass-transition cooperativity from heat-capacity spectroscopy: Two post-Adam-Gibbs variants. *Phys. Rev. B* **2000**, *61*, 15092–15101. [CrossRef]
87. Schneider, K.; Donth, E. Unterschiedliche Meßsignale am Glasübergang amorpher Polymere. 1. Die Lage der charakteristischen Frequenzen quer zur Glasübergangszone. *Acta Polym.* **1986**, *37*, 333–335. [CrossRef]
88. Dhotel, A.; Rijal, B.; Delbreilh, L.; Dargent, E.; Saiter, A. Combining Flash DSC, DSC and broadband dielectric spectroscopy to determine fragility. *J. Anal. Calorim.* **2015**, *121*, 453–461. [CrossRef]
89. Napolitano, S.; Glynos, E.; Tito, N.B. Glass transition of polymers in bulk, confined geometries, and near interfaces. *Rep. Prog. Phys.* **2017**, *80*, 36602. [CrossRef]

Review

Influence of the Polarity of the Plasticizer on the Mechanical Stability of the Filler Network by Simultaneous Mechanical and Dielectric Analysis

Sahbi Aloui [1,*], Horst Deckmann [1], Jürgen Trimbach [2] and Jorge Lacayo-Pineda [3]

[1] NETZSCH-Gerätebau GmbH, Schulstraße 6, 29693 Ahlden, Germany; horst.deckmann@netzsch.com
[2] Hansen & Rosenthal KG, Am Sandtorkai 64, 20457 Hamburg, Germany; juergen.trimbach@hur.com
[3] Continental Reifen Deutschland GmbH, Jädekamp 30, 30419 Hannover, Germany; jorge.lacayo-pineda@conti.de
* Correspondence: sahbi.aloui@netzsch.com

Abstract: Four styrene butadiene rubber (SBR) compounds were prepared to investigate the influence of the plasticizer polarity on the mechanical stability of the filler network using simultaneous mechanical and dielectric analysis. One compound was prepared without plasticizer and serves as a reference. The other three compounds were expanded with different plasticizers that have different polarities. Compared with an SBR sample without plasticizer, the conductivity of mechanically unloaded oil-extended SBR samples decreases by an order of magnitude. The polarity of the plasticizer shows hardly any influence because the plasticizers only affect the distribution of the filler clusters. Under static load, the dielectric properties seem to be oil-dependent. However, this behavior also results from the new distribution of the filler clusters caused by the mechanical damage and supported by the polarity grade of the plasticizer used. The Cole–Cole equation affirms these observations. The Cole–Cole relaxation time τ and thus, the position of maximal dielectric loss increases as the polarity of the plasticizer used is also increased. This, in turn, decreases the broadness parameter α implying a broader response function.

Keywords: plasticizer; polarity; carbon black network; simultaneous mechanical and dielectric analysis; mechanical stability

1. Introduction

Plasticizers are a widely used additive in rubber compounds [1–4]. They are particularly important and, as the third-highest ingredient in terms of content level, come in right after rubber and fillers. As processing aids, the plasticizers are added in different concentrations in order to impart rubber products with the desired elastic properties in the operating temperature range [5–10].

As a fluid component, the plasticizer migrates in the rubber matrix and its macromolecules are integrated into the polymer chains through intermolecular interactions. Consequently, the intermolecular forces of the polymer chains and the number of free valences in the three-dimensional structure are reduced. The internal space between the polymer chains is thus larger, and the free volume that allows the polymer chains to flow above their glass transition temperature increases [11–15]. This new conformation of the polymer chains, in turn, increases their mobility and enhances the filler distribution in the rubber mixture [16–21]. Above a certain percolation threshold, a filler network is formed that reinforces the rubber compounds and provides the necessary mechanical stability [16,17]. This applies to both the carbon-based fillers such as carbon black and silica [18–21]. Indeed, the plasticizer type strongly affects the mechanical properties of rubber products due to a shift in the glass transition temperature. Consequently, the strain, the mechanical stress, the modulus of elasticity and the damping behavior change [22–24].

Furthermore, the dielectric properties of rubber samples filled with electrically conductive filler depend on the structure of its filler network [25–32]. This applies to filler networks made of electrically conductive fillers such as carbon-based carbon black or hybrid filler networks, provided that at least one electrically conductive filler is present [25–28]. The non-conductive component is mainly used because of its excellent mechanical reinforcement, as is the case with silica used in dynamic systems such as car tires [29–32]. Aloui et al. have shown that mechanically induced changes in the structure of the electrically conductive filler network have a direct impact on dielectric mechanisms such as charge transport and polarization [33,34]. These, in turn, have consequences for the dielectric constant and the dielectric conductivity of rubber samples [35–39].

The direct relationship between mechanical and dielectric properties makes simultaneous mechanical and dielectric analysis of rubber samples filled with electrically conductive filler an outstanding technique for opening up new horizons in evaluating the microstructure dynamics of rubber materials under mechanical load and hence reproducing authentic situations from operation modes [40–43]. In addition to quality measurements on test samples, examinations on installed end products can also be guaranteed if sensors are installed to record the current material properties during use and to monitor them in the subsequent step. Mainly the dielectric properties are used as a response to the mechanical load [44].

In this study, the influence of the polarity of the plasticizer on the mechanical behavior of carbon black filled SBR under static loading is investigated. It is about the dynamics of the reinforcing filler network under mechanical loading and how this behavior can be described with simple material models. This study serves to understand the usage process and opens up the possibility of describing dynamic systems such as tires or seals during application. The electrical response is used as the display variable.

In this study, the influence of the plasticizer on the dielectric response of carbon black filled SBR under static load is examined, particularly with respect to the polarity of the plasticizer. The simultaneous dynamic-mechanical and dielectric analyzer DiPLEXOR® 500 N from NETZSCH-Gerätebau in Ahlden, Germany is used for this purpose.

2. Excursus: Dielectric Relaxation in Elastomers

Dielectric relaxation describes the build-up of the electric polarization of a dielectric medium after application of an external electric field. The characterization of the dielectric relaxation is based on the measurement of the variation of the permittivity as a function of frequency. The permittivity stems from dipole orientation and transport of free charge carriers under the action of an electric field. The measuring method uses capacitance measurements as a function of frequency for a sample placed between two electrodes. An extensive explanation of the phenomenon and the measurement technology can be found in [45].

The permittivity ε^* is a complex function with the real part ε' and the imaginary part ε'', also known as dielectric loss. As is typical for elastomers, not all dipoles have the same relaxation time, but different relaxation times, which exhibit a distribution with a relaxation peak. In order to describe these types of relaxation correctly, there are various empirical models derived from the Debye equation. In the case of symmetrical frequency response, the Cole–Cole approach is mainly used for amorphous dielectrics [46]. According to the Cole–Cole equation,

$$\varepsilon^*(\omega) = \varepsilon_{\inf} + \frac{\Delta\varepsilon}{1+(i\omega\tau)^\alpha} \text{ with } 0 < \alpha \leq 1 \tag{1}$$

where $\varepsilon_{\inf}$ is the infinite frequency dielectric permittivity, $\Delta\varepsilon$ is the relaxation strength, α is the broadness parameter and τ is the Cole–Cole relaxation time. $\omega = 2\pi f_{el}$ is the angular frequency and f_{el} is the electrical frequency. The expressions of ε' and ε'' take the following form:

$$\varepsilon'(\omega) = \varepsilon_{\inf} + \Delta\varepsilon \cdot \frac{1+(\omega\tau)^\alpha \cos\left[\alpha\frac{\pi}{2}\right]}{1+2(\omega\tau)^\alpha \cos\left[\alpha\frac{\pi}{2}\right]+(\omega\tau)^{2\alpha}} \tag{2}$$

And

$$\varepsilon''(\omega) = \frac{\sigma_{dc}}{\omega \varepsilon_0} + \Delta\varepsilon \cdot \frac{(\omega\tau)^\alpha \sin[\alpha\frac{\pi}{2}]}{1 + 2(\omega\tau)^\alpha \cos[\alpha\frac{\pi}{2}] + (\omega\tau)^{2\alpha}} \qquad (3)$$

where σ_{dc} is the direct current conductivity or DC conductivity [33,34].

3. Materials

Four carbon black filled SBR based compounds were prepared at Hansen and Rosenthal KG in Hamburg, Germany. The carbon black N 330 was used at a filler concentration of 60 phr. For a reference sample, no plasticizer was added. The three other samples each contain 20 phr of one plasticizer grade, which differ by polarity. Of course, the good miscibility of the plasticizers in the rubber matrix must be taken into account. Therefore, the following plasticizers are used: The plasticizers used are a paraffinic base oil (SN400), mild extraction solvate (MES) and distillate aromatic extract (DAE). Figure 1 shows the structural formula of the plasticizers with different polarities used [47].

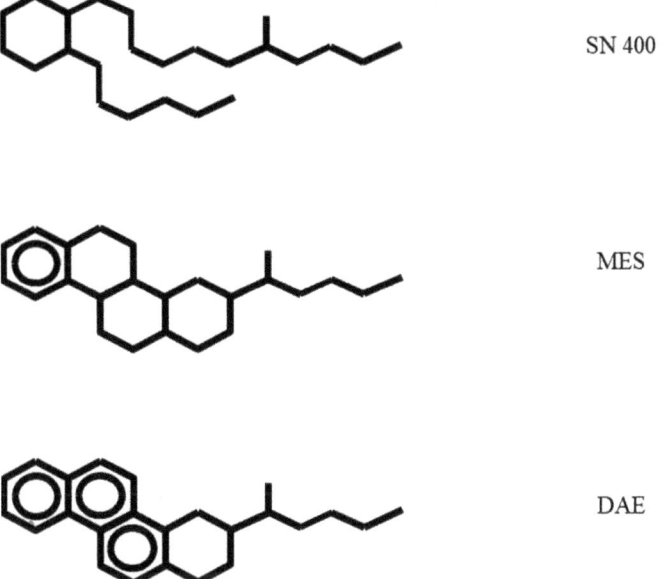

Figure 1. Structural formula of plasticizers SN400, MES and DAE.

Plasticizers SN400, MES and DAE have an aniline point in accordance with DIN ISO 2977 at 101 °C, 84 °C and 43 °C [47]. The aniline point is the temperature at which a homogeneous mixture of equal volumes of aniline and plasticizer separates into 2 phases during the cooling process. The degree of miscibility of aniline with the plasticizer estimates the aromatic content in the plasticizer. The lower the aniline point, the more polar the plasticizer.

The solubility parameter δ is an indicator of the miscibility quality of the various plasticizers within the SBR matrix. SN400, MES and DAE have a solubility parameter δ of 16 MPa$^{1/2}$, 16.7 Mpa$^{1/2}$ and 18.5 Mpa$^{1/2}$. With a value of 17.2 Mpa$^{1/2}$, the solubility parameter δ for SBR is in the same range as for the plasticizers, implying a good compatibility [47].

The aniline point and the solubility parameter are shown in Figure 2.

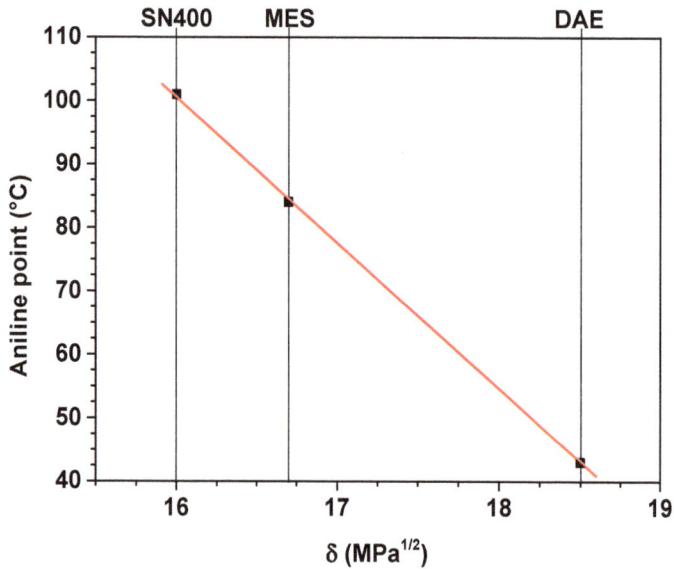

Figure 2. Aniline point and solubility parameter of the plasticizers.

The compound formulation is shown in Table 1.

Table 1. Compound formulation in phr.

	No Oil	SN400	MES	DAE
SBR 1502	100	100	100	100
N 330	60	60	60	60
SN400	-	20	-	-
MES	-	-	20	-
DAE	-	-	-	20
ZnO	2.5	2.5	2.5	2.5
Stearic acid	1	1	1	1
TMQ	1	1	1	1
6PPD	1	1	1	1
CBS	1.8	1.8	1.8	1.8
DDTD	0.2	0.2	0.2	0.2
Sulphur	1.5	1.5	1.5	1.5

The antioxidants 2,2,4-Trimethyl-1,2-dihydrochinolin (TMQ) and N-(1.3-Dimethylbutyl)-N′-phenyl-p-phenylenediamine (6PPD) were added at a concentration of 1 phr. The samples were sulfur-vulcanized. In addition to sulfur, the vulcanization accelerators N-cyclohexyl-2-benzothiazolesulfenamide (CBS) and Dimethyldiphenylthiuram disulfide (DDTD) were used.

4. Methods

4.1. Dielectric Analysis

Purely dielectric measurements were carried out at room temperature using the broadband dielectric spectrometer BDS from Novocontrol in Montabaur, Germany. The mechanical load was infinitesimally small, and it only served to maintain contact between the SBR samples and the electrodes. The coin-shape samples had a diameter of 30 mm and a thickness of 0.1 to 0.3 mm. The applied sinusoidal alternating voltage had an amplitude of 3V. The electrical frequency ranged between 1 Hz and 1 MHz.

4.2. Simultaneous Mechanical and Dielectric Analysis

Simultaneous mechanical and dielectric analysis were performed on the DiPLEXOR 500 N of NETZSCH-Gerätebau in Ahlden, Germany. The DiPLEXOR 500 N is the result of coupling the EPLEXOR 500 N dynamic-mechanical analyzer with the broadband dielectric spectrometer BDS from Novocontrol in Montabaur, Germany. Coin-shaped samples with a diameter of 10 mm and thickness of around 2 mm were used. The measurements were carried out at room temperature applying a static force of 10 N and a sinusoidal alternating voltage with an amplitude of 1 V. The electrical frequency ranged between 1 Hz and 1 MHz.

Three measurements were performed to confirm the validity of the results. However, only one measurement is shown to represent the overall result.

5. Results and Discussion

Carbon black filled elastomers have permanent dipoles and free charge carriers on the surface area of the carbon black clusters due to the graphitized surface area of carbon black particles.

The physical mechanisms behind the dielectric response of carbon black filled elastomers are based on the electrical frequency of the electrical alternating field applied. In the frequency range between 1 Hz and 1 MHz, the free charge carriers can be transported along the electric field lines. This conduction mechanism, described by the dielectric conductivity σ^*, is initially frequency-independent and reaches a constant plateau value, known as direct current conductivity or DC conductivity, abbreviated σ_{dc}. This is the result of phase-equal change in the electric field and the sample polarization. From a material-dependent frequency threshold, σ^* becomes frequency-dependent because the change in electric field and the change in sample polarization become time-delayed. This dielectric dispersion is caused by additional relaxation processes which come into play. This part is known as AC conductivity.

Furthermore, the present dipoles in the SBR materials are oriented along the electric field lines. Orientation polarization arises. Depending on the sample thickness, the applied electric field can also lead to accumulation of dipoles at the interfaces, also known as interface polarization. Both polarization mechanisms are described by the permittivity ε^*.

5.1. Dielectric Analysis

Purely dielectric measurements on the SBR samples are performed at room temperature without mechanical load. This serves first to determine the contribution of the different components to the dielectric response without the influence of static load. The measurements are performed on SBR samples with a thickness of 0.1 to 0.3 mm. Figure 3 shows the frequency-dependent change in the real part of the conductivity σ' of the SBR samples with and without plasticizers.

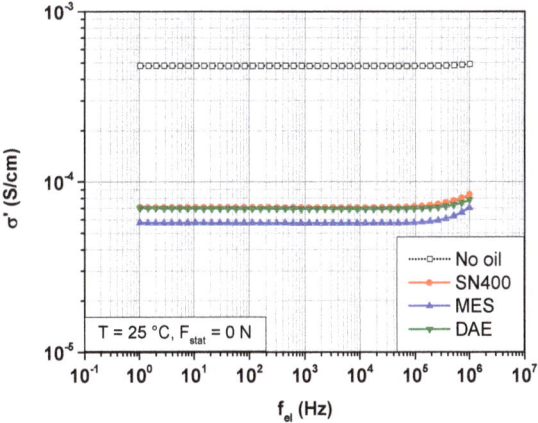

Figure 3. Frequency-dependent change of the real part of the conductivity σ' of SBR samples with different plasticizers at 25 °C.

Figure 3 indicates that the real part of the conductivity σ' of the SBR sample without plasticizer hardly shows any changes within the frequency range of the measurements. A direct current conductivity σ_{dc} of 4.8×10^{-4} S/cm is recorded. It is to note that this high conductivity in carbon black filled rubber is clearly related to the carbon black network. At a filler concentration of 60 phr, the mechanical percolation threshold is exceeded, resulting in a network of interconnected filler clusters. The contribution of polymer chains to the conductivity of SBR samples is approximately ten orders of magnitude less.

The oil-extended SBR samples show at least two disparities compared with the SBR sample without plasticizer. First, the real part of the conductivity σ' is from an electrical frequency of 100 kHz frequency-dependent, regardless of the polarity of the plasticizer used. σ' increases with increasing frequency. Second, the direct current conductivity σ_{dc} becomes smaller. It is 7×10^{-5} S/cm for the SBR samples with 20 phr SN400 and DAE. For the SBR samples with 20 phr MES, σ_{dc} is 5.8×10^{-5} S/cm.

In this context, it is worthy to note that the polarity of the plasticizer has no big influence on the conductivity. The addition of plasticizer has only influenced the distribution of the filler clusters within the rubber matrix. As a result, the distances covered by the free charge carriers along the electric field lines become longer.

5.2. Simultaneous Mechanical and Dielectric Analysis

Dielectric measurements on the SBR samples are carried out at room temperature under a static force of 10 N. The latter corresponds to a mechanical stress of 0.127 MPa. Figure 4 shows the frequency-dependent change in the real part of the conductivity σ' of the SBR samples with and without plasticizers.

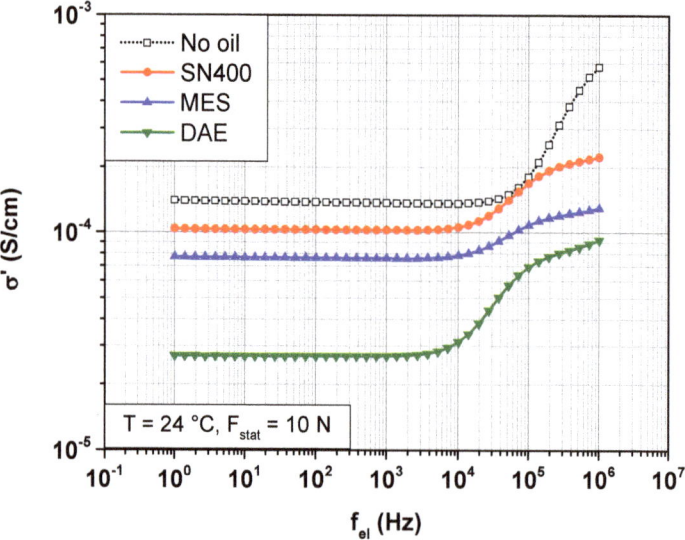

Figure 4. Frequency-dependent change in the real part of the conductivity σ' of SBR samples with different plasticizers at 24 °C under a static force of 10 N.

Figure 4 shows a completely different picture than Figure 3. This has to do with the different geometries of the test specimens and the applied contact forces. The simultaneous mechanical and dielectric analyses were performed on 2-mm-thick SBR samples with a contact force of 10 N. The purely dielectric measurements were performed on SBR samples with a thickness of 0.1 to 0.3 mm and a contact force in the mN range. It is therefore not possible to readily perform a direct comparison of the absolute measured values.

The real part of the conductivity σ' becomes frequency-dependent for all SBR samples independently of the plasticizer content. The characteristic frequency between the DC and AC conductivity shifts towards lower frequencies with the addition of the plasticizer.

For the SBR sample without plasticizer, the application of a static force of 10 N reduces σ_{dc} from 4.8×10^{-4} S/cm to 1.4×10^{-4} S/cm. This is due to the mechanical damage that reduces the density of the conduction paths within the carbon black network, and thus the SBR samples.

For the SBR samples with plasticizer, the polarity of the plasticizer used strongly influences the conductivity. σ_{dc} is 10^{-4} S/cm for the SBR sample with 20 phr SN400, 7.7×10^{-5} S/cm for the SBR sample with 20 phr MES and 2.7×10^{-5} S/cm for the SBR sample with 20 phr DAE. It is obvious that increasing the polarity of the plasticizer used decreases σ_{dc}.

The AC conductivity, abbreviated σ_{ac}, is the frequency-dependent part of σ' at which the change in the electric field and the change in sample polarization are time-delayed. The curve shape of σ_{ac} for the different SBR samples also suggests that the polarity of the plasticizer used has a huge influence on the relaxation processes caused by this dielectric dispersion.

However, as mentioned in the previous section, the polarity has no influence on the conductivity. It strongly affects the distribution of the filler clusters within the rubber matrix.

The difference in DC conductivity σ_{dc} between the mechanically undamaged (without mechanical load) and damaged (static force of 10 N) SBR samples is illustrated in Figure 5.

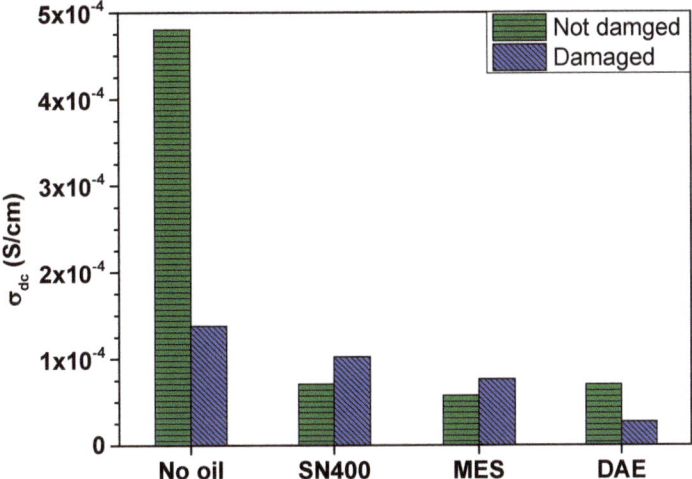

Figure 5. Difference in DC conductivity σ_{dc} between the mechanically not damaged (without static load) and damaged (static force of 10 N) SBR samples.

Figure 5 shows that σ_{dc} of the undamaged SBR samples is only affected by the presence of plasticizer, and not by the polarity of the plasticizer used. In contrast, σ_{dc} of the mechanically loaded and hence damaged SBR samples seems to be indirectly influenced by the polarity of the plasticizer used. In fact, the use of plasticizer influences the inner structure of the SBR samples by generating new dispersion states following the mechanical load.

In order to examine the impact of the plasticizer and its polarity on the dispersive part of the conductivity σ_{ac}, it is more convenient to consider the global dielectric response of the SBR samples, expressed in terms of permittivity. In addition to conductivity, relaxation processes are taken into account. Figures 6 and 7 show the change in the real and imaginary part of the permittivity, ε' and ε'' of SBR samples with different plasticizers at 24 °C under a static force of 10 N.

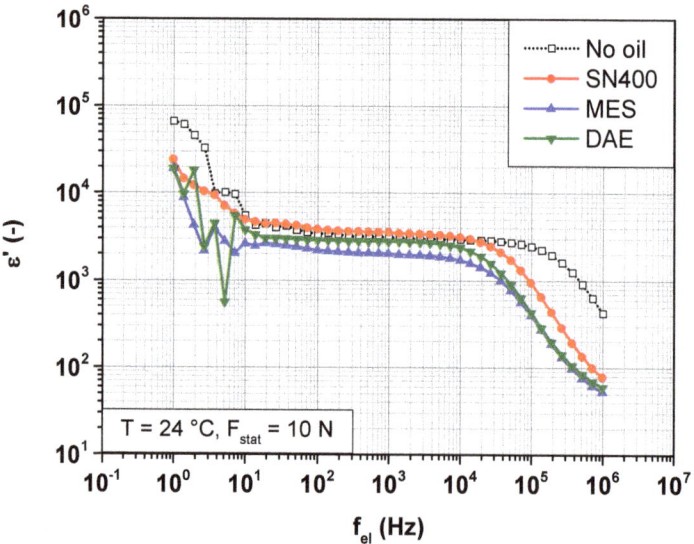

Figure 6. Frequency-dependent change in the real part of the permittivity ε' of SBR samples with different plasticizers at 24 °C under a static force of 10 N.

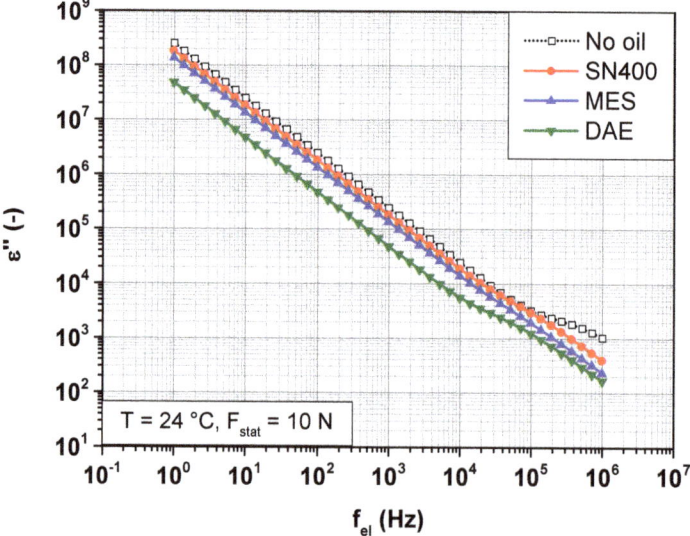

Figure 7. Frequency-dependent change in the imaginary part of the permittivity ε'' of SBR samples with different plasticizers at 24 °C under a static force of 10 N.

5.3. Cole–Cole Relaxation Model

To determine the dielectric relaxation in the SBR samples, the real and imaginary part of the permittivity ε' and ε'' are simultaneously fitted with the Equations (2) and (3) according to the Cole–Cole approach. With these basic equations, an own fitting program was developed, which simultaneously fits the real and imaginary part of the permittivity. Figure 8 displays the behavior of ε' and ε'' at 24 °C under a static force of 10 N. The SBR sample filled with 20 phr MES is shown as representative for all other SBR samples.

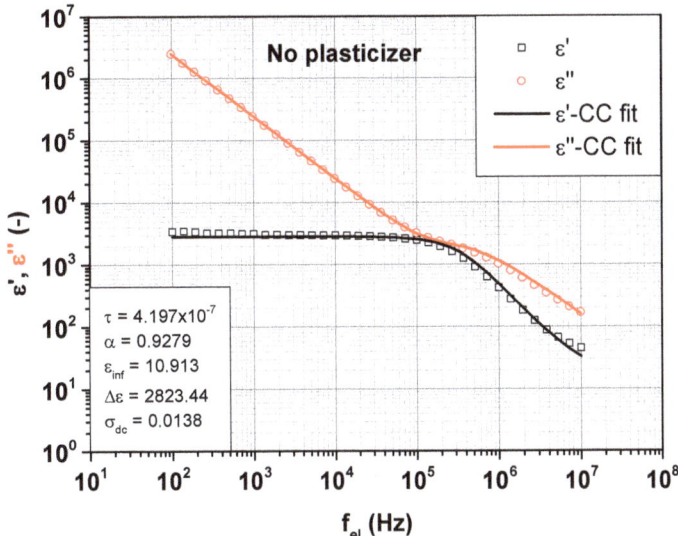

Figure 8. Simultaneous Cole–Cole fitting of real and imaginary part of the permittivity ε' and ε'' of SBR sample filled with 20 phr MES at 24 °C under a static force of 10 N.

Figure 8 suggests a good fitting quality for the measurement data of the SBR sample filled with 20 phr MES at 24 °C under a static force of 10 N. The fitting parameters are displayed on the inset. Hereinafter, the individual fitting parameters related to all SBR samples are presented. Figure 9 first shows the Cole–Cole relaxation time τ.

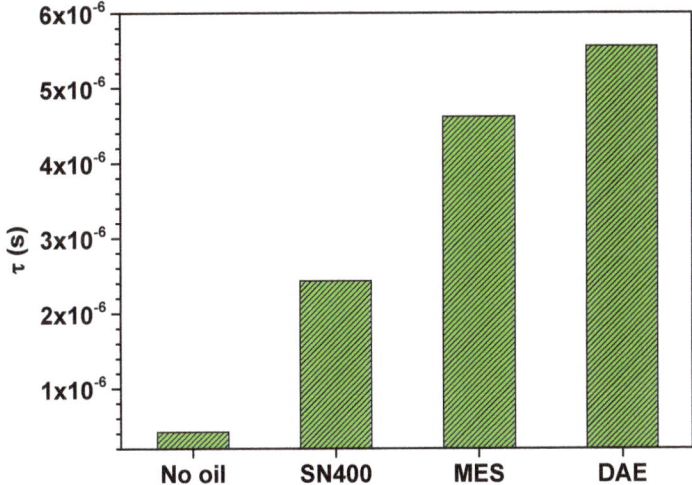

Figure 9. Cole–Cole relaxation time τ of the SBR samples determined by means of simultaneous mechanical and dielectric measurements at 24 °C under a static force of 10 N.

The Cole–Cole relaxation time τ gives the position of maximal dielectric loss. Figure 9 shows that τ increases as the polarity of the plasticizer used increases. τ is related to a characteristic electrical frequency of maximal loss f_{el} according to $\tau = 1/(2\pi f_{el})$. Increasing τ means a decrease in the characteristic electrical frequency of maximal loss f_{el}. This resembles the results shown in Figure 4, in which the frequency limit between DC and AC conductivity shifts to lower frequencies with higher polarity.

In the following, Figure 10 illustrates the broadness parameter α of the SBR samples determined by means of simultaneous mechanical and dielectric measurements at 24 °C under a static force of 10 N.

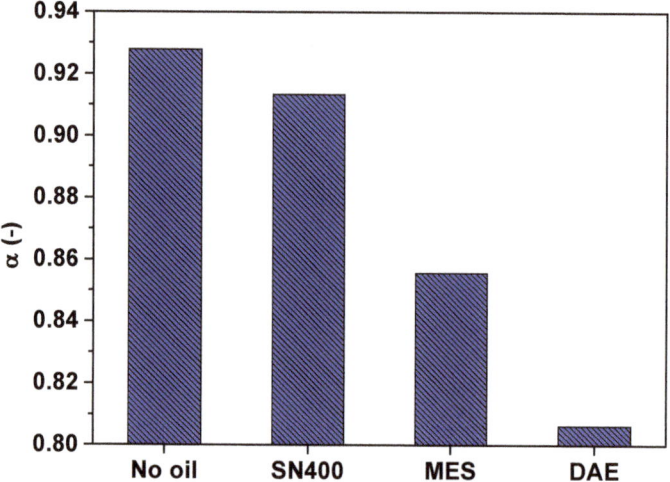

Figure 10. Broadness parameter α of the SBR samples determined by means of simultaneous mechanical and dielectric measurements at 24 °C under a static force of 10 N.

In contrast to the Debye model where the broadness parameter α equals one, Figure 10 shows that all α values for SBR samples are less than one. This indicates a symmetrically broad loss peak. Furthermore, α decreases as the polarity of the plasticizer used is increased, implying a broader response function. This behavior is caused by the interaction of the dipoles with each other, inducing a dispersion of the relaxation time τ. In physical terms, the dipoles participating in the relaxation phenomenon do not have the same relaxation time. Most probably this behavior is promoted by the polarity of the plasticizer, since the more polar the plasticizer, the more dipoles are available.

Figure 11 depicts the infinite-frequency dielectric permittivity $\varepsilon_{\inf}$ of the SBR samples determined by means of simultaneous mechanical and dielectric measurements at 24 °C under a static force of 10 N. $\varepsilon_{\inf} = \lim_{\omega \to \infty} \varepsilon'(\omega)$ is also known as the high frequency permittivity.

Figure 11 shows that $\varepsilon_{\inf}$ seems to depend on the presence of plasticizer within the SBR samples rather than the type of plasticizer. The doubling of the high frequency permittivity $\varepsilon_{\inf}$ for the oil-extended SBR samples as compared to the SBR sample without plasticizer can be attributed to the increase in dead ends caused by the mechanical damage of the filler network and the presence of additional polar molecules. Locally, more polar regions arise and permittivity increases.

The last parameter according to the Cole–Cole Equation (1) is the relaxation strength $\Delta\varepsilon$ and it is shown in Figure 12.

The relaxation strength $\Delta\varepsilon$ shown in Equation (1) is the difference between the static dielectric permittivity $\varepsilon_S = \lim_{\omega \to 0} \varepsilon'(\omega)$, also known as low frequency permittivity, and $\varepsilon_{\inf}$. $\Delta\varepsilon = \varepsilon_S - \varepsilon_{\inf}$ gives the contribution of the orientation polarization to the dielectric function and evaluates the mean molecular dipole moment on the conditions that the present dipoles do not interact with each other and shielding effects are insignificant [19]. Since this is not the case for the SBR samples, no reliable conclusions can be drawn in this regard.

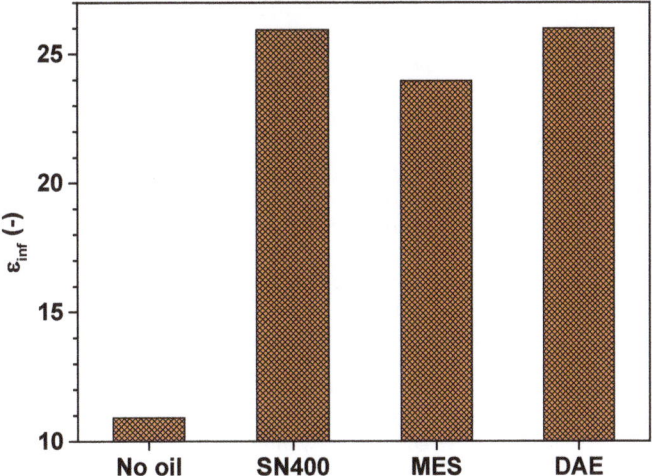

Figure 11. Infinite-frequency dielectric permittivity ε_{inf} of the SBR samples determined by means of simultaneous mechanical and dielectric measurements at 24 °C under a static force of 10 N.

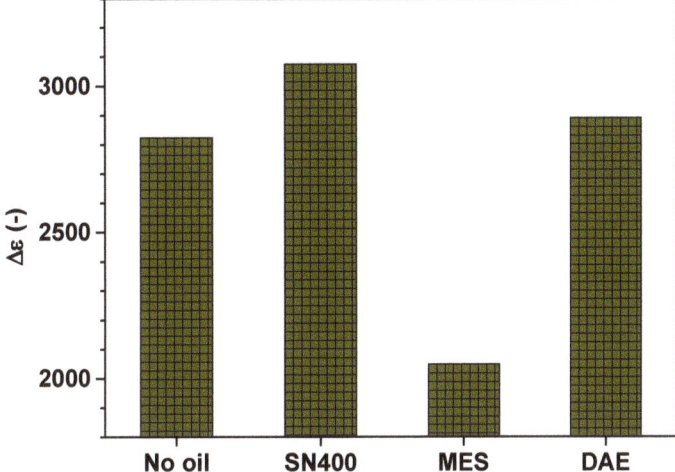

Figure 12. Relaxation strength $\Delta\varepsilon$ of the SBR samples determined by means of simultaneous mechanical and dielectric measurements at 24 °C under a static force of 10 N.

6. Summary and Conclusions

Four SBR compounds were prepared to investigate the influence of plasticizer polarity on the mechanical stability of the filler network using simultaneous mechanical and dielectric analysis.

The SBR compound without plasticizer serves as a reference. In a mechanically unloaded state, a constant conductivity of 4.8×10^{-4} S/cm was measured, showing no influence exerted by the electrical frequency. Under a static force of 10 N, which corresponds to a mechanical load of 0.127 MPa, the conductivity becomes frequency-dependent. The direct current conductivity σ_{dc} decreases from 4.8×10^{-4} S/cm to 1.4×10^{-4} S/cm due to the generated mechanical damage that reduces the density of the conduction paths within the carbon black network, and thus the SBR samples.

The other three compounds are expanded with oils that have different polarities.

The picture looks different for the three oil-expanded SBR samples. From an electrical frequency of 100 kHz, the real part of the conductivity σ' shows a dispersive part and increases as the frequency increases, regardless of the polarity of the plasticizer used. In addition, σ_{dc} is lower by almost an order of magnitude. It is 7×10^{-5} S/cm for the SBR samples with 20 phr SN400 and DAE. For the SBR samples with 20 phr MES, σ_{dc} is 5.8×10^{-5} S/cm.

The polarity of the plasticizers used does not directly contribute to the conductivity of the SBR samples. It solely influences the distribution of the filler particles within the matrix by increasing the distances covered by the free charge carriers along the electric field lines.

Under mechanical stress, the conductivity of all SBR samples becomes frequency-dependent, independent of the plasticizer content and type. Furthermore, applying a static force of 10 N reduces σ_{dc} from 4.8×10^{-4} S/cm to 1.4×10^{-4} S/cm for the SBR sample without plasticizer. This also applies to the SBR sample with 20 phr DAE. σ_{dc} decreases from 7×10^{-5} S/cm to 2.7×10^{-5} S/cm. For the SBR samples with 20 phr SN400 and MES, a small increase in σ_{dc} is observed. σ_{dc} is 10^{-4} S/cm for the SBR sample with 20 phr SN400 and 7.7×10^{-5} S/cm for the SBR sample with 20 phr MES. These experimental findings can be explained by the new internal structure, which is characterized not only by damage to the filler clusters, but also by a new distribution of the filler clusters within the matrix. It is also worth noting that σ_{dc} for the oil-expanded SBR samples decreases as polarity of the plasticizer increases.

The dielectric relaxation was analyzed to describe the behavior of the SBR samples under mechanical stress. The real and imaginary part of the permittivity, ε' and ε'', were simultaneously fitted according to the Cole–Cole approach. The Cole–Cole relaxation time τ and thus also the position of maximal dielectric loss both increase as the polarity of the plasticizer used increases. This, in turn, decreases the broadness parameter α, implying a broader response function. The high frequency permittivity $\varepsilon_{\inf}$ depends on the presence of plasticizer within the SBR samples rather than the type of plasticizer. $\varepsilon_{\inf}$ for the oil-expanded SBR samples is double that of the SBR sample without plasticizer. This can be attributed to the increase in dead ends, implying more polar regions in the SBR matrix. Finally, no clear trend can be seen for the relaxation strength $\Delta\varepsilon$. No reliable conclusions can be drawn in this regard.

This study is primarily application-oriented. These investigations are intended to describe and evaluate the influence of mechanical stress on a rubber compound in use as simply as possible. Rubber mixtures are mechanically loaded with the dynamic-mechanical analyzer in order to transfer the real operating conditions from the field to the laboratory and then to characterize the material behavior based on the dielectric properties.

In future work, the mechanical loads will be increased in order to reach the non-linear range of the rubber compounds in order to create conditions that are as realistic as possible. The study of rubber compounds with hybrid filler systems is also planned in order to characterize end products such as tires while driving, since the tire treads usually have a carbon black/silica filler system.

Author Contributions: Conceptualization, S.A., H.D., J.T. and J.L.-P.; methodology, S.A., H.D., J.T. and J.L.-P.; investigation, S.A., J.T. and J.L.-P.; writing—original draft preparation, S.A.; writing—review and editing, S.A., H.D., J.T. and J.L.-P. All authors have read and agreed to the published version of the manuscript.

Funding: This research received no external funding.

Institutional Review Board Statement: Not applicable.

Acknowledgments: The DiPLEXOR 500 N dynamic-mechanical and dielectric analyzer was developed within the ZIM project KF2473302DF0 in cooperation with M. Wilhelm from the Karlsruhe Institute of Technology.

Conflicts of Interest: The authors declare no conflict of interest.

References

1. Ni, Y.; Yang, D.; Wei, Q.; Yu, L.; Ai, J.; Zhang, L. Plasticizer-induced enhanced electromechanical performance of natural rubber dielectric elastomer composites. *Compos. Sci. Technol.* **2020**, *195*, 108202. [CrossRef]
2. Okamoto, K.; Toh, M.; Liang, X.; Nakajima, K. Influence of mastication on the microstructure and physical properties of rubber. *Rubber Chem. Technol.* **2021**, *94*, 533–548. [CrossRef]
3. Rahman, M.M.; Oßwald, K.; Reincke, K.; Langer, B. Influence of Bio-Based Plasticizers on the Properties of NBR Materials. *Materials* **2020**, *13*, 2095. [CrossRef]
4. Sokolova, M.D.; Fedorova, A.F.; Pavlova, V.V. Research of Influence of Plasticizers on the Low-Temperature and Mechanical Properties of Rubbers. *Mater. Sci. Forum* **2019**, *945*, 459–464. [CrossRef]
5. Kaliyathan, A.V.; Rane, A.V.; Huskic, M.; Kanny, K.; Kunaver, M.; Kalarikkal, N.; Thomas, S. The effect of adding carbon black to natural rubber/butadiene rubber blends on curing, morphological, and mechanical characteristics. *J. Appl. Polym.* **2021**, *139*, 51967. [CrossRef]
6. Kyei-Manu, W.A.; Herd, C.R.; Chowdhury, M.; Busfield, J.J.C.; Tunnicliffe, L.B. Influence of Colloidal Properties of Carbon Black on Static and Dynamic Mechanical Properties of Natural Rubber. *Polymers* **2022**, *14*, 1194. [CrossRef]
7. Shi, X.; Sun, S.; Zhao, A.; Zhang, H.; Zuo, M.; Song, Y.; Zheng, Q. Influence of carbon black on the Payne effect of filled natural rubber compounds. *Compos. Sci. Technol.* **2021**, *203*, 108586. [CrossRef]
8. Antoev, K.P.; Shadrinov, N.V. Effect of Conductive Carbon Black Concentration on Tyre Regenerate Properties. *IOP Conf. Ser. Mater. Sci. Eng.* **2021**, *1079*, 042025. [CrossRef]
9. Kim, I.J.; Ahn, B.; Kim, D.; Lee, H.J.; Kim, H.J.; Kim, W. Vulcanizate structures and mechanical properties of rubber compounds with silica and carbon black binary filler systems. *Rubber Chem. Technol.* **2021**, *94*, 339–354. [CrossRef]
10. Warasitthinon, N.; Robertson, C.G. Interpretation of the tand peak height for particle-filled rubber and polymer nanocomposites with relevance to tire tread performance balance. *Rubber Chem. Technol.* **2018**, *91*, 577–594. [CrossRef]
11. Hodge, R.M.; Bastow, T.J.; Edward, G.H.; Simon, G.P.; Hill, A.J. Free Volume and the Mechanism of Plasticization in Water-Swollen Poly (vinyl alcohol). *Macromolecules* **1996**, *25*, 8137–8143. [CrossRef]
12. Machin, D.; Rogers, C.E. Free volume theories for penetrant diffusion in polymers. *Macromol. Chem. Phys.* **1972**, *155*, 269–281. [CrossRef]
13. Gomes, A.C.O.; Soares, B.G.; Oliveira, M.G.; Machado, J.C.; Windmöller, D.; Paranhos, C.M. Characterization of crystalline structure and free volume of polyamide 6/nitrile rubber elastomer thermoplastic vulcanizates: Effect of the processing additives. *J. Appl. Polym. Sci.* **2017**, *134*, 45576. [CrossRef]
14. Mohamed, H.F.M.; Taha, H.G.; Alaa, H.B. Electrical conductivity and mechanical properties, free volume, and γ-ray transmission of ethylene propylene diene monomer/butadiene rubber composites. *Polym. Compos.* **2020**, *41*, 1405–1417. [CrossRef]
15. Švajdlenková, H.; Šauša, O.; Maťko, I.; Koch, T.; Gorsche, C. Investigating the Free-Volume Characteristics of Regulated Dimethacrylate Networks Below and Above Glass Transition Temperature. *Macromol. Chem. Phys.* **2018**, *219*, 1800119. [CrossRef]
16. Liu, J.; Li, B.; Jiang, Y.; Zhang, X.; Yu, G.; Sun, C.; Zhao, S. Investigation of filler network percolation in carbon black (CB) filled hydrogenated butadiene-acrylonitrile rubber (HNBR). *Polym. Bull.* **2022**, *79*, 87–96. [CrossRef]
17. Vas, J.V.; Thomas, M.J. Monte Carlo modelling of percolation and conductivity in carbon filled polymer nanocomposites. *IET Sci. Meas.* **2018**, *12*, 98–105. [CrossRef]
18. Nagaraja, S.M.; Henning, S.; Ilisch, S.; Beiner, M. Common Origin of Filler Network Related Contributions to Reinforcement and Dissipation in Rubber Composites. *Polymers* **2021**, *13*, 2534. [CrossRef]
19. Kumar, A.; Dalmiya, M.S.; Goswami, M.; Bansal, V.; Goyal, S.; Nair, S.; Hossain, S.J.; Chattopadhyay, S. Entangled network influenced by carbon black in solution SBR vulcanizates revealed by theory and experiment. *Rubber Chem. Technol.* **2021**, *94*, 324–338. [CrossRef]
20. Wen, S.; Zhang, R.; Xu, Z.; Zheng, L.; Liu, L. Effect of the Topology of Carbon-Based Nanofillers on the Filler Networks and Gas Barrier Properties of Rubber Composites. *Materials* **2020**, *13*, 5416. [CrossRef]
21. Shui, Y.; Huang, L.; Wei, C.; Sun, G.; Chen, J.; Lu, A.; Sun, L.; Liu, D. How the silica determines properties of filled silicone rubber by the formation of filler networking and bound rubber. *Compos. Sci. Technol.* **2021**, *215*, 109024. [CrossRef]
22. Ren, Y.; Zhao, S.; Yao, Q.; Li, Q.; Zhang, X.; Zhang, L. Effects of Plasticizers on the Strain-Induced Crystallization and Mechanical Properties of Natural Rubber and Synthetic Polyisoprene. *RSC Adv.* **2015**, *15*, 11317–11324. [CrossRef]
23. Sharma, P.; Roy, S.; Karimi-Varzaneh, H.A. Impact of Plasticizer Addition on Molecular Properties of Polybutadiene Rubber and its Manifestations to Glass Transition Temperature. *Macromol. Theory Simul.* **2019**, *28*, 4. [CrossRef]
24. Simon, P.P.; Ploehn, H.J. Modeling the effect of plasticizer on the viscoelastic response of crosslinked polymers using the tube-junction model. *J. Rheol.* **2000**, *44*, 169. [CrossRef]
25. Thaptong, P.; Jittham, P.; Sae-oui, P. Effect of conductive carbon black on electrical conductivity and performance of tire tread compounds filled with carbon black/silica hybrid filler. *J. Appl. Polym.* **2021**, *10*, 50855. [CrossRef]
26. Alves, A.M.; Cavalcanti, S.N.; da Silva, M.P.; Freitas, D.M.G.; Agrawal, P.; de Mélo, T.J.A. Electrical, rheological, and mechanical properties copolymer/carbon black composites. *J. Vinyl Addit. Technol.* **2021**, *27*, 445–458. [CrossRef]
27. Utrera-Barrios, S.; Manzanares, R.V.; Araujo-Morera, J.; González, S.; Verdejo, R.; López-Manchado, M.A.; Santana, M.H. Understanding the Molecular Dynamics of Dual Crosslinked Networks by Dielectric Spectroscopy. *Polymers* **2021**, *13*, 3234. [CrossRef]

28. Kropotin, O.V.; Nesov, S.N.; Polonyankin, D.A.; Drozdova, E.A. Structure and phase composition of electrically conductive carbon black. *J. Phys. Conf. Ser.* **2022**, *2182*, 012076. [CrossRef]
29. Salaeh, S.; Kao-ian, P. Conductive epoxidized natural rubber nanocomposite with mechanical and electrical performance boosted by hybrid network structures. *Polym. Test.* **2022**, *108*, 107493. [CrossRef]
30. Qian, M.; Zou, B.; Shi, Y.; Zhang, Y.; Wang, X.; Huang, W.; Zhu, Y. Enhanced mechanical and dielectric properties of natural rubber using sustainable natural hybrid filler. *Appl. Surf. Sci. Adv.* **2021**, *6*, 100171. [CrossRef]
31. Mathias, K.A.; Shivashankar, H.; Shankar, B.S.M.; Kulkarni, S.M. Influence of filler on dielectric properties of silicone rubber particulate composite material. *Mater. Today Proc.* **2020**, *33*, 5623–5627. [CrossRef]
32. Allah, M.M.D.; Ali, Z.M.; Raslan, M.A. Dielectric, thermal and morphological characteristics of Nitrile butadiene rubber under effect filler/hybrid filler. *Measures* **2019**, *131*, 13–18.
33. Aloui, S.; Lang, A.; Deckmann, H.; Klüppel, M.; Giese, U. Simultaneous characterization of dielectric and dynamic-mechanical properties of elastomeric materials under static and dynamic loads. *Polymeters* **2021**, *215*, 123413. [CrossRef]
34. Aloui, S.; Lang, A.; Deckmann, H.; Klüppel, M.; Giese, U. Corrigendum to 'Simultaneous characterization of dielectric and dynamic-mechanical properties of elastomeric materials under static and dynamic loads. *Polymeters* **2021**, *223*, 123686.
35. Abaci, U.; Guney, H.Y.; Yilmazoglu, M. Plasticizer effect on dielectric properties of poly (methyl methacrylate)/titanium dioxide composites. *Polym. Compos.* **2021**, *29*, S565–S574. [CrossRef]
36. Sengwa, R.J.; Dhatarwal, P.; Choudhary, S. Role of preparation methods on the structural and dielectric properties of plasticized polymer blend electrolytes: Correlation between ionic conductivity and dielectric parameters. *Electrochim. Acta* **2014**, *142*, 359–370. [CrossRef]
37. Maya, M.G.; George, S.C.; Jose, T.; Kailas, L.; Thomas, S. Development of a flexible and conductive elastomeric composite based on chloroprene rubber. *Polym. Test.* **2018**, *65*, 256–263. [CrossRef]
38. Chen, J.; Li, H.; Yu, Q.; Hu, Y.; Cui, X.; Zhu, Y.; Jiang, W. Strain sensing behaviors of stretchable conductive polymer composites loaded with different dimensional conductive fillers. *Compos. Sci. Technol.* **2018**, *168*, 388–396. [CrossRef]
39. Tonkov, D.N.; Kobylyatskaya, M.I.; Vasilyeva, E.S.; Semencha, A.V.; Gasumyants, V.E. Conductive properties of flexible polymer composites with different carbon-based fillers. *J. Phys. Conf. Ser.* **2022**, *2227*, 012022. [CrossRef]
40. Figuli, R.; Schwab, L.; Wilhelm, M.; Lacayo-Pineda, J.; Deckmann, H. Combined Dielectric (DEA) and Dynamic Mechanical Thermal Analysis (DMTA) in Compression Mode. *KGK-Kautsch. Gummi Kunstst.* **2016**, *4*, 22–27.
41. Aloui, S.; Wurpts, W.; Deckmann, H. Methods for simultaneous dynamic-mechanical and dielectric analysis–Part 3: Dielectric investigation of elastomer composites under dynamic deformation. *RFP Rubber Fibres Plast. Int.* **2020**, *4*, 197–200.
42. Aloui, S.; Wurpts, W.; Deckmann, H. Methods for simultaneous dynamic-mechanical and dielectric analysis–Part 2: Temperature dependence of the dielectric behavior of elastomer composites under static deformation. *RFP Rubber Fibres Plast. Int.* **2020**, *2*, 72–75.
43. Aloui, S.; Wurpts, W.; Deckmann, H. Methods for simultaneous dynamic-mechanical and dielectric analysis–Part 1: Dielectric investigation of elastomer composites under static deformation. *RFP Rubber Fibres Plast. Int.* **2020**, *1*, 26–31.
44. Aloui, S.; Deckmann, H. Static seals provide information about their own wear–Monitoring damage development by dynamic-mechanical and dielectric analyzer. *KGK-Kautsch. Gummi Kunstst.* **2019**, *4*, 14–19.
45. Kremer, F.; Schönhals, A. (Eds.) *Broadband Dielectric Spectroscopy*; Springer: Berlin/Heidelberg, Germany, 2003. [CrossRef]
46. Cole, K.S.; Cole, R.H. Dispersion and absorption in dielectrics I. Alternating current characteristics. *J. Chem. Phys.* **1941**, *9*, 341–351. [CrossRef]
47. Bergmann, C.; Trimbach, J. Influence of plasticizers on the properties of natural rubber based compounds. *KGK-Kautsch. Gummi Kunstst.* **2014**, *7*, 40–49.

Article

The Influence of Local Strain Distribution on the Effective Electrical Resistance of Carbon Black Filled Natural Rubber

E. Harea *, S. Datta, M. Stěnička, J. Maloch and R. Stoček

Centre of Polymer Systems, Tomas Bata University in Zlín, Tř. Tomáše Bati 5678, 76001 Zlín, Czech Republic; sdatta@utb.cz (S.D.); stenicka@utb.cz (M.S.); maloch@utb.cz (J.M.); stocek@utb.cz (R.S.)
* Correspondence: harea@utb.cz

Abstract: A monotonous relation between strain and measured electric resistance is highly appreciated in stretchable elastomer sensors. In real-life application the voids or technological holes of strained samples often induce non-homogeneous local strain. The present article focused on studying the effect of non-homogeneous local strain on measured direct current (DC) effective electric resistance (EER) on samples of natural rubber (NR), reinforced with 50, 60 and 70 phr of carbon black (CB). Samples were imparted geometrical inhomogeneities to obtain varied local strains. The resulting strain distribution was analyzed using Digital Image Correlation (DIC). EER exhibited a well-detectable influence of locations of inhomogeneities. Expectedly, the EER globally decreased with an increase in CB loading, but showed a steady increase as a function of strain for 50 and 60 phr over the complete testing protocol. Interestingly, for 70 phr of CB, under the same testing conditions, an alternating trend in EER was encountered. This newly observed behavior was explained through a novel hypothesis—"current propagation mode switching phenomenon". Finally, experimentally measured EERs were compared with the calculated ones, obtained by summing the global current flow through a diversity of strain dependent resistive domains.

Keywords: effective electrical resistance; elastomer sensors; natural rubber; local strain; conductive filler; digital image correlation

Citation: Harea, E.; Datta, S.; Stěnička, M.; Maloch, J.; Stoček, R. The Influence of Local Strain Distribution on the Effective Electrical Resistance of Carbon Black Filled Natural Rubber. *Polymers* **2021**, *13*, 2411. https://doi.org/10.3390/polym13152411

Academic Editor: Adriana Kovalcik

Received: 11 July 2021
Accepted: 20 July 2021
Published: 22 July 2021

Publisher's Note: MDPI stays neutral with regard to jurisdictional claims in published maps and institutional affiliations.

Copyright: © 2021 by the authors. Licensee MDPI, Basel, Switzerland. This article is an open access article distributed under the terms and conditions of the Creative Commons Attribution (CC BY) license (https://creativecommons.org/licenses/by/4.0/).

1. Introduction

1.1. Background

The change in electrical resistance of filled rubbers under mechanical stimulus opens a large number of possibilities for practical applications. The ability of cured rubber composites to deform enormously without visible failure is highly appreciated in stretchable sensors. Such sensors, prepared by blending an insulating rubber matrix with conductive fillers have a great perspective for industrial production [1]. These composites are also promising materials for transducers and flexible electrodes due to their conductivity, even in a deformed state [2–4].

Carbon black (CB) reinforced rubber compound consists of two interpenetrated phases with very different electrical properties: rubber forms a resistive network while CB produces a conductive network [5]. As such, a two-phase compound has an electrical conductivity dependent on both the phases.

Generally, the effective conductivity/resistivity of inhomogeneous materials was a subject of serious research for various materials and their applications. Conventionally all these studies can be divided into two groups: (1) involving the effective medium theory (EMT), averaging the multiple values of the constituents [6] and (2) focused on the calculation of equivalent resistor network (ERN) [7].

The effective electrical conductivity/resistivity was found to be dependent on relative amount of constituent phases, conductivity of phases and their distribution [7,8]. The attempt to discretize the phases in material by their geometry [7] had its advantage in reducing the complexity of EER calculation to solving the Kirchhoff equations [9] for the currents in conductive network. The suggested discretization involves substitution of resistance of one phase domain by a series of resistors, linking the neighboring domains.

Over the continuously enlarging number of investigations on highly deformable sensors, based on reinforced elastomers, the influence of strain-induced local deformation on EER has rarely been discussed.

The macroscopic geometrical inhomogeneity (e.g., cavity, voids, cracks) naturally present in rubber materials tend to form surrounding macroscopic domains with particular local strain distribution. Moreover, the intentionally produced geometrical inhomogeneity in a rubber product, which is required for its proper functioning, concentrates strain in the vicinity of the inhomogeneity during service. Therefore, the local stress–strain behavior around the inhomogeneity is close to intrinsic strength, causing crack growth initiation, whereas future loading leads to its propagation up to total failure [10]. On the other hand, the geometry of the inhomogeneity and global loading conditions are the reason for local multiaxial deformation leading to re-arrangement of the filler network at the affected location [11,12]. Thus, an influence on EER behavior is expected objectively.

Choice of composite, based on CB-filled natural rubber (NR) for the present study was dictated by several reasons:

Natural rubber (NR) is a strategic industrial raw material for manufacturing a wide variety of products, due to its very high strains and the respective ultimate strength before total rupture compared to other synthetic rubbers. NR becomes "self-reinforcing" at high strain level and simultaneously the mechanical properties increase. In most applications of NR, fillers are added to increase modulus, toughness and wear resistance (e.g., [13]), whereas for the rubber products with antistatic properties, the carbon black (CB) mostly is the applied conductive material along with its reinforcing properties.

The morphology and properties of filled rubber are greatly influenced by filler networking [14,15]. When a certain concentration is attained, the filler forms a continuous network that can be described in the frame of percolation or cluster, known as the cluster aggregation theory [16]. The long-range connectivity of conductive fillers can cause significant field intensification in the rubber matrix, and greatly enhance the dielectric behavior and thus the overall permittivity of the rubber matrix [17,18]. Due to the high conductivity of the CB particles, the electric field is most significantly concentrated in the small gap between two connected clusters separated by an individual distance. Thus, the electric field between two clusters changes dramatically with the distance between the interconnecting clusters, based on the ability of the electrons to overcome the gaps by tunneling effect or trap-assisted tunneling effect [19].

1.2. Research Approach

The experimental investigation for determination of change in electric response of cured rubber in dependence on strain generally can be done under a simple uniaxial tensile loading using the samples of strip geometry. Due to the application of a thin sample, in which the thickness significantly is lower than the length or even the width, the deformation in the direction of the thickness can fully be neglected [20–22].

The aim of the presented work was to perform an experimental investigation of EER of deformed rubber samples of an identical rectangular shape with implemented annular geometrical inhomogeneity, located differently across the orthogonal axes of sample to the main strain. The samples based on natural rubber (NR) reinforced with CB far above the percolation threshold (50, 60 and 70 phr), were prepared and subjected to tensile loading up to strain 26.7%. In the case of NR, the dedicated strain is still far below the strain values in which strain induced crystallization (SIC) appears [23,24]. The deformation of the sample has been monitored and the strain distribution over the complete sample surface has been determined by the Digital Image Correlation (DIC) system. The DC EER behavior has been measured simultaneously during the loading. The DC measurements was implemented taking into account the well-known facts that the current density distribution driven by the alternating current (AC) is often not uniform throughout the cross-section of any conductor because of the skin and proximity effects [25]. Finally, the proposal for numerical calculation of EER related to deformation of samples in dependence on varied geometrical inhomogeneity and the corresponding local strain distortion of conductive phase was introduced for the first time, in effect supporting the novelty of the work.

2. Materials and Methods

2.1. Rubber Formulation

The complete formulation of rubber used within this study is listed in Table 1. Natural rubber was supplied by the Astlett Rubber Inc. (Astlett Rubber Inc., Oakville, ON, Canada) (type SMR 20 CV/BP1). Sulfur used as the curing agent, zinc oxide (ZnO) and stearic acid used as activators were supplied by Sigma-Aldrich®. The reinforcing filler used in all the compounds was high abrasion furnace (HAF)–N330 carbon black (CB) supplied by Cabot Corporation, Boston, MA, USA. Moreover, Naphthenic oil NYTEX® (Nynas AB, Nynashamn, Sweden) was used as a plasticizer and CBS (N–cyclohexyl–2-benzothiazolesulfenamide), was employed as the curing accelerator.

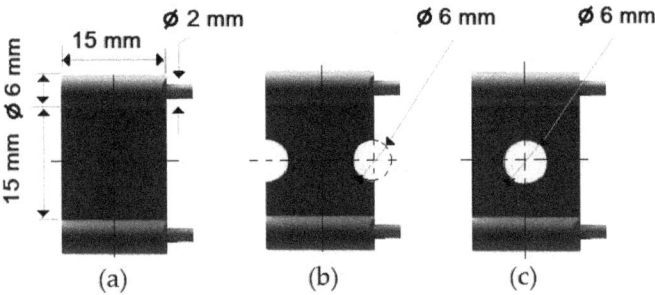

Figure 1. The geometry of studied samples: (**a**) basic configuration, (**b**) double side inhomogeneity, (**c**) central inhomogeneity.

Table 1. Rubber formulation.

	NR	Oil	Carbon Black	CBS	Sulfur	ZnO	Stearic Acid
	Content in phr *						
NR50_#			50.00				
NR60_#	100.00	10.00	60.00	1.00	2.50	5.00	2.00
NR70_#			70.00				

*—parts per hundred of rubber by weight. #—a, b or c depending on sample geometry (see Figure 1).

2.2. Rubber Compounding and Samples Preparation

The compounds were prepared in an internal mixer Brabender Plastograph (Brabender GmbH & Co., Duisburg, Germany) at 60 °C at a rotor speed of 50 rpm at a fill factor of 80%. The rubber and the compounding ingredients were successively added as follows: NR was masticated for 3 min followed by mixing of ZnO and stearic acid activators, both for 1 min successively. The filler was added in three stages alternating with the plasticizer and mixed for the next 3 min. Finally, CBS and sulfur were added and mixed for another 2 min. Thus, the total mixing time was 10 min. The optimum cure time at 160 °C for each batch was determined using a moving die rheometer (MDR 3000 MonTech, Buchen, Germany) according to ISO 3417. After 24 h conditioning at an ambient temperature of 23 °C, the compounds were molded using electrically heated hydraulic press (LabEcon, Delft, The Netherland) at 160 °C and 200 kN into samples of specific geometries defined generally with dimensions $15 \times 15 \times 2$ mm^3 with cylindrical shoulders of 6 mm in diameter at both ends. Each cylindrical shoulder contained a brass tube (2 mm external and 1.4 mm internal diameters) in the direction of the shoulder axis for realization of future electrical contacts. Finally, in two amongst the three different samples, shape inhomogeneities, characterized by top view of an annular circle (sample c) and two semi circles (sample b) having diameters of 6 mm were implemented. The detailed geometries of the investigated samples are shown in the Figure 1.

2.3. Electric Setup

The measurement of the DC EER in tensile mode was done in a servohydraulic testing equipment Instron 8871 (Instron, High Wycombe, UK) equipped with customized nonconductive clamps (Figure 2), which were used to fix the lateral cylindrical shoulders of the tested samples. Due to the application of cylindrical shoulders containing tubular brass contacts, the additional stress commonly induced by simple fixing system was efficiently avoided. The conductive wires were mechanically crimped into the brass tubes, avoiding meta–rubber interface overheating as encountered in the case of soldering.

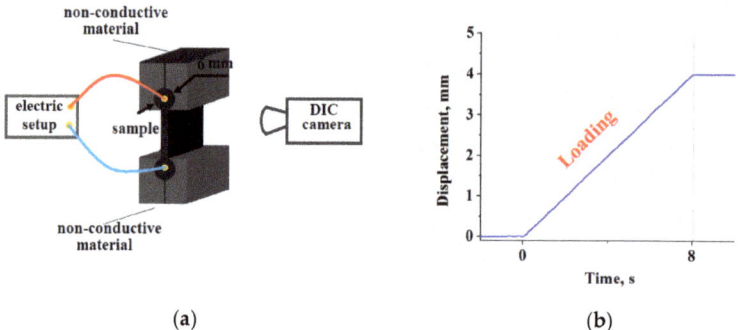

Figure 2. (a) Customized nonconductive clamps; (b) testing protocol.

The applied testing protocol, schematically visualized in Figure 2b was based on strain up to 4 mm (26.7% strain) at a constant rate of 0.5 mm/s.

The experimental setup for DC EER measurement is shown in Figure 3.

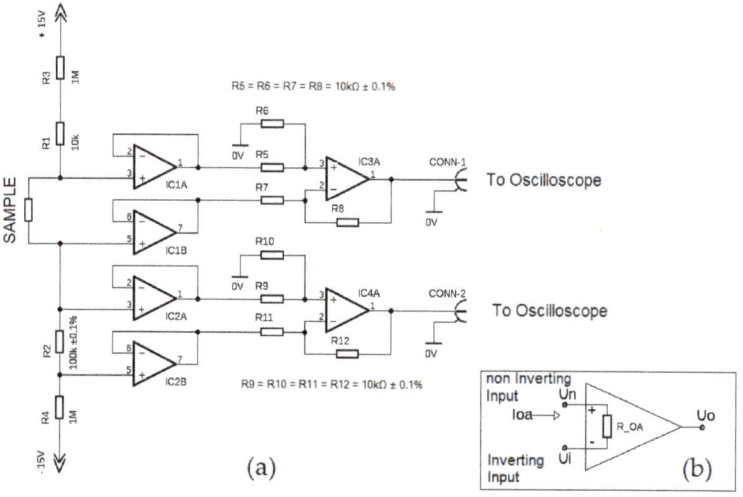

Figure 3. (a) The electric scheme of measurement; (b) operational amplifier.

The measuring was based on the indirect method, whereas the complete measuring setup was compiled and assembled for this study by the authors. The resistance was calculated from voltage drop on the measured sample and voltage drop on the serial high precision resistor R_2 (Figure 3a). Since all changes in the load were slow, a simple Ohm's Law was used in the form

$$R_S = U_S / I_S, \qquad (1)$$

where R_S is the resistance of the sample (Figure 3a), U_S is the voltage drop across the sample (measured directly) and I_S is the current going through the sample (measured indirectly from the voltage on the R_2).

Operational amplifiers (OA) assured the sensibility of measurements. Directly connected OA (denoted in the Figure 3a as IC1A, IC1B, IC2A, IC2B) were used as voltage followers. For OA functionality explanation, a simplified scheme is depicted in Figure 3b. Following the theory of the real OA output voltage, it can be assumed that

$$U_o = A_u (U_n - U_i), \qquad (2)$$

where, U_o is the output voltage, U_i is the inverting input voltage, U_n is the non-inverting input voltage, and A_u is the open loop voltage amplifying coefficient.

Theoretically, the difference $(U_n - U_i)$ can be assigned as U_{dif}. If the OA is connected as the voltage follower, then the inverting input voltage U_i is equal to the output voltage U_o. Equation (1) will then be modified with mathematical adjustment to the form

$$U_o = U_n A_u / (A_u + 1) \qquad (3)$$

where A_u is the open loop voltage amplifying coefficient. $A_u = 100000$ typically, for a used amplifier. Due to this condition, U_o can be taken as equal to the U_n. The input current passing through OA can be written as

$$I_{OA} = U_{dif} / R_{OA} \qquad (4)$$

and R_{OA} is internal resistance of OA.

Equation (4) can be modified with the substitution from Equation (2) and reads as follows:

$$I_{OA} = U_0 / (R_{OA} \times A_u) \qquad (5)$$

Substituting the value of U_0 from Equation (3) in Equation (5) leads to the framing of Equation (6):

$$I_{OA} = U_n / (R_{OA} \times (A_u + 1)) \qquad (6)$$

where the input resistance is transformed by multiplication with the term $(A_u + 1)$. So, taking into consideration the characteristic value of internal resistance for used OA (the minimal value R_{OA} = 30 kΩ), the customized installation is able to easily measure the resistances up to 750 MΩ and even higher. This fact guarantees that measuring method is suitable for tested samples.

Amplifiers IC3A, IC4A joined behind voltage followers were connected as typical differential amplifiers with voltage magnification [26,27]. Data recording was obtained using digital multichannel oscilloscope Rigol MS05104 (Rigol Technologies, Co Ltd., Suzhou, China).

Digital Image Correlation (DIC) was applied to determine the local strain fields in the studied samples. For this purpose, a stochastic pattern made by an anti-reflex spray, MR2000 Anti-Reflex L (MR Chemie GmbH, Unna, Germany) was applied on the surface of all the tested samples. The strains of the complete sample were recorded over the testing protocol via CCD monochrome camera Baeumer PXU 60 M Q (Bauemer, Frauenfeld, Switzerland) with a sampling frequency of 15 Hz. The DIC process was controlled over the software GOM Snap 2D, (GOM, Braunschweig, Germany). Subsequently, the captured pictures were processed and analyzed with DIC software (GOM Correlate, Braunschweig, Germany) for the strain field evaluation.

3. Results and Discussions

3.1. Local Strain Distribution and Measured EER

The measured strain contour image for all the studied samples under tensile loading and their initial shapes are visualized in Figure 4. Due to inappreciable differences in the determined local strains between the rubber samples loaded with different concentrations of CB, only the rubber samples compounded with 70 phr CB will be discussed in terms of DIC characterization. The DIC software monitored the deformation of complete sample during straining and evaluated the strain over the complete sample surface, as well as the local strain near to inhomogeneity and presented the data in colored map over the complete surface of the sample.

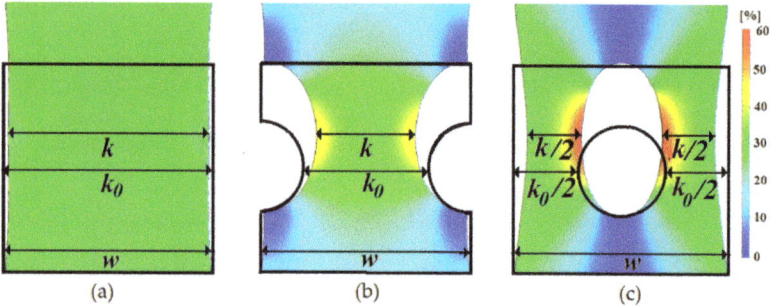

Figure 4. Strain maps (**a**) sample type a, (**b**) sample type b and (**c**) sample type c.

For all the analyzed samples, the contraction was observed in the horizontal axis, orthogonally to the main strain. The contraction gradually decreased in the vertical direction from the horizontal axis, whereas close to the clamping area, the contraction became minimal. It is obvious that the location of the inhomogeneity in the edges (sample type b) and central part (sample type c) provoked a very different strain distribution across the horizontal axis as well as over the complete affected sample area due to the stress concentrations focused on the sharp corners due to the abrupt change in the surface area [28].

The DIC of dynamic evolution of the contraction of the sample ($\Delta k = k_0 - k$) during a testing protocol was monitored continuously by measuring the width of the sample passing through the geometrical center (see Figure 4). To avoid any misunderstanding, here and in all following text, k represented the width of rubber composite in the geometrical center of sample, which is not always equal with samples width (w), due to intentionally create inhomogeneity. The vertical displacement of the sample (elongation, noted as Δl) was proportional to applied strain, and fully depended on the settings of the tensile equipment. However, the horizontal contraction of the monitored segment was a material and also the sample shape dependent term. The results experienced a similar trend for all CB concentration. Thus, only those obtained for the sample containing 70 phr CB are presented in Figure 5. Sample type (a) exhibited an absolutely logical trend of contraction, in close agreement with Poison's ratio value ($\nu = 0.5$ for CB filled natural rubber). For samples containing geometrical inhomogeneity, the elongation–contraction relationship exhibited an appreciable difference due to complicated non-homogeneous local strain distribution.

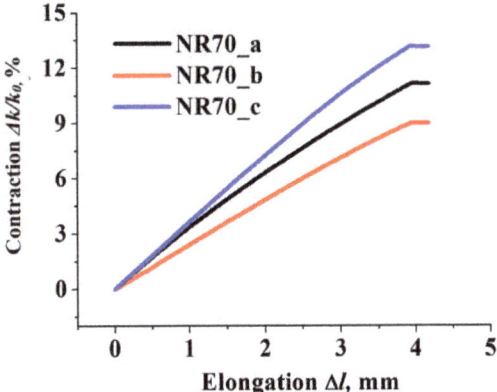

Figure 5. The evolution of elongation vs. contraction for sample containing 70 phr of CB (similar trends were observed for all tested CB concentrations).

The representative curves of resulting force and EER vs. elongation corresponding to applied strain, for all different filler concentrations and three different sample types are shown in Figure 6. From the mechanical point of view, the plot of force vs. elongation (Figure 6a,c,e) shows the well-known and expected trend of increasing force with an increase in the filler content [29] over the complete tensile loading.

The two samples containing different positions of geometrical inhomogeneity, although having the same total initial cross-sectional area in the horizontal axis, revealed a difference in the measured force under a certain strain over the complete straining process. The forces developed by tensile machine to achieve the maximum set strain are depicted in Figure 7. The sample type b possessing an undivided cross-section area exhibited higher force if compare to sample type c having a divided one. The cause of observed difference was the variation in the location of produced inhomogeneity, resulting in a different new surface area creation, preserving the same total cross-sectional area of the two different samples. Therefore, the sample with bulk cross-section area (type b) required more energy to be deformed compared to the divided one (type c) due to an excess in the number of internal molecular bonds.

Figure 6b,d,f exhibit the uniqueness of the results of EER for the used deformation settings. A gradual increase in CB concentration increased the non-monotonic runway of resistance variation. The results are presented in the form of a ratio between variation of resistance ($\Delta R = R - R_0$) and R_0, the latter being the resistance of the sample before deformation. Samples containing 50 phr of CB (Figure 6b) exhibited progressive, well-distinguished increase in resistance as a function of applied strain for all the geometries of the samples. Depending on sample geometry, increase in samples resistance ($\Delta R/R_0$) attained an impressive value close to 270 times at the maximum applied strain. This fact is highly appreciated for sensitivity of sensors.

Samples containing 60 phr CB (Figure 6d) showed an essential difference of $\Delta R/R_0$ in dependence of sample geometry, exhibited a modest increase in resistance for about seven times in the best case, while the samples containing 70 phr of CB (Figure 6f) showed a very low $\Delta R/R_0$ ranging between 0.02 and 0.2 times for maximum applied stress.

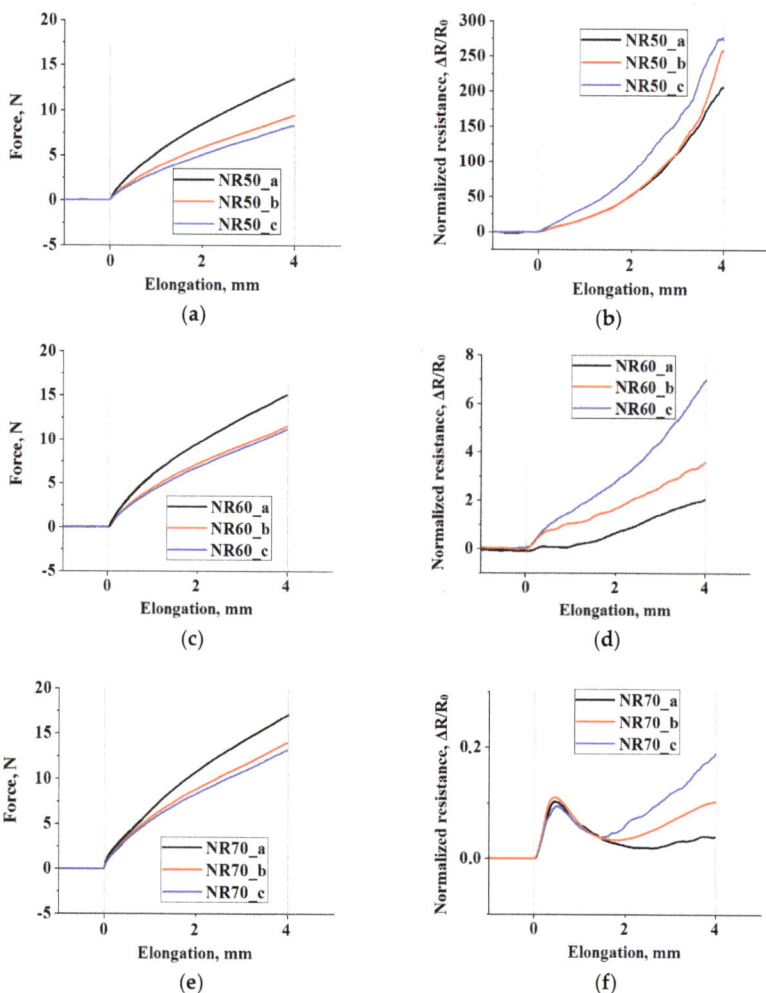

Figure 6. (**a**,**c**,**e**) force registered during the testing protocol; (**b**,**d**,**f**) electrical resistance of samples.

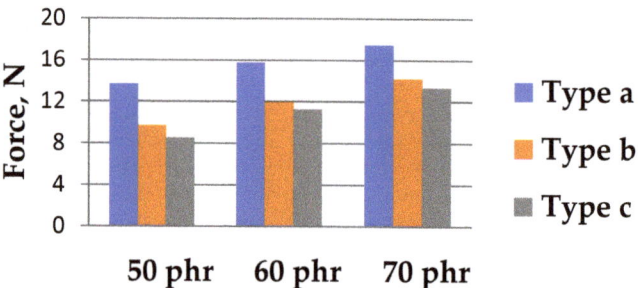

Figure 7. Force at maximum set elongation in strained samples.

3.2. Current Propagation Mode Switching Effect

The most unusual results were obtained for samples filled with 70 phr CB (Figure 6f). In just one tensile event, an initial increase in resistance was followed by a sudden decrease and then again, a continuous increase. Sequentially, such a strange behavior may be regarded as a negative and positive piezoresistive effect in filled polymers, discussed earlier, for example, in [30]. The uniqueness of the observed behavior was to have both these phenomena in the same material and under one loading process. This effect has not been discussed in any previous scientific studies from theoretical as well as an experimental point of view. In the opinion of the authors of the present article, the explanations given in the subsequent discussion describe the observed peculiarities in the effective resistance variation of the sample under deformation:

The contraction of the sample during the tensile test (Figure 4) confirmed the concurrence of two processes taking place in the conductive network. In the direction of applied tensile strain, the distance between the conductive particles increased and contrary to this happening, in the perpendicular direction of the applied strain, a hydrostatic pressure, made the particles approach each other. CB particles are usually considered as spheres coupled into aggregates. Taking into consideration the Hertzian contact theory, the contact area, A, of two spheres of identic radius, R, can be calculated from the equation:

$$A = \pi \frac{R}{2} \times x \tag{7}$$

where x is the penetration depth between spheres.

The Poisson coefficient, ν, for thin sample of square shape exhibits the ratio between contraction, Δk, and its elongation, Δl, and following expression can be written:

$$\Delta l = \frac{\Delta k}{\nu}. \tag{8}$$

Theoretically, the analysis of the evolution of contact areas, A_e and A_c which represent the contact area between two arbitrary carbon black spheres incorporated in rubber matrix, coaxially aligned, perpendicular to the direction of strain and to direction of contraction, respectively. Taking into consideration that Δl^i is the elongation between two particles forming couple i, and Δk^j is the contraction between particles forming couple j (see Figure 8a),

$$A_c = \pi \frac{R}{2} \times \left(x + \Delta l^i \right) \text{ and} \tag{9}$$

$$A_e = \pi \frac{R}{2} \times \left(x - \Delta k^j \right). \tag{10}$$

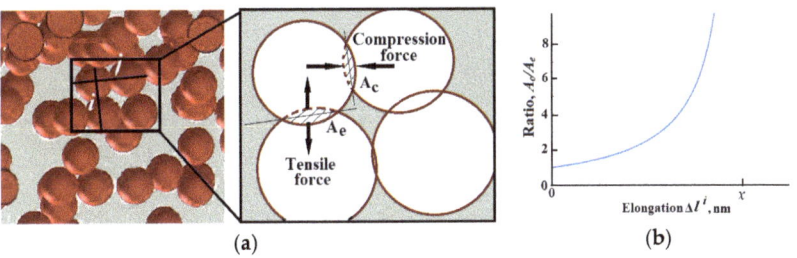

Figure 8. (a) Scheme of contact areas conductivity model; (b) Ratio between parallel and perpendicular contact area vs. interparticle displacement in tensile direction.

Thus, the ratio

$$\frac{A_c}{A_e} = \frac{x + \Delta l^i}{x - \Delta k^j} = \frac{x + \Delta l^i}{x - \nu \cdot \Delta l^i} \tag{11}$$

The solutions to this equation are depicted in Figure 8b.

Considering the framework of the present model, the relative motion of the CB particles in accordance to the Poisson's ratio of a rubber composite, the applied strain resulting in interparticle displacements Δl^i will generate, as well, an interparticle contraction Δk^j, proportional to ν. The aforementioned solution to Equation (11) clearly delimitates two regions: $\Delta l^i < x$ where the contact area of particles aligned in the direction of strain decrease much slower than the increase in the contact area of particles aligned in contraction direction, in effect, generating a global increase in interparticle contact area with possibility for easier carriers' flow (visualization sketch in Figure 8a). With further increase in strain, the increase of interparticle displacement becomes larger than the penetration depth ($\Delta l^i \geq x$), signaling the moment when the particles lose their contact and the current propagation change its mechanism, switching from Ohmic to Shottki or trap-assisted tunneling mechanisms. Since the EER is proportional to contact area, it was assumed that this phenomenon was the reason behind the sudden increase in the conductivity followed by a moderate decrease during the deformation of the sample. A reasonable question— "Why is this effect not observed for samples containing 50 or 60 phr CB?"—could be answered by assuming the initially predominant charge transfer mechanism to be the tunneling one.

3.3. The Deformed Samples EER Estimation

The aim of this paragraph was to check the possibility of EER calculation of non-homogeneously deformed samples by simulation, which will help to solve and thus replace a vital problem for otherwise finding the electrical properties of a product by cost and time demanding direct experimental testing. DIC converted the macroscopic domains of deformed samples into the colored maps, symbolizing the different magnitude of endured strain and consequently the difference in resistivity. As a base for present EER estimations it served the studies dedicated to effective resistance of homogeneously distributed two or three phase containing materials, these cases presenting with big approximations the conditions created in deformed materials. The main concept was founded on discrete networks of resistors, possessing one mutual node in each discrete domain and connecting the middle of the neighbor domain border [7]. Thus, the effective resistance calculation of deformed samples was reduced to calculation of equivalent circuit resistance (Figure 9). Taking into consideration the Kirchhoff's rules, and principles of symmetries, the contribution of maximally strained domains (yellow and red colored) was neglected assuming that current will not flow through these domains.

For simplification, the resistivity of each domain was considered homogenous and isotropic (contrary to the findings in the previous paragraph). Thus, the domain resistance was divided into equal resistances connecting the node of the domain with neighboring domains. Generally, the resistance of one rectangular domain may be calculated applying the formula:

$$R = \rho \frac{l}{d \cdot w}. \tag{12}$$

where R is the total resistance of the analyzed domain; ρ is its resistivity; l, d and w are the length, thickness and width respectively.

For the non-rectangular (arbitrary) shapes of samples (a usual characteristic of deformed samples), Equation (12) can be rewritten as:

$$R = \frac{\rho}{d} \sum_{i=1}^{\frac{l}{\Delta l}} \frac{\Delta l_i}{\Delta k_i} \tag{13}$$

where Δl_i is an arbitrarily chosen length of cell for an imaginary sample partition, i is the position of the cell along the tensile direction, d is the thickness of the sample considered in the present work as a constant, and Δw_i is the width of the sample in position i (see Figure 10). For successful calculation, it remained to find the resistivity of domains in accordance to experienced local strain depicted in different colors on the DIC map. It should be noted that only the reference samples (type a) exhibited a homogeneous strain during the whole testing protocol, giving reason to consider the resistivity of the deformed samples as well homogeneous.

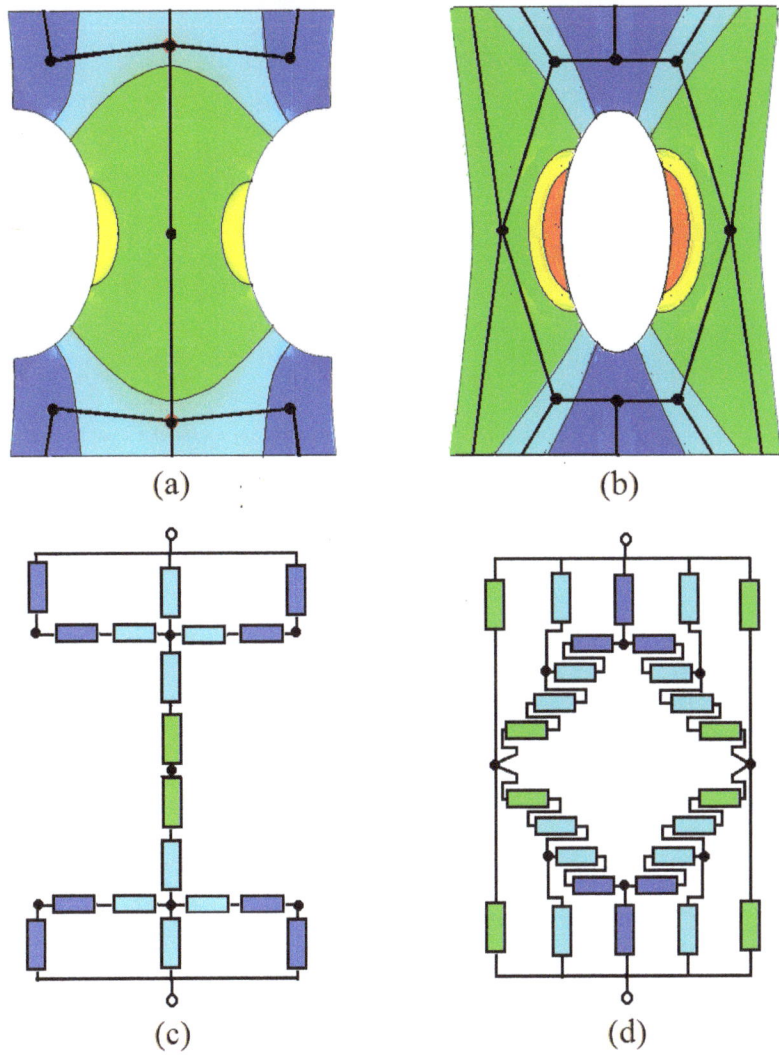

Figure 9. (**a**) Sketch defining the resistive domains with corresponding nodes for samples type b and (**b**) for samples type c, (**c**) The equivalent circuit for samples type b and (**d**) equivalent circuit for samples type c.

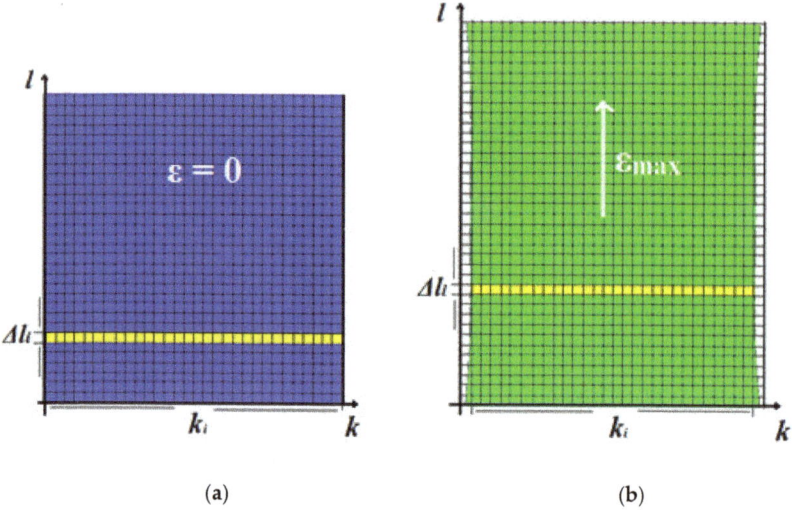

Figure 10. Illustration for resistance calculation for samples of arbitrary shape: (**a**) undeformed and (**b**) deformed state.

An approximate resistivity of domains (color delimited) was determined by extrapolation of fitted curve of referential samples EER (type a from Figure 6b,d,e) and subsequent conversion into resistivity. In the case of samples loaded with 70 phr of CB, the part of the curve containing the switching effect was omitted for the fitting process. The results of extrapolation are presented in Figure 10.

As can be observed, the sample containing 70 phr CB exhibited a reverse trend compared with the samples with lower filler content.

Finding the resistivity of each domain from Figure 11 according to color, dictated by local strain (shown in Figure 9) and applying Equation (13), resistance of the separate domains—the constituent parts of deformed at maximum strain samples types b and c—was calculated (see Tables 2 and 3).

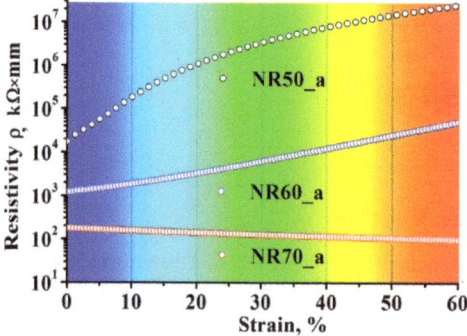

Figure 11. The resistivity of reference samples containing 50 phr CB, 60 phr CB and 70 phr CB vs. the sample's strain and corresponding DIC domain's color.

Table 2. The calculated resistivity and resistance of separate domains (samples type b).

		Shape 1	Shape 2	Shape 3	Shape 4
50 phr	ρ, Ωm	6.80×10^4	8.10×10^5	3.60×10^6	9.20×10^6
	R, Ω	9.50×10^7	3.10×10^8	3.10×10^9	3.10×10^{10}
60 phr	ρ, Ωm	1.40×10^3	2.80×10^3	6.00×10^3	1.40×10^4
	R, Ω	1.90×10^6	1.00×10^6	5.10×10^6	4.60×10^7
70 phr	ρ, Ωm	1.70×10^2	1.40×10^2	1.20×10^2	1.10×10^2
	R, Ω	2.40×10^2	5.30×10^1	1.10×10^2	3.60×10^2

Table 3. The calculated resistivity and resistance of separate domains (samples type c).

		Shape 1	Shape 2	Shape 3	Shape 4	Shape 5
50 phr	ρ, Ωm	6.80×10^4	8.10×10^5	3.60×10^6	9.20×10^6	2.10×10^7
	R, Ω	3.40×10^7	1.80×10^9	6.30×10^9	5.20×10^{10}	4.90×10^{10}
60 phr	ρ, Ωm	1.40×10^3	2.80×10^3	6.00×10^3	1.40×10^4	3.90×10^4
	R, Ω	7.00×10^5	6.10×10^6	1.10×10^7	7.90×10^7	9.30×10^7
70 phr	ρ, Ωm	1.70×10^2	1.40×10^2	1.20×10^2	1.10×10^2	9.80×10^1
	R, Ω	8.80×10^4	3.10×10^5	2.10×10^5	6.30×10^5	2.30×10^5

Finally, the calculated resistance of each domain was evenly divided to the corresponding number of imaginary resistors to fulfill the equivalent circuit (Figure 9), where the colors of the resistors denote their belonging to respective resistive domains. The rule of resistive domain partition into resistors was described earlier, in the introductory part. Thus, the complicated task to calculate the EER of non-homogeneously deformed sample was reduced to calculation of total resistance of an ordinary resistors network.

The results of calculated EER compared with the measured ones for sample types b and c determined at maximal applied strain, are presented in Figure 12. It is clearly visible that qualitatively the calculated data follows the identical trends observed during experimental investigation, for all sample types and CB loading.

Taking into consideration the accepted approximations the method of EER of non-homogeneously deformed samples by discrete resistors network approach, showed a reasonably good accuracy, and could be considered as a prospective one for related studies.

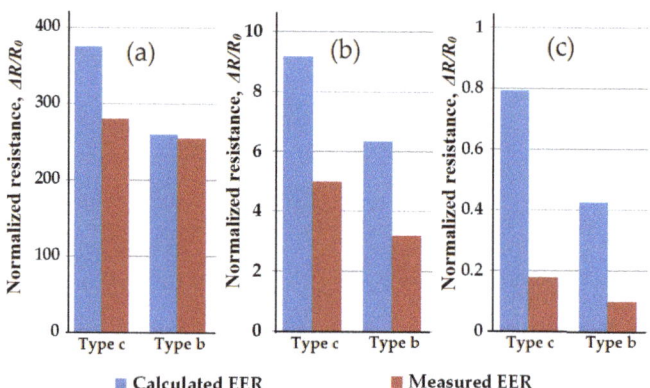

Figure 12. The relative effective electric resistance for samples type b and c experimentally measured and calculated: (**a**) samples containing 50 phr CB, (**b**) 60 phr CB and (**c**) 70 phr CB.

4. Conclusions

In the present study, the effects of filler loading at 50, 60 and 70 phr CB in NR matrix on variation in strain induced electric resistance was investigated. At such comparatively high filler loadings, a nonlinear strain-resistance variation was observed which in general is a huge impediment for stress–strain self-sensing applications. The obtained results showed a strong limitation of strain induced resistance growth pronounced with increasing filler concentration. As such, samples containing 60 phr and 70 phr of CB could be considered as non-efficient for strain gauge fabrication. However, this work was not only confined to prove this inefficiency and was further channelized to investigate some other interesting phenomena.

Thus, in continuation, a very first trial for theoretical EER calculation of non-homogeneously deformed samples was done. Initially, by fitting and extrapolation of experimental results for a rectangular reference samples (considered homogeneously strained) the strain-dependent resistivity of studied compounds was found. Using this strain-dependent resistivity and further, following the discrete resistors network approach with the simultaneous application of DIC technique to determine locally strained domains, reasonably good calculated EER results were obtained for non-homogeneously deformed samples. This finding opens a large possibility for practical application through simulation of EER for arbitrary shaped products under deformation.

Even with some differences between the magnitudes of the calculated and measured results, the trends were the same and thus, the proposed method may "feel" the geometrical inhomogeneities and their locations. For all concentrations of CB, it was found that the placement position of inhomogeneity has a tremendous impact on local strain distribution as well as on the EER. This relation could be the base for in-situ, nondestructive defect monitoring technology.

Finally, a unique current propagation mode switching phenomenon was observed and explained in a novel approach. The trustable explanation of this effect was done by analyzing the simultaneous decrease in the contact area of the conductive CB particles in the direction of strain and an increase in the perpendicular contraction direction. This effect, according to the knowledge of the authors of the present work was not previously reported.

Author Contributions: Conceptualization, E.H. and R.S.; methodology, E.H.; software, E.H.; validation, E.H., S.D. and R.S.; formal analysis, E.H.; investigation, E.H.; resources, R.S.; data curation, M.S. and J.M.; writing—original draft preparation, E.H.; writing—review and editing, R.S. and S.D.; visualization, E.H.; supervision, R.S.; project administration, R.S.; funding acquisition, R.S. All authors have read and agreed to the published version of the manuscript.

Funding: This research was funded by the Ministry of Education, Youth and Sports of the Czech Republic–DKRVO (RP/CPS/2020/004).

Institutional Review Board Statement: Not applicable.

Data Availability Statement: The data presented in this study are available on request from thecorresponding author.

Conflicts of Interest: The authors declare no conflict of interest.

Nomenclature

A_c (nm^2)	Contact area between CB particles, aligned perpendicular to the direction of sample contraction.
A_e (nm^2)	Contact area between CB particles, aligned perpendicular to the direction of sample strain
A_u (-)	Open loop voltage amplifying coefficient
d (mm)	Thickness of sample
Δk (mm)	Contraction of sample
Δk^j (nm)	The contraction between two particles forming couple j
Δl (mm)	Elongation of sample
Δl^i (nm)	The elongation between two particles forming couple i
Δl_i (mm)	Length of cell i for an imaginary sample partition
$\Delta R/R_0$ (-)	Normalized resistance
Δw_i (mm)	The width of the sample in position i
I_{OA} (A)	The input current passing through operational amplifier
I_s (A)	Current going through the sample
l (mm)	Length of sample
ν (-)	Poisson's ratio
R_s (Ω)	Resistance of the sample
R_{OA} (Ω)	Internal resistance of operational amplifier
ρ (Ωm)	Resistivity
U_s (V)	Voltage drop across the sample
U_i (V)	Inverting input voltage
U_o (V)	Output voltage
w (mm)	Width of sample
x (nm)	The penetration depth between two arbitrary CB particles

Abbreviations

AC	Alternating current
CB	Carbon black
CBS	N-cyclohexyl-2-benzothiazolesulfenamide
DIC	Digital Image Correlation
DC	Direct current
EER	Effective electric resistance
EMT	Effective medium theory
ERN	Equivalent resistor network
HAF	High abrasion furnace
NR	Natural rubber
OA	Operational amplifier
phr	Parts per hundred rubber
SIC	Strain induced crystallization
ZnO	Zinc oxide

References

1. Wang, S.L.; Wang, P.; Ding, T.H. Piezoresistivity of silicone-rubber/carbon black composites excited by AC electrical field. *J. Appl. Polym. Sci.* **2009**, *113*, 337–341. [CrossRef]
2. Natarajan, T.S.; Eshwaran, S.B.; Stöckelhuber, K.W.; Wießner, S.; Pötschke, P.; Heinrich, G.; Das, A. Strong Strain Sensing Performance of Natural Rubber Nanocomposites. *ACS Appl. Mater. Interfaces* **2017**, *9*, 4860–4872. [CrossRef] [PubMed]
3. Liu, P.; Liu, C.X.; Huang, Y.; Wang, W.H.; Fang, D.; Zhang, Y.G.; Ge, Y.J. Transfer function and working principle of a pressure/temperature sensor based on carbon black/silicone rubber composites. *J. Appl. Polym. Sci.* **2016**, *133*, 42979. [CrossRef]
4. Ciselli, P.; Lu, L.; Busfield, J.; Peijs, T. Piezoresistive polymer composites based on EPDM and MWNTs for strain sensing applications. *e-Polymers* **2010**, *10*, 1–13. [CrossRef]
5. Bakošová, D. The Study of the Distribution of Carbon Black Filler in Rubber Compounds by Measuring the Electrical Conductivity. *Manuf. Technol.* **2019**, *19*, 366–370. [CrossRef]
6. Myles, T.D.; Peracchio, A.A.; Chiu, W.K.S. Extension of anisotropic effective medium theory to account for an arbitrary number of inclusion types. *J. Appl. Phys.* **2017**, *117*, 025101. [CrossRef]
7. Söderberg, M.; Grimvall, G. Conductivity of inhomogeneous materials represented by discrete resistor networks. *J. Appl. Phys.* **1986**, *59*, 186–190. [CrossRef]
8. Niklasson, G.A.; Granqvist, C.G.; Hunderi, O. Effective medium models for the optical properties of inhomogeneous materials. *Appl. Opt.* **1981**, *20*, 26–30. [CrossRef]
9. Ruehli, A.; Antonini, A.; Jiang, L. *Circuit Oriented Electromagnetic Modeling Using the PEEC Techniques*, 1st ed.; Wiley-IEEE Press: New York City, NY, USA, 2017.
10. Robertson, C.G.; Stoček, R.; Mars, W.V. The Fatigue Threshold of Rubber and Its Characterization Using the Cutting Method. In *Fatigue Crack Growth in Rubber Materials. Advances in Polymer Science*; Heinrich, G., Kipscholl, R., Stoček, R., Eds.; Springer: Cham, Switzerland, 2020; pp. 19–38. [CrossRef]
11. Harea, E.; Datta, S.; Stěnička, M.; Stoček, R. Electrical conductivity degradation of fatigued carbon black reinforced natural rubber composites: Effects of carbon nanotubes and strain amplitudes. *Express Polym. Lett.* **2019**, *13*, 1116–1124. [CrossRef]
12. Harea, E.; Datta, S.; Stěnička, M.; Stoček, R. Undesirable Aspects of Fatigue on Stretchable Elastomer Sensors. In *Nanoscience and Nanotechnology in Security and Protection against CBRN Threats*; Petkov, P., Achour, M., Popov, C., Eds.; Springer: Dordrecht, The Netherlands, 2020; pp. 95–105. [CrossRef]
13. Harea, E.; Stoček, R.; Storozhuk, L.; Sementsov, Y.; Kartel, N. Study of tribological properties of natural rubber containing carbon nanotubes and carbon black as hybrid fillers. *Appl. Nanosci.* **2019**, *9*, 899–906. [CrossRef]
14. Bohm, G.G.A.; Nguyen, M.N. Flocculation of carbon black in filled rubber compounds. I. Flocculation occurring in unvulcanized compounds during annealing at elevated temperatures. *J. Appl. Polym. Sci.* **1995**, *55*, 1041–1050. [CrossRef]
15. Kraus, G. Reinforcement of elastomers by carbon black. *Macromol. Mater. Eng.* **1977**, *60*, 215–248. [CrossRef]
16. Stauffer, D.; Aharony, A. *Introduction to Percolation Theory*, 2nd ed.; Taylor and Francis: London, UK, 2017. [CrossRef]
17. Huang, M.; Tunnicliffe, L.B.; Zhuang, J.; Ren, W.; Yan, H.; Busfield, J.J.C. Strain dependent dielectric behavior of carbon black reinforced natural rubber. *Macromolecules* **2016**, *49*, 2339–2347. [CrossRef]
18. Huang, Y.; Schadler, L.S. Understanding the Strain-Dependent Dielectric Behavior of Carbon Black Reinforced Natural Rubber—An interfacial or bulk phenomenon? *Compos. Sci. Technol.* **2017**, *142*, 91–97. [CrossRef]
19. Gismatulin, A.A.; Kamaev, G.N.; Kruchinin, V.N.; Gritsenko, V.A.; Orlov, O.M.; Chin, A. Charge transport mechanism in the forming-free memristor based on silicon nitride. *Sci. Rep.* **2021**, *11*, 2417. [CrossRef]
20. Stoček, R.; Heinrich, G.; Gehde, M.; Rauschenbach, A. Investigations about notch length in pure-shear test specimen for exact analysis of crack propagation in elastomers. *J. Plast. Technol.* **2012**, *1*, 2–22.
21. Stoček, R.; Heinrich, G.; Gehde, M.; Kipscholl, R. Analysis of dynamic crack propagation in elastomers by simultaneous tensile- and pure-shear-mode testing. In *Fracture Mechanics and Statistical Mechanics of Reinforced Elastomeric Blends. Lecture Notes in Applied and Computational Mechanics*; Grellmann, W., Heinrich, G., Kaliske, M., Klüppel, M., Schneider, K., Vilgis, T., Eds.; Springer: Berlin, Germany, 2013; pp. 269–301. [CrossRef]
22. Stoček, R.; Stěnička, M.; Maloch, J. Determining Parametrical Functions Defining the Deformations of a Plane Strain Tensile Rubber Sample. In *Fatigue Crack Growth in Rubber Materials. Advances in Polymer Science*; Heinrich, G., Kipscholl, R., Stoček, R., Eds.; Springer: Cham, Switzerland, 2020; pp. 19–38. [CrossRef]
23. Trabelsi, S.; Albouy, P.A.; Rault, J. Effective Local Deformation in Stretched Filled Rubber. *Macromolecules* **2003**, *36*, 9093–9099. [CrossRef]
24. Biben, T.; Munch, E. Strain-Induced Crystallization of Natural Rubber and Cross-Link Densities Heterogeneities. *Macromolecules* **2014**, *47*, 5815–5824. [CrossRef]
25. Middleton, W.I.; Davis, E.W. Skin effect in large stranded conductors at low frequencies. *J. Am. Inst. Elect. Eng.* **1921**, *40*, 757–763. [CrossRef]
26. Franco, S. *Design with Operational Amplifiers and Analog Integrated Circuits*, 4th ed.; McGraw-Hill: New York, NY, USA, 2015; 672p.
27. Carter, B.; Mancini, R. *Op Amps for Everyone*, 5th ed.; Elsevier: Oxford, UK, 2018.

28. Pilkey, W.D.; Pilkey, D.F. *Peterson's Stress Concentration Factors*, 3rd ed.; John Wiley & Sons Inc.: Hoboken, NJ, USA, 2008; pp. 180–184.
29. Robertson, C.G.; Ned, J.; Hardman, N.J. Nature of Carbon Black Reinforcement of Rubber: Perspective on the Original Polymer Nanocomposite. *Polymers* **2021**, *13*, 538. [CrossRef]
30. Tang, Z.; Jia, S.; Zhou, C.; Li, B. 3D Printing of Highly Sensitive and Large-Measurement-Range Flexible Pressure Sensors with a Positive Piezoresistive Effect. *ACS Appl. Mater. Interfaces* **2020**, *12*, 28669–28680. [CrossRef] [PubMed]

Article

Characterization of Viscoelastic Poisson's Ratio of Engineering Elastomers via DIC-Based Creep Testing

Jonathan A. Sotomayor-del-Moral [1,†], Juan B. Pascual-Francisco [1,†], Orlando Susarrey-Huerta [2], Cesar D. Resendiz-Calderon [3], Ezequiel A. Gallardo-Hernández [2] and Leonardo I. Farfan-Cabrera [3,*]

[1] Departamento de Mecatrónica, Universidad Politécnica de Pachuca, Carretera Pachuca-Cd. Sahagún Km. 20, Ex-Hacienda de Santa Barbara, Zempoala 43830, HGO, Mexico; allan16@micorreo.upp.edu.mx (J.A.S.-d.-M.); jbpascualf@hotmail.com (J.B.P.-F.)

[2] SEPI-Escuela Superior de Ingeniería Mecánica y Eléctrica, Instituto Politécnico Nacional, Unidad Zacatenco, Col. Lindavista, Ciudad de México 07738, CDMX, Mexico; osusarrey@ipn.mx (O.S.-H.); egallardo@ipn.mx (E.A.G.-H.)

[3] Escuela de Ingeniería y Ciencias, Tecnologico de Monterrey, Ave. Eugenio Garza Sada 2501, Monterrey 64849, NL, Mexico; resendiz.cesar@tec.mx

* Correspondence: farfanl@hotmail.com

† These authors contributed equally to this work.

Abstract: New data of creep and viscoelastic Poisson's ratio, $\nu(t)$, of five engineering elastomers (Ethylene Propylene-Diene Monomer, Flouroelastomer (Viton®), nitrile butadiene rubber, silicone rubber and neoprene/chloroprene rubber) at different stress (200, 400 and 600 kPa) and temperature (25, 50 and 80 °C) are presented. The $\nu(t)$ was characterized through an experimental methodological approach based on creep testing (30 min) and strain (axial and transverse) measurements by digital image correlation. Initially, creep behavior in axial and transverse directions was characterized for each elastomer and condition, and then each creep curve was fitted to a four-element creep model to obtain the corresponding functions. The obtained functions were used to estimate $\nu(t)$ for prolonged times (300 h) through a convolution equation. Overall, the characterization was achieved for the five elastomers results exhibiting $\nu(t)$ increasing with temperature and time from about 0.3 (for short-term loading) to reach and stabilize at about 0.48 (for long-term loading).

Keywords: rubber; material testing; rheology; Poisson's ratio; viscoelasticity

1. Introduction

Elastomers are viscoelastic materials widely used in engineering applications. The performance of elastomeric mechanical components not only depends on the material's mechanical properties but also on viscoelastic properties such as creep compliance, stress relaxation and viscoelastic Poisson's ratio, $\nu(t)$, which are time- and temperature-dependent properties. The viscoelasticity of a material is represented by a combination of both elasticity and viscosity properties in different proportions, which is the reason that an elastomer exhibits a variable elastic modulus dependent of time, stress and temperature. The Poisson's ratio, ν, is defined as the negative constant ratio between transverse and axial strains in a uniaxial state of stress, which is applied along the axial direction for any elastic, homogenous and isotropic material. Practically, for polymers, this property is assumed as a constant, ν, instead of a time and temperature variable, $\nu(t)$, in design and performance simulation of engineering components due to the complexity for its experimental determination [1–3]. However, the progress of computing and software technology for modern engineering design has promoted the inclusion of more complex or nonlinear properties, namely, the viscoelastic properties (stress relaxation, creep and $\nu(t)$) of soft materials for obtaining more accurate predictions of performance and service life of engineering components via simulation [1–4]. For elastomers, the data of viscoelastic properties, particularly

$\nu(t)$, are scarce in the literature. There only few works reporting on the characterization of the time-dependent Poisson's ratio of some elastomers. For example, Kuggler et al. [4] determined the $\nu(t)$ of different Hypalon-based rubbers and hydroxyl terminated polybutadiene rubber by using an optoelectronic system constant strain rate and stress relaxation tests; they found an increase in Poisson's ratio with time for all cases. Saseendran et al. [5] determined the evolution of $\nu(t)$ of the commercial LY5052 epoxy resin at different cure states under uniaxial tension subject to constant deformation stress relaxation testing. They found that Poisson's ratio evolved from 0.32 to 0.44 over time depending on the cure state of the resin. Pandini and Pegoretti [6] investigated the phenomenology of the dependence of Poisson's ratio with temperature, time and strain of two crosslinked epoxy resins with different glass transition temperatures using contact extensometers and the simultaneous measurement of the axial and transverse deformations under two dissimilar tensile and relaxation testing. They found that $\nu(t)$ increased with strain rate, temperature, and time. Cui et al. [7] proposed a fully numerical framework based on a theory of stress relaxation for the determination of time-dependent Poisson's ratio for solid propellants (elastomer composites). The time-dependent Poisson's ratio was obtained under different cohesive parameters, namely, loading conditions (loading temperature, loading rate and fixed strain) and area fraction. They found that the numerical simulation revealed that time-dependent Poisson's ratio can be nonmonotonic or monotonic depending on different cohesive parameters. In addition, all time-dependent Poisson's ratios increased at the beginning of the relaxation stage because of cohesive contact. Then, once transverse and axial strains stop changing, all time-dependent Poisson's ratios achieved equilibrium values. In a more recent research work, Cui et al. [8] proposed constitutive models relating $\nu(t)$ with a classical creep constitutive model using a Laplace transform method and compared with stress relaxation models. They found that, in analytical analyses, creep and relaxation models solutions correlated well. It can be a reference that $\nu(t)$ can be obtained from creep or stress relaxation data.

According to theory of viscoelasticity [9–11], $\nu(t)$ can be directly obtained from stress relaxation tests by measuring the transversal strain with time, $\varepsilon_x(t)$, after applying a constant axial deformation, ε_0, as expressed by Equation (1).

$$\nu(t) = -\frac{\varepsilon_x(t)}{\varepsilon_0} \quad (1)$$

In this manner, $\nu(t)$ is difficult to obtain accurately due to the minimal transverse strain produced with time during a stress relaxation test even using sophisticated measurement technology with high resolution. This is one of the reasons that published data of $\nu(t)$ of elastomers are rarely reported. An alternative method to obtain $\nu(t)$ is through creep tests followed by a converse methodological approach [9,12]. Creep tests are advantageous because they allow the generation of larger strains in both axial and transverse directions with time under a constant tensile or compressive load, which can be measured more accurately and easily. This methodological approach and its rationalization have been recently published previously elsewhere [13]. It has been demonstrated to be effective for the evaluation of $\nu(t)$ of elastomers under different stress levels and temperatures by using digital image correlation (DIC) for strain measurement.

Generally, the measurement of creep strain has been achieved by using gripping extensometers or strain gauges adhered or gripped to the material sample in standard tensile creep testing devices. Nevertheless, the application of these strain measurement gauges can penetrate the sample producing disturbances in structural homogeneity and producing stress concentrators in the material, as well as, restricting the free strain produced in the sample. It is known to introduce some errors in the collected strain data. Hence, in order to avoid errors in the strain measurement, some additional data correction techniques [6] and techniques based on non-contact optical measurement such as Moire interferometry, electronic speckle pattern interferometry (ESPI), shearography, and digital image correlation (DIC) have been applied effectively [12,14,15]. The utilization of a non-contact strain measurement technique such as DIC allows accurate creep strain determination in elastomers without interfering with the sample deformation during the creep strain state. Other alternatives have been proposed and used by several research groups for evaluating the viscoelastic behavior of soft materials such as elastomers, particularly creep. For instance, the standard methods include ASTM-D2990 and ISO 899–1:2003, some non-standard tensile test methods [16], dynamic-mechanical analysis (DMA) [17], some methodologies based on nanoindentation [18,19] and micro- and macro-indentation with axi-symmetric indenters [20–23]. Nonetheless, DIC-based creep testing, in particular, has been demonstrated as a very suitable and accurate tool for mechanical and viscoelasticity characterization of a wide range of materials, including elastomers [24]. Moreover, DIC has been employed for purposes in elastomers. For example, it has been used for studying fatigue crack growth behavior of elastomers, in which plane strain tensile samples (thin and rectangular strips), also named as pure shear samples, are preferred for this testing [25–27].

DIC is a non-invasive optical full-field measurement technique based on the comparison of digital images of an image in different stages of change/deformation. For the comparison of the images, it recognizes patterns with different light intensity of an area. Usually, the light intensity pattern is represented by small and well-defined contrasting points detected in the images taken and processed. The points identified in the undeformed image is recognized by contrasting with the light intensity pattern from the surrounding area. Depending on the light intensity of each point, the points with identical light intensity are identified in the deformed image. About 256 levels of grayscale are used for the digitization of the light intensity in black and white images. Using a single camera-based DIC system is sufficient to measure in-plane (two directions) deformations simultaneously, which is sufficient to determine creep and $\nu(t)$.

Thus, the aim of this paper is to obtain and provide new data from an extensive novel characterization of the $\nu(t)$ of various common engineering elastomers under different tensile loads and temperatures through creep tests and using a single camera-based DIC for obtaining accurate axial and transverse strain measurements in accordance with the methodology reported in [13], which is described in the following section for purpose of this research.

2. Materials and Methods

The $\nu(t)$ of five commercial elastomers (Ethylene-Propylene-Diene Monomer (EPDM), Flouroelastomer (Viton®) (FKM), nitrile butadiene rubber (NBR), silicone rubber (VMQ) and neoprene/chloroprene rubber (CR)), which are contemporarily used in a wide range of engineering applications (static and dynamic seals, belts, support inserts, vibration insulators, etc.), was determined by a methodological approach using tensile creep tests and strain measurement by DIC. Overall, it comprises the methodological steps shown in Figure 1: (1) measurement of creep strains (generation of the strain map by DIC) in transverse, $\varepsilon_x(t)$, and axial, $\varepsilon_y(t)$, directions of an elastomeric sample during a determined creep test period at constant temperature; (2) determination of the creep strain functions in both directions by fitting the obtained data to a known viscoelastic model; (3) estimation of the $\nu(t)$ using a numerical solution of a convolution equation for each material and condition.

Both the axial and transverse creep strains of rectangle-shaped samples (60 mm × 5 mm and 2 mm thickness) cut from black sheets of each elastomer were obtained simultaneously by using a DIC equipment (Q-450: Dantec Dynamics, Skovlunde, Denmark) instrumented in a home-built creep test set-up, as shown in Figure 1. Commonly, carbon black is added to the elastomers during their manufacturing process to enhance their mechanical properties [28] and provide black pigmentation to elastomers, which is the case of tested elastomers. It is noteworthy that the strain measurement with DIC can be also applied successfully in the study of materials pigmented with other colors, or even colorless, as long as the required speckle pattern be achieved.

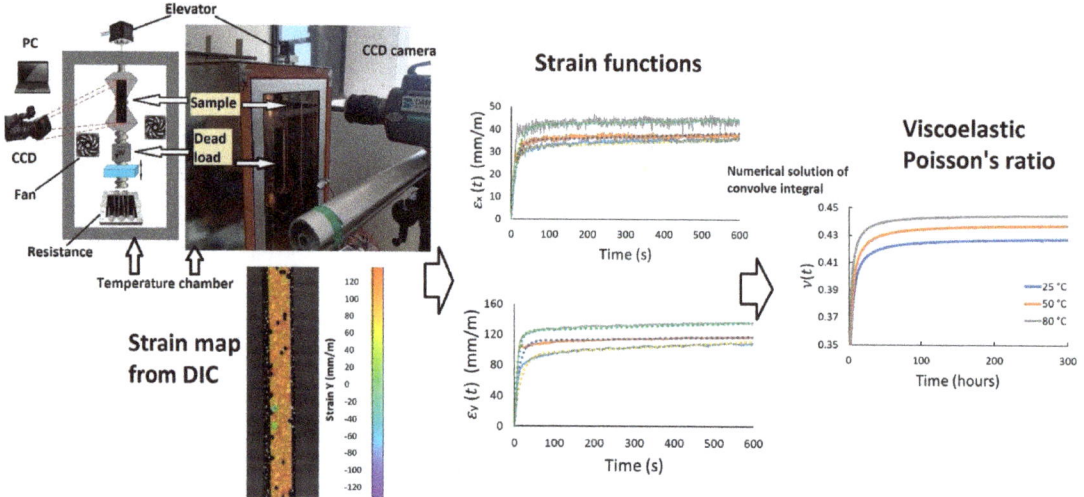

Figure 1. Experimental setup and methodological steps for determining viscoelastic Poisson's ratio of elastomers.

The mechanical properties of the elastomers and parameters of the creep test and DIC measurement are shown in Table 1. The tensile tests were conducted according to the method specified in ASTM-D412 using dumbbell shape specimens with a gauge length of 30 mm at a strain rate of 50 mm/min by using a tensile tester (UE22XX Digital Electronic Tensile Testing Machine, Laryee, Beijing, China) with a load cell of 1 kN. The hardness measurements were performed according to the method in ASTMD2240 in square specimens with dimension of 20 mm using a Shore A type digital durometer (DD-100 Digital Shore Durometer Tester: ABQ Industrial, The Woodlands, TX, USA). The surface roughness of each material was determined in an optical profilometer (Contour GT-K: Bruker, Billerica, MA, USA) with an objective of 5X. The mean roughness of area (Sa) was found to be in the range of 0.2–0.47 μm for all elastomers. The DIC parameters selected have been useful and effective for the creep characterization of elastomers, as reported in a previous research work [24]. The tested samples were finely speckled with white paint for enabling the suitable detection for DIC. The creep tests were run under different proportional tensile stresses and temperatures. They consisted of hanging a rectangular specimen on a frame inside a thermal chamber with temperature control and then applying the predefined tensile load through a dead weight lever system. Once the sample is heated up and maintained at a predefined temperature, the tensile load is applied while DIC measurements are run simultaneously. The sample temperature is measured and monitored by three infrared sensors positioned over different regions of the sample to confirm a quasi-homogenous temperature inside the chamber. To measure $\varepsilon_x(t)$ and $\varepsilon_y(t)$ produced in the sample, a CCD camera (SpeedSense 9070: Phantom, Wayne, NJ, USA) possessing a Zeiss Makro-Planar 50 mm f/2 ZF.2 lens and an image resolution of 1280 × 800 pixels was employed. The camera was connected to a computer loaded with software Istra 4D (Istra 4D: Dantec Dynamics, Skovlunde, Denmark) for the camera configuration, data collection, image processing, and strain computing. The Istra 4D software integrates an algorithm for the numerical calculation of displacements and strain in the x- and y-directions. This calculation is based on the tracking of every point of the image of the speckled sample at any time. Thus, the "true" strain is directly calculated by the software.

Table 1. Material properties and parameters of creep test and DIC measurement.

Material/Test	Property/Parameter	Value
Ethylene-Propylene-Diene Monomer (EPDM) (Manufacturer: Rodillos BMR®, Guadalajara, Jalisco, México)	Hardness, ASTM-D2240 (Shore A)	68.5 ± 2
	Tensile breaking strength, ASTM-D412 (MPa)	14 ± 0.6
Flouroelastomer, Viton® (FKM) (Manufacturer: Rodillos BMR®, Guadalajara, Jalisco, México)	Hardness, ASTM-D2240 (Shore A)	77.5 ± 2
	Tensile breaking strength, ASTM-D412 (MPa)	11 ± 0.7
Nitrile Butadiene Rubber (NBR) (Manufacturer: Rodillos BMR®, Guadalajara, Jalisco, México)	Hardness, ASTM-D2240 (Shore A)	73 ± 2
	Tensile breaking strength, ASTM-D412 (MPa)	6.9 ± 0.5
Silicone rubber/Vinyl-Methyl silicone (VMQ) (Manufacturer: Rodillos BMR®, Guadalajara, Jalisco, México)	Hardness, ASTM-D2240 (Shore A)	47.5 ± 1.5
	Tensile breaking strength, ASTM-D412 (MPa)	5 ± 0.8
Neoprene/Chloroprene Rubber (CR) (Manufacturer: Rodillos BMR®, Guadalajara, Jalisco, México)	Hardness, ASTM-D2240 (Shore A)	69 ± 2
	Tensile breaking strength, ASTM-D412 (MPa)	3.5 ± 0.5
Strain measurement/DIC parameters	Subset (pixels)	17
	Step (pixels)	3
	Field of view (mm × mm)	55 × 36
	Measurement points (points)	425
	Temporal resolution (fps)	1
	Camera distance (mm)	200
	Image resolution (pixels × pixels)	1280 × 800
	Spatial resolution (mm)	0.1
	Strain resolution (mm/m)	0.25
	Frame amount	1800
	Measurement time (minutes)	30
Creep test	Tensile load (N)	2, 4, 6
	Stress (kPa)	200, 400, 600
	Temperature (°C)	25 ± 1, 50 ± 2 and 80 ± 2
	Test time (minutes)	30

The tests were run in triplicate (using new specimens without load history), each for 30 min, for all elastomers and conditions. The time selected was enough to generate the first and second creep stages consistently in all elastomers, which is required to estimate $\nu(t)$ [9,12].

Once the creep data ($\varepsilon_x(t)$ and $\varepsilon_y(t)$) were obtained by DIC, the mean creep strain functions ($\varepsilon_x(t)$ and $\varepsilon_y(t)$) were determined from the three repeats. Afterwards, both mean creep functions were used to predict the $\nu(t)$ functions for each material and condition by Equation (2), which is a convolution integral equation based on a secure analytical foundation of the viscoelasticity theory [9].

$$\int_0^t \frac{\nu(t-u)}{du}\varepsilon_y(u)du = \varepsilon_x(t) + \nu_g \varepsilon_y(t) \qquad (2)$$

ν_g is the glassy (instantaneous) Poisson's ratio. Due to there not being an available analytical solution for Equation (1), $\nu(t)$ should be evaluated for any time (t_n) using the next recurrence formula [1,12]:

$$\nu(t_n) = \frac{-2\varepsilon_x(t_n) + \nu(t_{n-1})\bigl(\varepsilon_y(t_0) - \varepsilon_y(t_n - t_{n-1})\bigr) + X(t_n)}{\varepsilon_y(t_n) + \varepsilon_y(t_n - t_{n-1})} \qquad (3)$$

where

$$X(t_n) = \sum_{i=1}^{i=n-1}(\nu(t_i) + \nu(t_{i-1}))\bigl(\varepsilon_y(t_n - t_i) - \varepsilon_y(t_n - t_{i-1})\bigr) \qquad (4)$$

with

$$v(t_0) = v_g = -\frac{\varepsilon_x(t_0)}{\varepsilon_y(t_0)} \qquad (5)$$

and the following is the case.

$$v(t_1) = -\frac{-2\varepsilon_x(t_1) + v_g(\varepsilon_y(t_1) - \varepsilon_y(t_0))}{\varepsilon_y(t_1) + \varepsilon_y(t_0)} \qquad (6)$$

Equation (3) should be evaluated for $t_n \geq 2$, where t_n is the time to be evaluated, t_{n-1} is the immediate previous time and t_0 stands for $t = 0$. In this manner, $v(t)$ can be estimated and predicted for the time required.

3. Results and Discussions

The data obtained from the three repeats were considered for the creep characterization of all materials and conditions. As an example of the creep raw data results obtained by DIC, the dispersion of the results of transverse and axial creep from the three repeats obtained for FKM at 600 kPa and different temperatures is presented in Figures 2 and 3, respectively. The maximum and minimum creep values, as well as the corresponding average behavior of creep strain of both transverse and axial creep data, are plotted. Considering the dispersion of the three repeats (gray shaded area), an average curve was generated for each case. It was observed that three tests were sufficient to obtain significant repeatability. The standard deviations (% error) obtained from the three repeats were in the range of 1.5–5.1 and 2.2–6.5 mm/m for axial and transverse creep strain, respectively. For most of the materials, it was observed that the higher standard deviations correspond to the measurements of creep in transverse direction (x). This is ascribed to the magnitude of the transverse strains generated; they are very small in contrast to those obtained in the axial direction. The DIC technique is known to be less effective for small strains. When the strain is lower, accuracy and effectiveness become lower. The range of measurement of the DIC system is from 100 micro-strains up to several 100% strain. In addition, other sources causing error in the measurement are rigid body motion when load is applied to the specimen, material structural defects, the non-uniformity in the geometry of the specimens, and the speckle pattern painted over the sample. Rigid body motion is produced when load is applied to the specimen. The electromechanical lever generates small vibrations, especially in the transverse direction, which generates slight oscillations to the specimen. Defects or inconsistencies in the homogeneity, continuity, and properties of the material promote different creep behavior, generating a wider creep strain dispersion. The specimens' preparation, particularly the cutting of the samples, can produce some irregularities such as non-uniform geometry. Variations in the geometry, in particular the cross-section of the sample, causes higher or less stress to the sample and, therefore, higher or lower strains varying the creep results. On the other hand, in a small extent, another source of the error in the results is the variation of the light intensity captured by the CCD camera together with the non-uniformity of the speckle pattern generated in the surface of the specimens. The ideal speckle pattern for DIC should be composed of many as possible points with similar geometry and light intensity and separated by a well contrasting area in order to an accurate identification of the light patterns by DIC. However, it is very difficult to achieve this speckle pattern homogeneity. Finally, despite the data dispersion obtained by the creep tests and repeats, a clear trend of the creep curves was observed for all materials and test conditions.

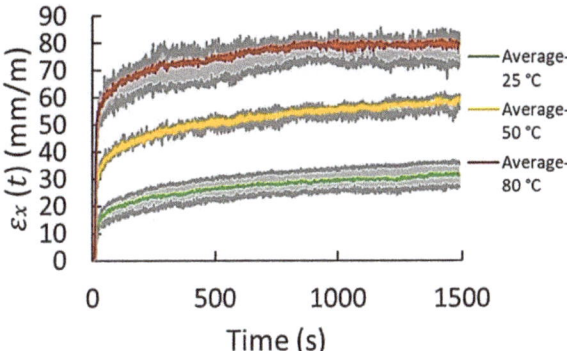

Figure 2. Dispersion of transverse creep data ("x" direction) obtained from three test repeats for FKM at 600 kPa and different temperatures.

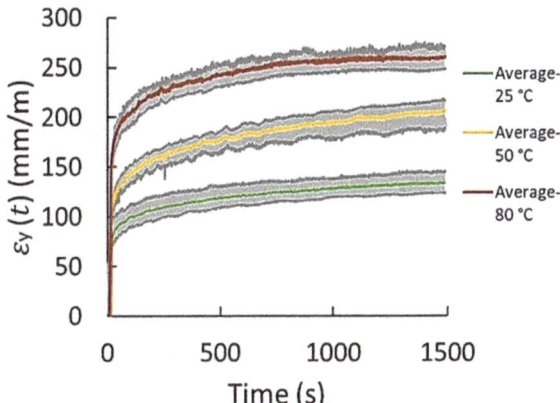

Figure 3. Dispersion of axial creep data ("y" direction) obtained from three test repeats for FKM at 600 kPa and different temperatures.

For all cases, including all the repeatability tests, the first and second stages of creep were clearly generated in both the axial, ε_x, and transverse, ε_y, directions. The first stage is characterized by the instantaneous (elastic) strain and an abrupt strain rate decrease while the second creep stage is recognized by a behavior approaching a nearly constant strain rate. Thus, according to the creep strain behavior obtained in both axial and transverse directions for all elastomers and conditions, it was found that all ε_x and ε_y average curves correlated well with a four-element creep model. The model involves the connection in series of the Maxwell and Kelvin–Voigt models [29], as expressed by the following:

$$\varepsilon(t) = \frac{\sigma_0}{R_1} + \frac{\sigma_0}{\eta_1}t + \frac{\sigma_0}{R_2}\left(1 - e^{-\frac{R_2 t}{\eta_2}}\right) \quad (7)$$

where σ_0 is the imposed constant stress, R_1 and R_2 are the elastic constants in the Maxwell and Kelvin–Voigt models, respectively, and η_1 and η_2 are the viscous constants in each model. Hence, the data of both mean transverse and axial creep strains were fitted to the model in Equation (7), obtaining the corresponding creep strain function and elastic and viscous constants for each elastomer and condition tested. The average creep functions and the errors (standard deviations) of the fitted models obtained from the three test repetitions conducted for all materials and conditions are summarized in Table 2.

Table 2. Axial and transverse creep strain functions for the elastomers and conditions tested.

Material	Stress (kPa)	Temperature (°C)	Axial Creep Strain Function, $\varepsilon_y(t)$	Error (%)	Transverse Creep Strain Function, $\varepsilon_x(t)$	Error (%)
EPDM	200	25	$\varepsilon_y(t) = 20 + 0.0078t + 35(1 - e^{-0.0492\,t})$	3.7	$\varepsilon_x(t) = 5 + 0.0035\,t + 10(1 - e^{-0.0835t})$	9.5
		50	$\varepsilon_y(t) = 30 + 0.0111\,t + 30(1 - e^{-0.0912\,t})$	5.5	$\varepsilon_x(t) = 5 + 0.0051\,t + 13(1 - e^{-0.0912\,t})$	4.4
		80	$\varepsilon_y(t) = 40 + 0.01754t + 85(1 - e^{-0.0321\,t})$	1.8	$\varepsilon_x(t) = 10 + 0.008\,t + 25(1 - e^{-0.0227\,t})$	3.1
	400	25	$\varepsilon_y(t) = 35 + 0.0069\,t + 75(1 - e^{-0.023\,t})$	1.6	$\varepsilon_x(t) = 10 + 0.0031\,t + 28(1 - e^{-0.0298\,t})$	3.5
		50	$\varepsilon_y(t) = 100 + 0.0061\,t + 57(1 - e^{-0.0121\,t})$	2.0	$\varepsilon_x(t) = 35 + 0.0028\,t + 19(1 - e^{-0.0244\,t})$	2.7
		80	$\varepsilon_y(t) = 120 + 0.0226\,t + 120(1 - e^{-0.0097\,t})$	3.6	$\varepsilon_x(t) = 40 + 0.0104t + 30(1 - e^{-0.0117\,t})$	3.6
	600	25	$\varepsilon_y(t) = 150 + 0.0061\,t + 70(1 - e^{-0.0051\,t})$	3.0	$\varepsilon_x(t) = 40 + 0.0026\,t + 30(1 - e^{-0.0088\,t})$	2.0
		50	$\varepsilon_y(t) = 170 + 0.0139\,t + 90(1 - e^{-0.0062\,t})$	4.0	$\varepsilon_x(t) = 60 + 0.0061\,t + 23(1 - e^{-0.0074\,t})$	3.3
		80	$\varepsilon_y(t) = 200 + 0.0611\,t + 350(1 - e^{-0.0047\,t})$	4.8	$\varepsilon_x(t) = 80 + 0.0104\,t + 30(1 - e^{-0.0055t})$	2.5
CR	200	25	$\varepsilon_y(t) = 30 + 0.0061\,t + 18(1 - e^{-0.0317\,t})$	3.1	$\varepsilon_x(t) = 8 + 0.0026\,t + 4(1 - e^{-0.0212\,t})$	6.3
		50	$\varepsilon_y(t) = 40 + 0.0099\,t + 32(1 - e^{-0.0215\,t})$	1.4	$\varepsilon_x(t) = 10 + 0.0044\,t + 7(1 - e^{-0.0514\,t})$	3.4
		80	$\varepsilon_y(t) = 70 + 0.0192\,t + 40(1 - e^{-0.0204\,t})$	2.4	$\varepsilon_x(t) = 15 + 0.0085\,t + 14(1 - e^{-0.0119\,t})$	4.1
	400	25	$\varepsilon_y(t) = 80 + 0.0069\,t + 20(1 - e^{-0.0084\,t})$	1.9	$\varepsilon_x(t) = 10 + 0.0031\,t + 28(1 - e^{-0.0298\,t})$	4.8
		50	$\varepsilon_y(t) = 100 + 0.0113\,t + 54(1 - e^{-0.0127\,t})$	3.0	$\varepsilon_x(t) = 33 + 0.0052\,t + 12(1 - e^{-0.0218\,t})$	3.8
		80	$\varepsilon_y(t) = 140 + 0.0226\,t + 100(1 - e^{-0.0055\,t})$	2.3	$\varepsilon_x(t) = 35 + 0.0106\,t + 35(1 - e^{-0.01\,t})$	3.2
	600	25	$\varepsilon_y(t) = 100 + 0.0069\,t + 56(1 - e^{-0.0101\,t})$	2.3	$\varepsilon_x(t) = 35 + 0.0031\,t + 18(1 - e^{-0.0147\,t})$	4.6
		50	$\varepsilon_y(t) = 170 + 0.0174\,t + 150(1 - e^{-0.0029\,t})$	2.2	$\varepsilon_x(t) = 60 + 0.0073\,t + 12(1 - e^{-0.0066\,t})$	4.5
		80	$\varepsilon_y(t) = 400 + 0.0874\,t + 350(1 - e^{-0.0031\,t})$	3.1	$\varepsilon_x(t) = 100 + 0.0104\,t + 50(1 - e^{-0.0113\,t})$	4.2
NBR	200	25	$\varepsilon_y(t) = 10 + 0.0036\,t + 18(1 - e^{-0.0660\,t})$	2.0	$\varepsilon_x(t) = 2 + 0.0015\,t + 7(1 - e^{-0.0822\,t})$	11
		50	$\varepsilon_y(t) = 25 + 0.0038\,t + 22(1 - e^{-0.0379\,t})$	3.0	$\varepsilon_x(t) = 8 + 0.0017\,t + 3(1 - e^{-0.1207\,t})$	11
		80	$\varepsilon_y(t) = 30 + 0.0052\,t + 22(1 - e^{-0.0379\,t})$	1.6	$\varepsilon_x(t) = 10 + 0.0024\,t + 1(1 - e^{-0.3615\,t})$	4.6
	400	25	$\varepsilon_y(t) = 60 + 0.0026\,t + 15(1 - e^{-0.024\,t})$	1.0	$\varepsilon_x(t) = 10 + 0.0012\,t + 9(1 - e^{-0.064\,t})$	3.1
		50	$\varepsilon_y(t) = 70 + 0.0066t + 23(1 - e^{-0.0301\,t})$	3.0	$\varepsilon_x(t) = 20 + 0.0031\,t + 3(1 - e^{-0.0572\,t})$	5.5
		80	$\varepsilon_y(t) = 90 + 0.0071\,t + 33(1 - e^{-0.0172\,t})$	2.3	$\varepsilon_x(t) = 20 + 0.0034\,t + 10(1 - e^{-0.036\,t})$	3.6
	600	25	$\varepsilon_y(t) = 100 + 0.0043\,t + 33(1 - e^{-0.0173\,t})$	2.0	$\varepsilon_x(t) = 30 + 0.0019\,t + 12(1 - e^{-0.0145\,t})$	2.4
		50	$\varepsilon_y(t) = 120 + 0.0061\,t + 55(1 - e^{-0.0103\,t})$	2.7	$\varepsilon_x(t) = 40 + 0.0027\,t + 12(1 - e^{-0.007\,t})$	2.6
		80	$\varepsilon_y(t) = 150 + 0.0157\,t + 80(1 - e^{-0.0056\,t})$	2.0	$\varepsilon_x(t) = 50 + 0.0073\,t + 17(1 - e^{-0.0099\,t})$	3.9
VMQ	200	25	$\varepsilon_y(t) = 60 + 0.0015\,t + 20(1 - e^{-0.0232\,t})$	3.2	$\varepsilon_x(t) = 10 + 0.0006\,t + 25(1 - e^{-0.023\,t})$	4.2
		50	$\varepsilon_y(t) = 80 + 0.0034\,t + 8(1 - e^{-0.0578\,t})$	1.9	$\varepsilon_x(t) = 20 + 0.0015t + 12(1 - e^{-0.0479\,t})$	3.6
		80	$\varepsilon_y(t) = 80 + 0.0036\,t + 8(1 - e^{-0.1044\,t})$	1.3	$\varepsilon_x(t) = 20 + 0.0017\,t + 5(1 - e^{-0.0724\,t})$	5.5
	400	25	$\varepsilon_y(t) = 140 + 0.0026\,t + 28(1 - e^{-0.0129t})$	2.6	$\varepsilon_x(t) = 40 + 0.0012\,t + 38(1 - e^{-0.0151\,t})$	3.3
		50	$\varepsilon_y(t) = 180 + 0.0036\,t + 45(1 - e^{-0.0154\,t})$	1.7	$\varepsilon_x(t) = 60 + 0.0017\,t + 17(1 - e^{-0.0102\,t})$	2.4
		80	$\varepsilon_y(t) = 200 + 0.0054\,t + 45(1 - e^{-0.0127t})$	1.5	$\varepsilon_x(t) = 70 + 0.0026\,t + 5(1 - e^{-0.0722\,t})$	2.0
	600	25	$\varepsilon_y(t) = 400 + 0.0069\,t + 150(1 - e^{-0.0115\,t})$	2.0	$\varepsilon_x(t) = 120 + 0.0031\,t + 30(1 - e^{-0.0278\,t})$	3.0
		50	$\varepsilon_y(t) = 400 + 0.0034\,t + 95(1 - e^{-0.0181\,t})$	1.1	$\varepsilon_x(t) = 120 + 0.0016\,t + 17(1 - e^{-0.0492\,t})$	2.3
		80	$\varepsilon_y(t) = 400 + 0.0034\,t + 30(1 - e^{-0.0154\,t})$	2.2	$\varepsilon_x(t) = 100 + 0.0017\,t + 20(1 - e^{-0.0087\,t})$	2.2
FKM	200	25	$\varepsilon_y(t) = 15 + 0.0038\,t + 18(1 - e^{-0.0256\,t})$	4.3	$\varepsilon_x(t) = 3 + 0.0015\,t + 5(1 - e^{-0.1151\,t})$	22
		50	$\varepsilon_y(t) = 40 + 0.0083\,t + 15(1 - e^{-0.0305\,t})$	2.4	$\varepsilon_x(t) = 4 + 0.0034\,t + 5(1 - e^{-0.0167\,t})$	5.5
		80	$\varepsilon_y(t) = 50 + 0.0062\,t + 28(1 - e^{-0.0203\,t})$	1.4	$\varepsilon_x(t) = 10 + 0.0026\,t + 8(1 - e^{-0.0451\,t})$	5.6

Table 2. Cont.

Material	Stress (kPa)	Temperature (°C)	Axial Creep Strain Function, $\varepsilon_y(t)$	Error (%)	Transverse Creep Strain Function, $\varepsilon_x(t)$	Error (%)
	400	25	$\varepsilon_y(t) = 40 + 0.0052\,t + 15(1 - e^{-0.0114\,t})$	2.4	$\varepsilon_x(t) = 10 + 0.002\,t + 8(1 - e^{-0.0217\,t})$	3.6
		50	$\varepsilon_y(t) = 40 + 0.0076\,t + 50(1 - e^{-0.0166\,t})$	3.8	$\varepsilon_x(t) = 10 + 0.0031\,t + 8(1 - e^{-0.0105\,t})$	3.7
		80	$\varepsilon_y(t) = 90 + 0.0075\,t + 45(1 - e^{-0.0126t})$	2.3	$\varepsilon_x(t) = 20 + 0.0031\,t + 15(1 - e^{-0.024\,t})$	2.8
	600	25	$\varepsilon_y(t) = 50 + 0.0061\,t + 55(1 - e^{-0.0103\,t})$	3.2	$\varepsilon_x(t) = 10 + 0.0024\,t + 15(1 - e^{-0.0115\,t})$	3.9
		50	$\varepsilon_y(t) = 80 + 0.0095\,t + 80(1 - e^{-0.007\,t})$	2.7	$\varepsilon_x(t) = 15 + 0.004\,t + 28(1 - e^{-0.0094\,t})$	2.9
		80	$\varepsilon_y(t) = 150 + 0.0069\,t + 60(1 - e^{-0.0059\,t})$	3.3	$\varepsilon_x(t) = 40 + 0.0029\,t + 27(1 - e^{-0.0064\,t})$	3.1

Using the mean axial and transverse creep functions, $\nu(t)$ was determined for 300 h for the different elastomers and test conditions. 300 h were selected as a considerable long-term use period. However, using the same axial and transverse creep function reported in Table 2 and Equations (2)–(6), longer predictions of $\nu(t)$ can be estimated. The $\nu(t)$ results for each tested elastomer at the different stress and temperature for 300 h are shown in Figures 4–8. t_0 was assumed to occur at 1–2 s during stress application since the frame rate employed was 1 fps. Hence, the first strains (used to obtain glassy value) detected by DIC were obtained by the correlation of the two first frames taken. In all cases, except EPDM and CR at the highest temperature (80 °C) and stress (600 kPa), $\nu(t)$ increased with time and temperature reaching a stable behavior. The increasing $\nu(t)$ with temperature has been demonstrated also for other elastomers, e.g., hydroxyl-terminated poly-butadiene (HTPB) propellant [30]. In the cases of EPDM and CR at the highest temperature and stress, $\nu(t)$ decreases with time. It is because these materials are not resistant to those temperatures. They exhibit very high creep rates reaching the third creep stage at high temperatures promoting very large axial strains and minimal transverse strains, which reduces $\nu(t)$, especially at high stress. The increase in $\nu(t)$ of the elastomers with temperature is associated to the approach of a liquid-like behavior with increasing temperature, which tends to reach the Poisson's ratio of an incompressible material ($\nu \approx 0.5$) [31]. Stress had also influence on $\nu(t)$ for all the elastomers; however, it did not exhibit a clear trend. Overall, $\nu(t)$ of all the elastomers increased from about 0.3 to 0.48 with time; the last being near to the constant values frequently used for characterizing elastomers (≈ 0.45–0.5). Thus, it can be stated that $\nu(t)$ is low (about 0.3) at short-term loading, but it increases and stabilizes to about 0.48 with the long-term loading for the elastomers tested. This growth of $\nu(t)$ with time until reaching an almost stable value has been also reported by other research groups for other viscoelastic materials [4–6,30–33]. It is noteworthy that this behavior is also similar to the ν(t) behavior for stress relaxation for linear viscoelastic materials, as reported in [32,33]. Considering the foundations of linear viscoelasticity, Aili et al. [32] and Charpin and Sanahuja [33] postulated that the instantaneous and the stable ν(t) are similar for creep and stress relaxation.

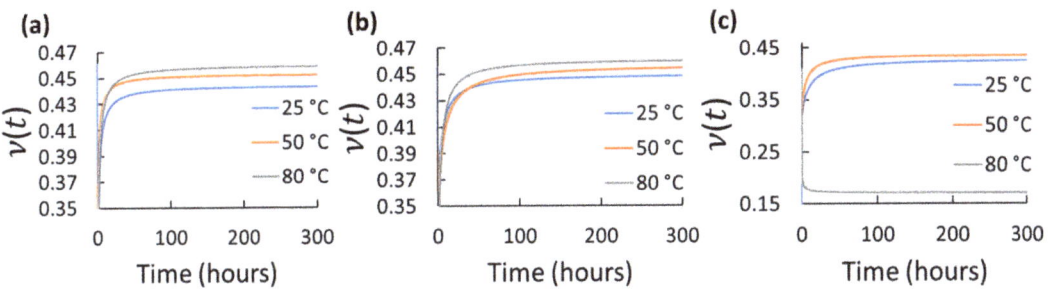

Figure 4. Viscoelastic Poisson's ratio of EPDM at different temperatures and stress: (a) 200 kPa; (b) 400 kPa; (c) 600 kPa.

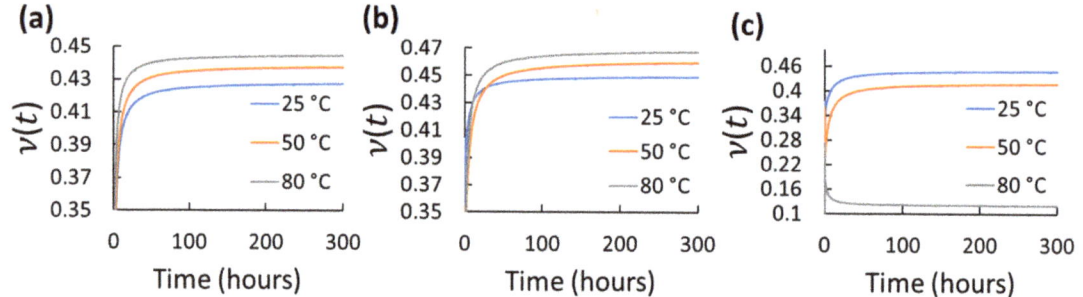

Figure 5. Viscoelastic Poisson's ratio of CR at different temperatures and stress: (**a**) 200 kPa; (**b**) 400 kPa; (**c**) 600 kPa.

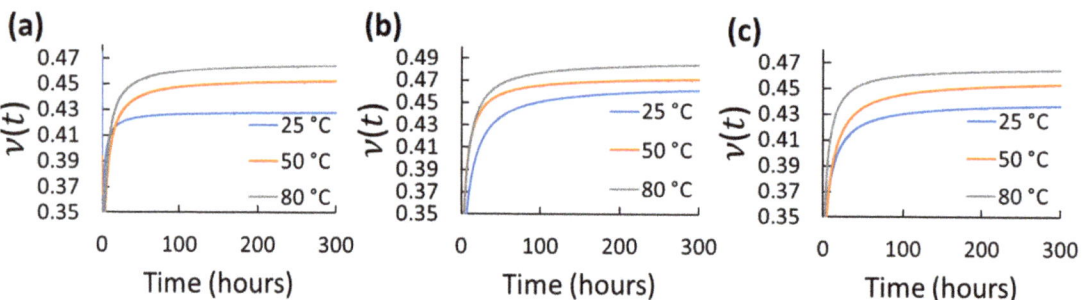

Figure 6. Viscoelastic Poisson's ratio of NBR at different temperatures and stress: (**a**) 200 kPa; (**b**) 400 kPa; (**c**) 600 kPa.

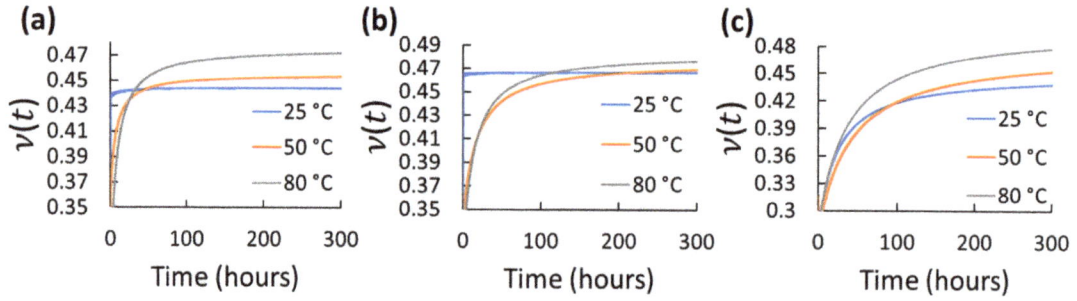

Figure 7. Viscoelastic Poisson's ratio of VMQ at different temperatures and stress: (**a**) 200 kPa; (**b**) 400 kPa; (**c**) 600 kPa.

Finally, the viscoelastic behaviors of the tested elastomers somehow depend on their molecular weight, crosslinking strength and reinforcements (silica, carbon black, graphene, carbon nanotubes, etc.). However, a deeper analysis of the relationship between chemical composition/structure and $\nu(t)$ is out of the scope of the present work. It requires extensive further research. The new behavior data of $\nu(t)$ for different elastomers at different conditions obtained by this characterization work can be useful for modern design of a wide range of elements with different engineering applications in which Poisson's ratio plays an important role on their performance with short- and long-term use. The experimental method followed in this work for characterizing $\nu(t)$ was demonstrated to be suitable for evaluating different elastomers with relative ease. Moreover, the implementation of DIC for creep measurement allowed an accurate (with acceptable error) measurement of creep without restricting the strain produced in the material opposite to those techniques employing strain gauges gripped on the specimens.

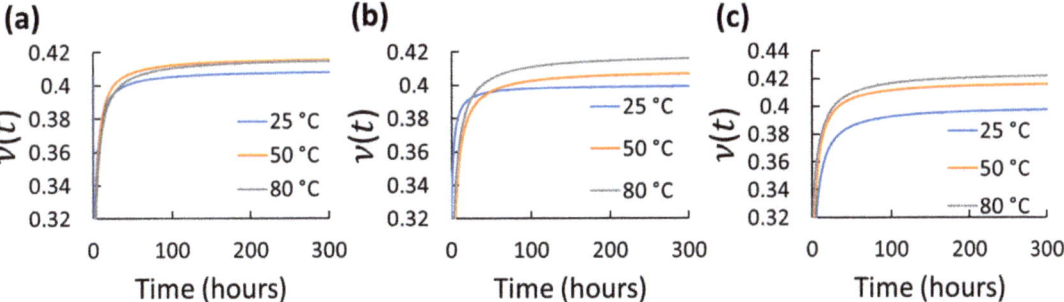

Figure 8. Viscoelastic Poisson's ratio of FKM at different temperatures and stress: (**a**) 200 kPa; (**b**) 400 kPa; (**c**) 600 kPa.

4. Conclusions

The viscoelastic properties ($\varepsilon_x(t)$, $\varepsilon_y(t)$ and $\nu(t)$) of ethylene-propylene-diene monomer, flouroelastomer (Viton®), nitrile butadiene rubber, silicone rubber and neoprene/chloroprene rubber at different stress (200, 400 and 600 kPa) and temperatures (25, 50 and 80 °C) were successfully characterized through an experimental methodological method based on creep testing and strain measurements via digital image correlation (DIC). The entire field of DIC techniques was successfully employed for the simultaneous measurement of strain in axial (ε_y) and transverse (ε_x) directions with time. The tests were effective to obtain the first and second creep stages for all elastomers and conditions. The creep behaviors obtained in both the transverse and axial directions for all the materials and conditions were found to correlate well with a four-element creep model. Thus, average creep curves and models for transverse and axial directions for each material and condition were obtained and used for estimating the corresponding $\nu(t)$. The reported $\nu(t)$ of the five elastomers were estimated through the convolve equation for 300 h. However, it can be determined for more prolonged times by using the axial and transverse creep functions presented in this work and the solution of the convolve equation. Overall, the new data reported in this work suggest $\nu(t)$ to increase with temperature and time, raising from about 0.3 (for short-term loading) to reach and stabilize a value to about 0.48 (for long-term loading) for all tested elastomers. Finally, these results can be potentially applied for more accurate analytical or numerical strain and stress analyses of the components made of elastomers for both short- and long-term uses. In addition, the method implemented in this work will facilitate the characterization of complex viscoelastic properties, particularly, creep and $\nu(t)$, of conventional and new soft materials (e.g., elastomers), which could be a potential tool for screening materials produced by different manufacturing technologies.

Author Contributions: J.A.S.-d.-M.: Investigation; validation; writing—original draft. J.B.P.-F.: Conceptualization; data curation; formal analysis; investigation; writing—original draft; methodology; supervision. O.S.-H.: Resources; visualization; data curation. C.D.R.-C.: Resources; visualization; data curation. E.A.G.-H.: Visualization; data curation. L.I.F.-C.: Conceptualization; writing—review and editing; formal analysis; resources; data curation; project administration; supervision. All authors have read and agreed to the published version of the manuscript.

Funding: The APC was funded by Tecnologico de Monterrey.

Institutional Review Board Statement: Not applicable.

Informed Consent Statement: Not applicable.

Data Availability Statement: Not applicable.

Acknowledgments: The authors would like to acknowledge the financial support of Tecnologico de Monterrey in the production and publication of this work. Moreover, the authors would like to thank Consejo Nacional de Ciencia y Tecnología (CONACyT) of the Government of Mexico for funding and scholarship grants.

Conflicts of Interest: The authors declare no conflict of interest.

References

1. Ernst, L.J.; Zhang, G.Q.; Jansen, K.M.B.; Bressers, H.J.L. Time- and Temperature-Dependent Thermo-Mechanical Modeling of a Packaging Molding Compound and its Effect on Packaging Process Stresses. *J. Electron. Packag.* **2003**, *125*, 539–548. [CrossRef]
2. Świeszkowski, W.; Ku, D.N.; Bersee, H.E.; Kurzydlowski, K.J. An elastic material for cartilage replacement in an arthritic shoulder joint. *Biomaterials* **2006**, *27*, 1534–1541. [CrossRef] [PubMed]
3. Al-Hiddabi, S.A.; Pervez, T.; Qamar, S.; Al-Jahwari, F.; Marketz, F.; Al-Houqani, S.; van de Velden, M. Analytical model of elastomer seal performance in oil wells. *Appl. Math. Model.* **2015**, *39*, 2836–2848. [CrossRef]
4. Kugler, H.P.; Stacer, R.G.; Steimle, C. Direct Measurement of Poisson's Ratio in Elastomers. *Rubber Chem. Technol.* **1990**, *63*, 473–487. [CrossRef]
5. Saseendran, S.; Wysocki, M.; Varna, J. Cure-state dependent viscoelastic Poisson's ratio of LY5052 epoxy resin. *Adv. Manuf. Polym. Compos. Sci.* **2017**, *3*, 92–100. [CrossRef]
6. Pandini, S.; Pegoretti, A. Time, temperature, and strain effects on viscoelastic Poisson's ratio of epoxy resins. *Polym. Eng. Sci.* **2008**, *48*, 1434–1441. [CrossRef]
7. Cui, H.; Ma, W.; Lv, X.; Li, C.; Ding, Y. Investigation on Viscoelastic Poisson's Ratio of Composite Materials considering the Effects of Dewetting. *Int. J. Aerosp. Eng.* **2022**, *2022*, 3696330. [CrossRef]
8. Cui, H.-R.; Shen, Z.-B. An investigation on a new creep constitutive model and its implementation with the effects of time- and temperature-dependent Poisson's ratio. *Acta Mech.* **2018**, *229*, 4605–4621. [CrossRef]
9. Tschoegl, N.W.; Knauss, W.G.; Emri, I. Poisson's Ratio in Linear Viscoelasticity—A Critical Review. *Mech. Time-Depend. Mater.* **2002**, *6*, 3–51. [CrossRef]
10. Van Der Varst, P.G.T.; Kortsmit, W.W. Notes on the lateral contraction of linear isotropic visco-elastic materials. *Arch. Appl. Mech.* **1992**, *62*, 338–346. [CrossRef]
11. Ashrafi, H.; Shariyat, M. A mathematical approach for describing the time-dependent Poisson's ratio of viscoelastic ligaments mechanical characteristics of biological tissues. In Proceedings of the 17th Iranian Conference of Biomedical Engineering (ICBME), Isfahan, Iran, 3–4 November 2010. [CrossRef]
12. O'Brien, D.J.; Sottos, N.R.; White, S.R. Cure-dependent Viscoelastic Poisson's Ratio of Epoxy. *Exp. Mech.* **2007**, *47*, 237–249. [CrossRef]
13. Farfan-Cabrera, L.I.; Pascual-Francisco, J.B. An Experimental Methodological Approach for Obtaining Viscoelastic Poisson's Ratio of Elastomers from Creep Strain DIC-Based Measurements. *Exp. Mech.* **2022**, *62*, 287–297. [CrossRef]
14. Pan, B.; Qian, K.; Xie, H.; Asundi, A. Two-dimensional digital image correlation for in-plane displacement and strain measurement: A review. *Meas. Sci. Technol.* **2009**, *20*, 062001. [CrossRef]
15. Pascual-Francisco, J.B.; Barragán-Pérez, O.; Susarrey-Huerta, O.; Michtchenko, A.; Martínez-García, A.; Farfán-Cabrera, L.I. The effectiveness of shearography and digital image correlation for the study of creep in elastomers. *Mater. Res. Express* **2017**, *4*, 115301. [CrossRef]
16. Sahu, R.; Patra, K.; Szpunar, J. Experimental Study and Numerical Modelling of Creep and Stress Relaxation of Dielectric Elastomers. *Strain* **2014**, *51*, 43–54. [CrossRef]
17. Degrange, J.-M.; Thomine, M.; Kapsa, P.; Pelletier, J.M.; Chazeau, L.; Vigier, G.; Dudragne, G.; Guerbé, L. Influence of viscoelasticity on the tribological behaviour of carbon black filled nitrile rubber (NBR) for lip seal application. *Wear* **2005**, *259*, 684–692. [CrossRef]
18. Herbert, E.G.; Phani, P.S.; Johanns, K.E. Nanoindentation of viscoelastic solids: A critical assessment of experimental methods. *Curr. Opin. Solid State Mater. Sci.* **2015**, *19*, 334–339. [CrossRef]
19. Tweedie, C.A.; Van Vliet, K.J. Contact creep compliance of viscoelastic materials via nanoindentation. *J. Mater. Res.* **2006**, *21*, 1576–1589. [CrossRef]
20. Yoneyama, S.; Takashi, M.; Gotoh, J. Photoviscoelastic Stress Analysis near Contact Regions under Complex Loads. *Mech. Time-Depend. Mater.* **1997**, *1*, 51–65. [CrossRef]
21. Darlix, B.; Montmitonnet, P.; Monasse, B. Creep of polymers under ball indentation: A theoretical and experimental study. *Polym. Test.* **1986**, *6*, 189–203. [CrossRef]
22. Oyen, M.L. Spherical Indentation Creep Following Ramp Loading. *J. Mater. Res.* **2005**, *20*, 2094–2100. [CrossRef]
23. Rubin, A.; Favier, D.; Danieau, P.; Giraudel, M.; Chambard, J.-P.; Gauthier, C. Direct observation of contact on non transparent viscoelastic polymers surfaces: A new way to study creep and recovery. *Prog. Org. Coat.* **2016**, *99*, 134–139. [CrossRef]
24. Pascual-Francisco, J.B.; Farfan-Cabrera, L.I.; Susarrey-Huerta, O. Characterization of tension set behavior of a silicone rubber at different loads and temperatures via digital image correlation. *Polym. Test.* **2020**, *81*, 106226. [CrossRef]

25. Stoček, R.; Stěnička, M.; Maloch, J. Determining Parametrical Functions Defining the Deformations of a Plane Strain Tensile Rubber Sample. In *Fatigue Crack Growth in Rubber Materials. Advances in Polymer Science*; Heinrich, G., Kipscholl, R., Stoček, R., Eds.; Springer: Cham, Germany, 2020; Volume 286, pp. 19–38. [CrossRef]
26. Stoček, R.; Heinrich, G.; Gehde, M.; Kipscholl, R. A new testing concept for determination of dynamic crack propagation in rubber materials. *KGK Rubberpoint* **2012**, *65*, 49–53.
27. Stoček, R.; Heinrich, G.; Gehde, M.; Kipscholl, R. Analysis of dynamic crack propagation in elastomers by simultaneous tensile- and pure-shear-mode testing. In *Fracture Mechanics and Statistical Mechanics of Reinforced Elastomeric Blends (Lecture Notes in Applied and Computational Mechanics)*; Grellmann, W., Heinrich, G., Kaliske, M., Klüppel, M., Schneider, K., Vilgis, T., Eds.; Springer: Berlin/Heidelberg, Germany, 2013; Volume 70, pp. 269–301. [CrossRef]
28. Robertson, C.G.; Hardman, N.J. Nature of Carbon Black Reinforcement of Rubber: Perspective on the Original Polymer Nanocomposite. *Polymers* **2021**, *13*, 538. [CrossRef] [PubMed]
29. Findley, W.N.; Lai, J.S.; Onaran, K. *Creep and Relaxation of Nonlinear Viscoelastic Materials*; Dover Publications: New York, NY, USA, 1976; pp. 50–70.
30. Cui, H.R.; Tang, G.J.; Shen, Z.B. Study on the Viscoelastic Poisson's Ratio of Solid Propellants Using Digital Image Correlation Method. *Propellants Explos. Pyrotech.* **2016**, *41*, 835. [CrossRef]
31. Grassia, L.; D'Amore, A.; Simon, S.L. On the viscoelastic Poisson's ratio in amorphous polymers. *J. Rheol.* **2010**, *54*, 1009–1022. [CrossRef]
32. Aili, A.; Vandamme, M.; Torrenti, J.-M.; Masson, B. Theoretical and practical differences between creep and relaxation Poisson's ratios in linear viscoelasticity. *Mech. Time-Depend. Mater.* **2015**, *19*, 537–555. [CrossRef]
33. Charpin, L.; Sanahuja, J. Creep and relaxation Poisson's ratio: Back to the foundations of linear viscoelasticity. Application to concrete. *Int. J. Solids Struct.* **2017**, *110*, 2–14. [CrossRef]

Article

Fast Evaluation and Comparison of the Energy Performances of Elastomers from Relative Energy Stored Identification under Mechanical Loadings

Jean-Benoît Le Cam

Institut de Physique UMR 6251 CNRS de Rennes 1, Campus de Beaulieu, Université de Rennes 1, Bât. 10B, CEDEX, 35042 Rennes, France; jean-benoit.lecam@univ-rennes1.fr

Abstract: The way in which elastomers use mechanical energy to deform provides information about their mechanical performance in situations that require substantial characterization in terms of test time and cost. This is especially true since it is usually necessary to explore many chemical compositions to obtain the most relevant one. This paper presents a simple and fast approach to characterizing the mechanical and energy behavior of elastomers, that is, how they use the mechanical energy brought to them. The methodology consists of performing one uniaxial cyclic tensile test with a simultaneous temperature measurement. The temperature measurement at the specimen surface is processed with the heat diffusion equation to reconstruct the heat source fields, which in fact amounts to surface calorimetry. Then, the part of the energy involved in the mechanical hysteresis loop that is not converted into heat can be identified and a quantity γ_{se} is introduced for evaluating the energy performance of the materials. This quantity is defined as an energy ratio and assesses the ability of the material to store and release a certain amount of mechanical energy through reversible microstructure changes. Therefore, it quantifies the relative energy that is not used to damage the material, for example to propagate cracks, and that is not dissipated as heat. In this paper, different crystallizable materials have been considered, filled and unfilled. This approach opens many perspectives to discriminate, in an accelerated way, the factors affecting these energetic performances of elastomers, at the first order are obviously the formulation, the aging and the mechanical loading. In addition, such an approach is well adapted to better characterize the elastocaloric effects in elastomeric materials.

Keywords: elastomer; fast characterization; energy stored and released; heat source reconstruction; intrinsic dissipation; infrared thermography

Citation: Le Cam, J.-B. Fast Evaluation and Comparison of the Energy Performances of Elastomers from Relative Energy Stored Identification under Mechanical Loadings. *Polymers* 2022, 14, 412. https://doi.org/10.3390/polym14030412

Academic Editors: Radek Stoček, Gert Heinrich and Reinhold Kipscholl

Received: 30 November 2021
Accepted: 17 January 2022
Published: 20 January 2022

Publisher's Note: MDPI stays neutral with regard to jurisdictional claims in published maps and institutional affiliations.

Copyright: © 2022 by the authors. Licensee MDPI, Basel, Switzerland. This article is an open access article distributed under the terms and conditions of the Creative Commons Attribution (CC BY) license (https://creativecommons.org/licenses/by/4.0/).

1. Introduction

Elastomers are widely used in many industries, such as automotive, nuclear or civil engineering, for their high deformability, high damping and, for some of them, their high fatigue resistance. As for most of the engineering materials, their design requires substantial characterization time and cost, typically for fatigue [1–7], crack propagation [8–10], aging [11–13] and damping [14], non-exhaustively. At a time when many sectors that use elastomeric materials are undergoing deep technological changes, such as health with implantable medical devices or mobility with the desire to decarbonize transportation, it is necessary to develop new materials and new technological solutions that must be evaluated in an increasingly limited time and at a low cost. In addition, crystallizing elastomers are increasingly studied for their elastocaloric properties [15–17] and criteria are needed to compare their energy performance. This is why the development of fast characterization methods for rubbers, allowing a classification of the desired performances, however basic, is today a major challenge in most industries.

What makes elastomers attractive for many applications is that their mechanical response exhibits a hysteresis loop. The mechanical hysteresis is generally obtained by adding fillers in the rubber matrix and/or strain-induced crystallization (SIC) and melting. Classically, the mechanical energy involved in the hysteresis loop is assumed to be mainly dissipated into heat. Nevertheless, several observations question this assumption:

- The mechanical response of some elastomers exhibits a hysteresis loop only when strain-induced crystallization (SIC) occurs, typically in case of unfilled natural rubber (NR) [18]. In this case, no self-heating accompanies the mechanical cycles, which indicates that SIC does not induce or has very little viscosity. This was first intuited by Clark [19] and confirmed by [20–22];
- The mechanical hysteresis can be not time-dependent [23–26]. This is not expected for viscous materials;
- If all the energy contained in the hysteresis loop were due to viscosity, the self-heating would be much higher than that observed experimentally, especially under repeated cycles.

One can therefore wonder about the nature and the time dependency of the phenomena involved in the formation of the hysteresis loop and about the real contribution of the intrinsic dissipation to it. To go further in the discussion, it is necessary to recall that in addition to the intrinsic dissipation two other factors can affect to the mechanical hysteresis:

(a) The thermal dissipation (under non adiabatic test conditions). If heat is exchanged with the specimen's outside, then a hysteresis loop in the stretch-stress relationship can theoretically form, the current temperature appearing in the elastic coupling. In most of the homogeneous tests (in terms of heat source field, see Section 2) performed, considering the thermal properties of elastomers and the loading rate relatively high, the thermal dissipation does not contribute significantly to the mechanical hysteresis;

(b) The change in microstructure. In this case, all the work done to the system is not measured as an apparent temperature change (see for instance the recent studies by [27] on polyurea who concluded that a significant part of the mechanical energy is used by the material to reorganize or by Le Cam [22] who demonstrated that the mechanical hysteresis of the unfilled natural rubber he studied was entirely due to the difference in kinetics between crystallization and crystallite melting).

Thus, of the many phenomena that can contribute to the hysteresis loop, some do not appear to dissipate energy as heat. This is in fact what is assessed with self-heating tests for evaluating the fatigue properties [28]. Therefore, the level of self-heating or more precisely intrinsic dissipation must be put into perspective in relation to the energy rate involved in the hysteresis loop. Even more interesting is the part of the mechanical energy brought that is stored and released by reversible microstructure changes, which is not dissipated as heat or used to damage the material, for example to propagate cracks. In the extreme, the mechanical response of an unfilled natural rubber exhibits a hysteresis loop as soon as it is crystallizing, while no (or minimally) energy is converted as heat [22].

Considering that contributions to the hysteresis loop change when one chemical formulation is substituted to another one (see for example [29]), or when one or more ingredients of a chemical formulation is changed to improve a given property (see, for example, [30–33]), the present study assumes that identifying the energy contributions and associated phenomena involved in the mechanical hysteresis is a reliable indicator of the material's performances.

Such analysis echoes the pioneering work by [34,35] who measured the latent energy remaining after cold working in a metal under quasi-static monotonous loadings. Today, the fraction of the anelastic deformation energy rate irreversibly converted into heat is studied through the Taylor–Quinney ratio [36–40]. Polymers have then benefited from this approach in the Rittel's group [41,42] and the Chrysochoos's group [43,44]. Concerning elastomers, only four recent studies investigate the energetic behavior and the energy storage during deformation [22,45–47]. From these studies, the energy storage in different types of elastomers, filled and unfilled, crystallizing or not, can be discussed. The main difference with the pre-mentioned materials is that the energy stored elastically with a given kinetics is released within the same mechanical cycle, but with a different kinetics. This energy serves for reversible changes in microstructure, typically when the crystallization—crystallite melting process occurs in natural rubbers [20,48]. Section 2 presents the thermodynamics framework for calorimetry under stretch from surface temperature measurements, the methodology for energy balances and the definition of the characteristic quantity giving the energy performance. Section 3 presents a typical experimental setup. Section 4 provides illustrations with different materials, inspired from three recent studies carried out in our group [22,45,47], which provides the first comparison of elastomers according to their energy performances. Concluding remarks close the paper.

2. Thermodynamic Framework

Identifying the phenomena involved in the mechanical hysteresis requires the calculation of several continuum quantities that are recalled in this section. The calculations are carried out in the case of the uniaxial tension.

2.1. Total Strain Energy Density and Energy Rate Involved in the Hysteresis Loop

The strain energy density W_{strain} (in J/m^3) is the energy brought mechanically to deform the material. It corresponds to the area under the load (unload) strain–stress curve and is calculated as follows:

$$W_{strain}^{load} = \int_{load} \pi \, d\lambda \text{ and } W_{strain}^{unload} = \int_{unload} \pi \, d\lambda, \quad (1)$$

where λ is the stretch defined as the ratio between current and initial lengths. π is the nominal stress, defined as the force per unit of initial (undeformed) surface. If the material's behavior is purely elastic and if the test is carried out under adiabatic loading conditions, then the mechanical response obtained during a load-unload cycle is such that no hysteresis loop forms ($W_{strain}^{load} = W_{strain}^{unload}$). If a hysteresis loop forms, the mechanical energy dissipated over one cycle W_{hyst}^{cycle} is defined as follows:

$$W_{hyst}^{cycle} = W_{strain}^{load} - W_{strain}^{unload}. \quad (2)$$

From this energy, a quantity P_{hyst}^{cycle} is calculated in W/m^3. It is obtained by dividing W_{hyst}^{cycle} by the cycle duration. It is therefore an energy density per time unit or an energy rate. It should be noted that when the mechanical cycle is also a thermodynamic cycle, the energy contained in the hysteresis loops is only due to intrinsic dissipation and thermomechanical coupling effects as long as the specific heat is assumed to be constant (further details are provided in [49,50]). This will be used in the following for carrying out the energy balances and to identify the intrinsic dissipation.

2.2. Heat Sources

During the mechanical cycle, the material produces and absorbs heat. Under non adiabatic conditions, corresponding temperature variations are influenced by heat diffusion effects. Therefore, temperature is not the relevant quantity to investigate the thermomechanical behavior of materials under non-adiabatic conditions, as most of classical mechanical tests are. This is the reason why the heat source (or the heat power density) is calculated from the heat diffusion equation and the temperature measurement (see [51,52] for further details). The heat source does not depend on the heat diffusion effects and is thus intrinsic to the thermomechanical behavior of materials. Considering the heat source field as homogeneous during uniaxial tensile loading, the formulation of the heat diffusion equation can be simplified as follows:

$$\rho C \left(\dot{\theta} + \frac{\theta}{\tau} \right) = S, \quad (3)$$

where ρ is the initial density, C is the specific heat, θ is the temperature variation with respect to the equilibrium temperature T^{ref} in the reference state, corresponding here to the undeformed state. The heat sources S can be divided into two terms that differ in nature:
- the intrinsic dissipation D_{int}: this positive quantity corresponds to the heat production due to mechanical irreversibilities during the deformation process, for instance viscosity or damage;
- the thermomechanical couplings S_{tmc}: they correspond to the couplings between the temperature and the state variables, and describe reversible deformation processes. Consequently, their integration with respect to time over one thermodynamical cycle is null.

Equation (3) is formulated in the case where the ambient temperature T_{amb} is constant during the test. In case where changes in ambient temperature occur, the term $\frac{\theta}{\tau}$ has to be corrected accordingly as $\frac{T-T_{amb}}{\tau}$. τ is a parameter characterizing the heat exchanges between the specimen and its surroundings. It can be easily identified from a natural return to room temperature after a heating (or a cooling) for each testing configuration (machine used, environment, stretch level, etc). For instance, in cases where the material is beforehand heated, the exponential formulation of the temperature variation $\theta(t) = \theta_0 e^{-\frac{(t-t_0)}{\tau}}$ is used to determine τ, where t is the time in s and $\theta_0 = \theta(t = t_0)$. Further details are provided in [18]. It should be noted that under uniaxial tension, τ depends on the stretch only ($\tau = \tau(\lambda)$) (It should be noted that in the case of heterogeneous strain fields, τ spatially varies according to the heterogeneous change in the thickness, that is, to the stretch and to the biaxiality ratio B ($\tau = \tau(\lambda, B)$. Further information on the identification of a τ field is provided in [53]). Several strategies can be applied for determining $\tau(\lambda)$. Either it is evaluated for different stretch levels or by considering the material to be incompressible and the thickness changes to be proportional to $\lambda^{-\frac{1}{2}}$, so that $\tau = \tau(\lambda) = \tau_0 \lambda^{-\frac{1}{2}}$. It should be noted that when the material is crystallizing, the second strategy is preferred. Indeed, a change in temperature of the material (heating or cooling) changes the crystallinity. Therefore, an additional heat production or absorption is obtained during the return at the initial crystallinity, which affects the temperature variation during the return at ambient temperature and, consequently, the identification of τ.

2.3. Identifying the Mean Intrinsic Dissipation

Integrating the heat source, that is, the left member of Equation (3), with respect to time over one thermodynamical cycle gives the energy density due to intrinsic dissipation, the integration of the thermomechanical couplings being null. This energy is then divided by the cycle duration to obtain the mean intrinsic dissipation $\tilde{D}_{int}$. This amounts to applying the following formula:

$$\tilde{D}_{int} = \frac{1}{t_{cycle}} \int_{cycle} S \, dt = \frac{1}{t_{cycle}} \int_{cycle} (D_{int} + S_{tmc}) \, dt = \frac{1}{t_{cycle}} \int_{cycle} D_{int} \, dt. \quad (4)$$

2.4. Energy Balance

Energy balance is carried out from both the mechanical and the calorific responses. The mechanical response provides the strain energy density and its rate P_{hyst}^{cycle} involved in the hysteresis loop. The difference between P_{hyst}^{cycle} and $\tilde{D}_{int}$ gives the rate of the energy stored and released due to microstructure changes at each cycle:

$$P_{stored}^{cycle} = P_{hyst}^{cycle} - \tilde{D}_{int}. \quad (5)$$

To further discuss on the relative contribution of the energy stored in the hysteresis loop of rubbers, a ratio γ_{se} has been proposed in our group [45,46]. It is written in terms of energy as follows:

$$\gamma_{se} = \frac{W_{stored}^{cycle}}{W_{hyst}^{cycle}} \tag{6}$$

- if γ_{se} tends to 0, no energy is stored during the deformation. The whole hysteresis loop is due to the intrinsic dissipation,
- if γ_{se} tends to 1, the whole hysteresis loop is due to energy stored and no intrinsic dissipation is detected. This is typically the case in unfilled natural rubber [18], for which the energy stored in the crystallites is released with a different kinetics during their melting.

3. Overview of the Experimental Setups

Three different materials have been considered:

- an unfilled natural rubber, denoted U-NR and studied in [21]. Its chemical composition is given in Table 1;
- a carbon black filled natural rubber studied in [47], denoted F-NR. Its chemical composition is given in Table 1;
- an unfilled thermoplastic polyurethane, also crystallizable under tension, denoted TPU and studied in [45]. It is referred to as Irogran® A87H4615 TPU from the Huntsman corporation (The Woodlands, TX, USA). It is elaborated by reacting together a diisocyanate, a macro diol (long chain diol), which is a polyester in the present case, and a small molecule chain-extender (butane diol) [54].

Table 1. Chemical composition in parts per hundred rubber (phr).

Ingredient	U-NR [21]	F-NR [47]
Natural rubber NR	100	100
Carbon black	0	20–30
Antioxidant	1.9	2–4
Stearic acid	2	2
Zinc oxide ZnO	2.5	10
Accelerator	1.6	2–4
Sulfur	1.6	1.5

The experimental setups are briefly recalled in Table 2. It should be noted that the loading conditions differ from one study to another, depending on the considered application, but lead to thermodynamical cycles from which energy balance can be carried out. Generally, the stretch is calculated from the initial length of the virgin specimens and is not corrected while a permanent deformation can appear and grow as the cycles follow each other. The temperature was measured with infrared cameras, which were switched on several hours before testing in order to ensure their internal temperature to be stabilized. The calibration of camera detectors was performed with a black body using a Non-Uniformity Correction (NUC) procedure. A typical experiment setup is given in Figure 1, which enabled us to stretch the specimens symmetrically. All the specimens tested had cylindrical ends that avoid any slippage with the grips and makes accurate the determination of the initial length.

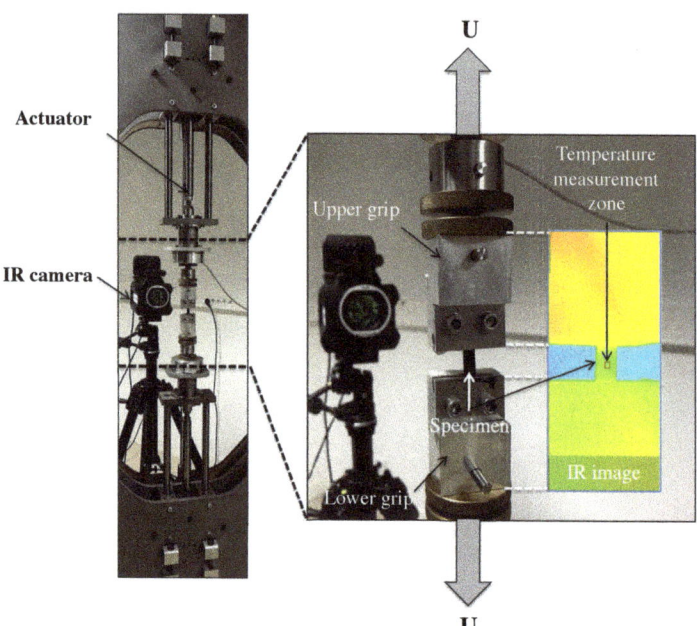

Figure 1. Typical experimental setup: the specimen is stretched symmetrically with a home-made biaxial testing machine, the temperature field is measured with an infrared camera.

Table 2. Summary of the experiments performed.

	Materials		
Reference	U-NR [21]	F-NR [47]	TPU [45]
Filler type and amount (phr)	- 0	CB 20–30	-
Crystallizable under strain	Yes	Yes	Yes
Specimen geometry (mm): Width × Length × thickness	5 × 10 × 1.4	10 × 24 × 2	9 × 20 × 5
Testing machine	Instron 5543, one moving grip	Homemade biaxial tensile machine, symmetric loading	Instron 5543, one moving grip
Mechanical loading	3 cycles at λ = 2, 5, 6 and 7.5	3 cycles at λ = 2.5, 4 and 6	5 cycles at λ = 1.5, 2, 2.5 and 3
Constant loading rate (mm/min)/strain rate (s^{-1})	±100 and ±300/±0.17 and ±0.51	±300/±0.21	±100 and ±300/±0.08 and ±0.25
Infrared camera, resolution	Cedip Jade III, 320 × 240 px	FLIR X6540sc, 640 × 512 px	FLIR X6540sc, 640 × 512 px
Motion compensation technique	Yes	No	Yes

Remark on the Temperature Measurement

If the specimen is stretched by the displacement of only one jaw, which means that the initial temperature measurement area is shifted in the tensile direction, the temperature field must be initially homogeneous in order to calculate the heat sources. However, most of the testing machines and in particular the hydraulic ones lead to a thermal gradient in the specimen because the jaws are not at the same temperature. Therefore, a thermal camera has to be used in order to track the measurement area in the thermal images (see for instance [21]). If the specimen is stretched by moving the two jaws symmetrically, the initial temperature measurement area does not move and only one spot measurement with, for example, a pyrometer is possible, which reduces drastically the thermal measurement cost.

4. Results and Discussion

Figure 2 presents the mechanical responses of the unfilled NR obtained at ± 0.17 s^{-1} (on the left hand side) and ± 0.51 s^{-1} (on the right hand side) in terms of the nominal stress versus the stretch. They are extracted from [21]. For the two loading rates, no stress softening is observed between cycles and the loading rate has no effect on the stiffness expect for the highest stretch at $\lambda_4 = 7.5$ that induces the highest crystallinity. In this case, the highest loading rate leads to the highest stiffness. For cycles at $\lambda_1 = 2$ and $\lambda_2 = 5$, no hysteresis loop is observed. For cycles at $\lambda_3 = 6$, a hysteresis loop forms, which closes at a stretch equal to 3, which is close to the stretch at which crystallite melting is assumed to be complete. It should be noted that the crystallization onset is about 4, meaning that the crystallinity is not high enough to form a hysteresis loop. For cycles at $\lambda_4 = 7.5$, a plateau is observed from $\lambda = 6$ on, followed by a stress increase that is higher at the highest loading rate, as previously observed and explained in [55,56]. As for the previous cycles at $\lambda_3 = 6$, the hysteresis loop closes at around $\lambda = 3$.

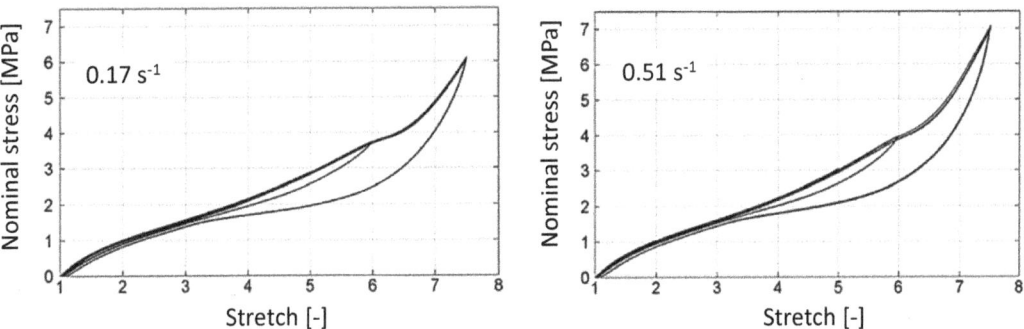

Figure 2. Mechanical responses of the unfilled NR obtained at ± 0.17 s^{-1} (on the left hand side) and ± 0.51 s^{-1} (on the right hand side), extracted from [21].

Figure 3 presents the mechanical responses of the TPU obtained at ± 0.08 s^{-1} (on the left hand side) and ± 0.25 s^{-1} (on the right hand side), extracted from [45]. The mechanical responses exhibit softening, residual stretch and hysteresis, which was widely investigated in the literature by [57–60] and is similar to that observed in filled rubbers. It should be noted that the loading rate does not affect significantly the stiffness, which is a less intuitive result when considering that a hysteresis loop is mainly due to viscosity. This is further discussed in [45].

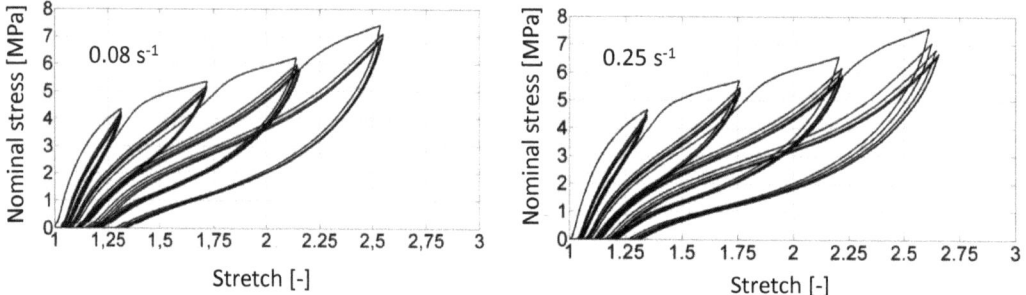

Figure 3. Mechanical responses of the TPU obtained at ±0.08 s^{-1} (on the left hand side) and ±0.25 s^{-1} (on the right hand side), extracted from [45].

Figure 4 presents the mechanical response of the F-NR obtained at ±0.21 s^{-1}, extracted from [47]. As expected, adding fillers to a natural rubber induces a hysteresis loop, this is the reason why it is observed in the mechanical response even at a stretch inferior to the crystallization onset. Furthermore, a stress softening is observed as well as a permanent set. In the case of a filled crystallizing rubber, a typical point is to define what is the part of the energy converted into heat and the one stored reversibly in one thermodynamical cycle. Obviously, this has a consequence on the self-heating and on the energy available for damaging the material, typically discussed in studies dealing with fatigue and fracture mechanics in elastomers.

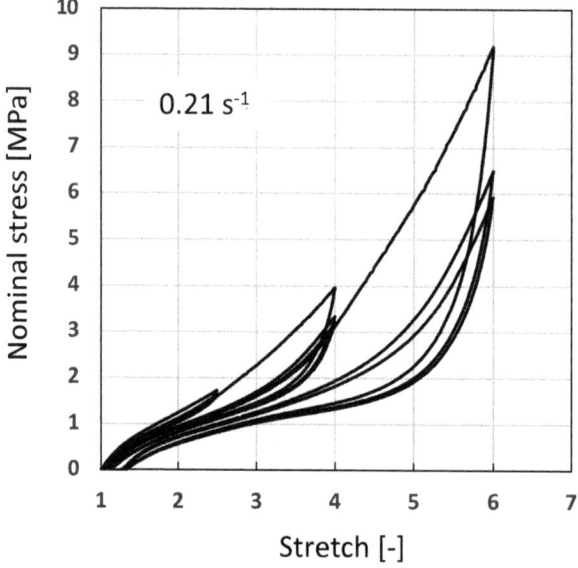

Figure 4. Mechanical response of the F-NR obtained at ±0.21 s^{-1}, extracted from [47].

The hysteresis loop observed in the mechanical responses of these three materials is due to the different possible contributions recalled in the introduction section. The aim of the following is to identify these contributions. For that purpose, the methodology described in Section 2 has been applied to each material:

- First, the heat source S has been calculated from Equation (3) for each stabilized cycle at different maximum stretches (these cycles are considered as thermodynamical ones),

- Second, P_{hyst}^{cycle} and $\tilde{\mathcal{D}}_{int}$ have been determined from Equations (2) and (4) respectively,
- Third, the γ_{se} ratio has been calculated from Equation (6).

In the following, γ_{se} is first plotted versus stretch for the unfilled NR and the TPU in Figure 5, which provides the first comparison of rubber compounds with respect to their energy behavior. We recall here that the higher the value of γ_{se}, the higher the energy storage capacity of the material. If $\gamma_{se} = 1$, then no intrinsic dissipation is involved in the hysteresis loop, that is, no mechanical energy is converted into heat and 100% of the normalized area of the hysteresis loop is energy stored and released during the cycle. The γ_{se}—stretch relationship for the unfilled NR is given by the green curve in the diagram and was found close to 1, whatever the loading rate and the stretch level once the material is crystallizing ($\lambda > \lambda_c \approx 4$). For stretches inferior to the crystallization onset, no hysteresis loop forms. This illustrates, as demonstrated in [22], that all the energy involved in the hysteresis loop is stored by the crystallization processes and fully released during the crystallite melting. It can be therefore seen as a "cold energy". For the TPU material at the strain rate of ± 0.25 s^{-1}, γ_{se} is about 0.8 at a stretch of 1.5 and increases with stretch. It is close to 1 for stretches from 2.5 on. The energy involved in the hysteresis loop is therefore almost not converted into heat and the higher the stretch level the higher the relative energy stored in the material. This strong similitude with the unfilled natural rubber is explained by the fact that the TPU considered here crystallizes under tension, but also by the fact that the multi-phase nature of TPU induces strong self-organization as suggested in [27] and regardless of whether it crystallizes or not as shown in [61]. It should be noted that for a strain rate of four times lower, the ratio is close to 1 whatever the stretch applied.

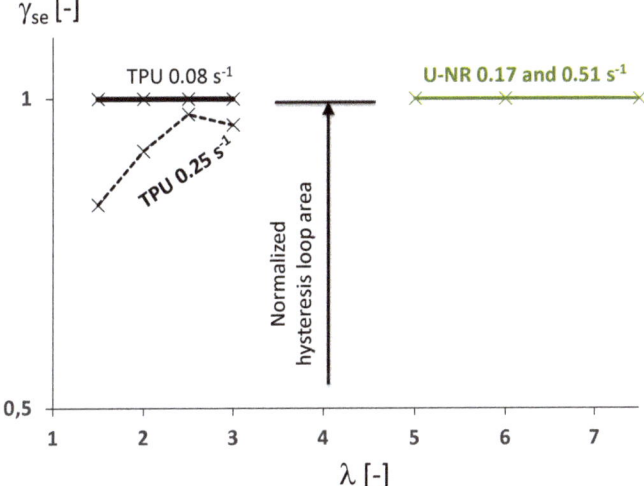

Figure 5. Energy characterization in terms of γ_{se} versus stretch for the U-NR at 0.17 and 0.51 s^{-1} (in green) and the TPU (in black).

Figure 6 gives γ_{se} versus stretch for the filled NR stretched at the rate of the same order of magnitude as the lowest rate applied to the unfilled NR. It was observed that about 80% of the energy involved in the hysteresis loop is dissipated as heat at the lowest stretches applied (maximum 4, which induces a very low crystallinity). This means that adding fillers does not only induce viscosity but enables the material to store a part of the mechanical energy without converting it into heat, mechanical energy that is not used to damage and self-heat the material (this energy storage effect of the filler network was also highlighted in the case of carbon black filled acrylonitrile rubber (see [46] for further information)). When the material is crystallizing (when stretch at $\lambda = 6$), the relative energy stored strongly increases from 20% of the hysteresis loop area at $\lambda = 4$ to 50% ($\gamma_{se} = 0.5$) at $\lambda = 6$. As industrial natural rubbers are filled, comparing their energy behavior amounts to comparing the area under the red curve in the diagram. Typically, the lower this area in the deformation domain defined by the industrial application, the higher the self-heating and energy used to damage the material and the lower the elastic energy stored reversibly. From an elastocaloric point of view, identifying the relative contribution of the intrinsic dissipation during one cycle is of paramount importance as it opposes the heat absorption during the unloading and consequently it reduces the cooling.

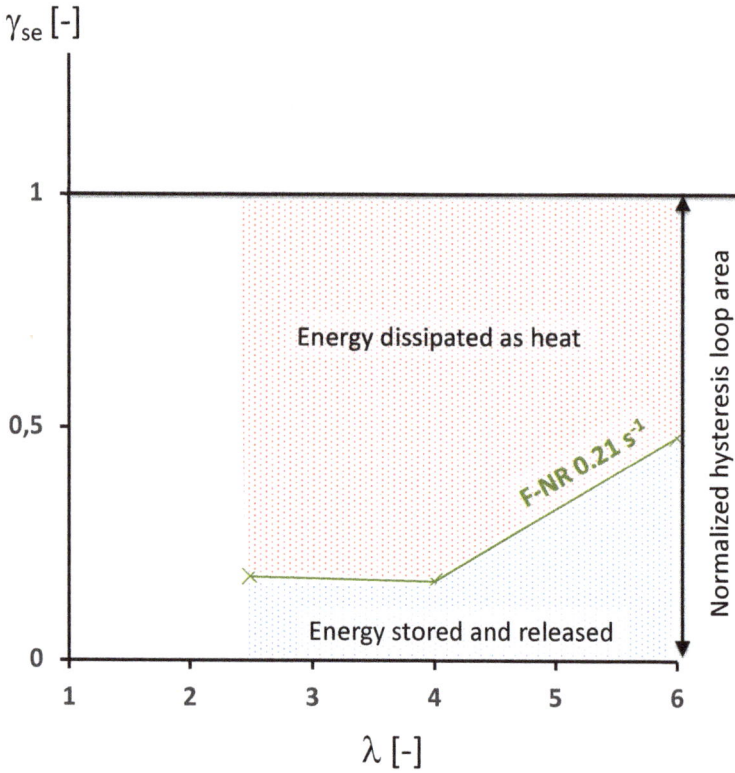

Figure 6. Energy characterization in terms of γ_{se} versus stretch for the F-NR at 0.21 s^{-1}.

5. Conclusions

This paper presents a simple and fast approach to determining the energy behavior of elastomers from cyclic uniaxial tensile loadings and temperature measurements. The temperature measurement is used to reconstruct the heat source, that is, the calorimetric response, from the 0D formulation of the heat diffusion equation. Based on the mechanical and calorimetric responses, a ratio γ_{se} is calculated to relativise the energy stored and released at each cycle. Results show that viscosity is not systematically the preponderant contribution to the hysteresis loop, whatever the elastomer considered, filled and unfilled, crystallizing or not: the mechanical energy brought to the material is not entirely dissipated into heat and can be mainly used by the material to change its microstructure. Therefore, γ_{se} assesses the ability of the material to store mechanical energy through reversible microstructure changes, energy that is not dissipated as heat and used to damage the material, for example, to propagate cracks. The results obtained enable us to characterize and to compare different materials according to their energetic performance. This study thus proposes an approach well adapted to the design of elastomer parts and opens many perspectives to discriminate, in an accelerated way, the factors affecting these energetic performances, at the first order are obviously the formulation, the aging and the mechanical loading. This makes it a very promising tool for further investigating the thermomechanical behavior of rubbers and for validating and enriching physical thermodynamical models.

Funding: This research received no external funding.

Institutional Review Board Statement: Not applicable.

Informed Consent Statement: Not applicable.

Data Availability Statement: Data can be found in [21,45,47].

Acknowledgments: We renew the acknowledgements in [21,45,47] to our industrial and academic partners for their financial support and the fruitful discussions. The National Center for Scientific Research (MRCT-CNRS and MI-CNRS), Rennes Metropole and Region Bretagne are also thank for their financial support.

Conflicts of Interest: The authors declare no conflict of interest.

Abbreviations

The following abbreviations are used in this manuscript:

CB	Carbon Black
F-NR	Filled Natural Rubber
F-NBR	Filled Nitrile Butadiene Rubber (acrylonitrile rubber)
phr	part per hundred of rubber in weight
NR	Natural Rubber
SIC	Strain-Induced Crystallization
TPU	Thermoplastic PolyUrethane
U-NR	Unfilled Natural Rubber

References

1. Cadwell, S.M.; Merril, R.A.; Sloman, C.M.; Yost, F.L. Dynamic fatigue life of rubber. *Ind. Eng. Chem.* **1940**, *12*, 19–23.
2. Beatty, J.R. Fatigue of rubber. *Rubber Chem. Technol.* **1964**, *37*, 1341–1364. [CrossRef]
3. Mars, W.V.; Fatemi, A. A literature survey on fatigue analysis approaches for rubber. *Int. J. Fatigue* **2002**, *24*, 949–961. [CrossRef]
4. Saintier, N.; Cailletaud, G.; Piques, R. Multiaxial fatigue life prediction for a natural rubber. *Int. J. Fatigue* **2006**, *28*, 530–539. [CrossRef]
5. Le Cam, J.B.; Verron, E.; Huneau, B. Description of fatigue damage in carbon black filled natural rubber. *Fatigue Fract. Eng. Mater. Struct.* **2008**, *31*, 1031–1038. [CrossRef]
6. Masquelier, I. Influence of Formulation on the Fatigue Properties of Elastomeric Materials. Ph.D. Thesis, Université de Bretagne Occidentale, Brest, France, 2014.
7. Ruellan, B.; Le Cam, J.B.; Jeanneau, I.; Canévet, F.; Mortier, F.; Robin, E. Fatigue of natural rubber under different temperatures. *Int. J. Fatigue* **2019**, *124*, 544–557. [CrossRef]
8. Greensmith, H.W.; Thomas, A.G. Rupture of rubber-III-Determination of tear properties. *J. Polym. Sci.* **1955**, *18*, 189–200. [CrossRef]
9. Greensmith, H.W. Rupture of rubber-IV-Tear properties of vulcanized containing carbon black. *J. Polym. Sci.* **1956**, *21*, 175–187. [CrossRef]
10. Lindley, P. Non-relaxing crack growth and fatigue in a non-crystallizing rubber. *Rubber Chem. Technol.* **1974**, *47*, 1253–1264. [CrossRef]
11. Lake, G.J.; Lindley, P.B. Role of ozone in dynamic cut growth of rubber. *J. Appl. Polym. Sci.* **1965**, *9*, 2031–2045. [CrossRef]
12. Gent, A.N.; Liu, G.L.; Sueyasu, T. Effect of temperature and oxygen on the strength of elastomers. *Rubber Chem. Technol.* **1991**, *64*, 96–107. [CrossRef]
13. Le Saux, V. Fatigue and Ageing of Rubbers under Marine and Thermal Environments: From Accelerated Tests to Structure Numerical Simulations. Ph.D. Thesis, Université de Bretagne Occidentale, Brest, France, 2010.
14. Emminger, C.; Çakmak, U.D.; Preuer, R.; Graz, I.; Major, Z. Hyperelastic Material Parameter Determination and Numerical Study of TPU and PDMS Dampers. *Materials* **2021**, *14*, 7639. [CrossRef]
15. Xie, Z.; Sebald, G.; Guyomar, D. Comparison of elastocaloric effect of natural rubber with other caloric effects on different-scale cooling application cases. *Appl. Therm. Eng.* **2017**, *111*, 914–926. [CrossRef]

16. Coativy, G.; Haissoune, H.; Seveyrat, L.; Sebald, G.; Chazeau, L.; Chenal, J.M.; Lebrun, L. Elastocaloric properties of thermoplastic polyurethane. *Appl. Phys. Lett.* **2020**, *117*, 193903. [CrossRef]
17. Candau, N.; Vives, E.; Fernández, A.I.; Maspoch, M.L. Elastocaloric effect in vulcanized natural rubber and natural/wastes rubber blends. *Polymer* **2021**, *236*, 124309. [CrossRef]
18. Samaca Martinez, J.R.; Le Cam, J.B.; Balandraud, X.; Toussaint, E.; Caillard, J. Mechanisms of deformation in crystallizable natural rubber. Part 2: Quantitative calorimetric analysis. *Polymer* **2013**, *54*, 2727–2736. [CrossRef]
19. Clark, G.L.; Kabler, M.; Blaker, E.; Ball, J.M. Hysteresis in crystallization of stretched vulcanized rubber from X-ray data. *Ind. Eng. Chem.* **1940**, *32*, 1474–1477. [CrossRef]
20. Trabelsi, S. Etude Statique et Dynamique de la Cristallisation des Élastomères Sous Tension. Ph.D. Thesis, University of Paris 11, Paris, France, 2002.
21. Samaca Martinez, J.R.; Le Cam, J.B.; Balandraud, X.; Toussaint, E.; Caillard, J. Mechanisms of deformation in crystallizable natural rubber. Part 1: Thermal characterization. *Polymer* **2013**, *54*, 2717–2726. [CrossRef]
22. Le Cam, J.B. Energy storage due to strain-induced crystallization in natural rubber: The physical origin of the mechanical hysteresis. *Polymer* **2017**, *127*, 166–173. [CrossRef]
23. D'Ambrosio, P.; De Tommasi, D.; Ferri, D.; Puglisi, G. A phenomenological model for healing and hysteresis in rubber-like materials. *Int. J. Eng. Sci.* **2008**, *46*, 293–305. [CrossRef]
24. Dorfmann, K.N.G.F.A.; Ogden, R.W. Modelling dilatational stress softening of rubber. In *Constitutive Models for Rubber III*; Busfield, J., Muhr, A., Eds.; Balkema: London, UK, 2003; pp. 253–261.
25. Rey, T.; Chagnon, G.; Favier, D.; Le Cam, J.B. Hyperelasticity with rate-independent microsphere hysteresis model for rubberlike materials. *Comput. Mater. Sci.* **2014**, *90*, 89–98. [CrossRef]
26. Vandenbroucke, A.; Laurent, H.; Hocine, N.A.; Rio, G. A Hyperelasto-Visco-Hysteresis model for an elastomeric behaviour: Experimental and numerical investigations. *Comput. Mater. Sci.* **2010**, *48*, 495–503. [CrossRef]
27. Mott, P.; Giller, C.; Fragiadakis, D.; Rosenberg, D.; Roland, C. Deformation of polyurea: Where does the energy go? *Polymer* **2016**, *105*, 227–233. [CrossRef]
28. Marco, Y.; Huneau, B.; Masquelier, I.; Le Saux, V.; Charrier, P. Prediction of fatigue properties of natural rubber based on the descriptions of the cracks population and of the dissipated energy. *Polym. Test.* **2017**, *59*, 67–74. [CrossRef]
29. Li, L.; Ji, H.; Yang, H.; Zhang, L.; Zhou, X.; Wang, R. Itaconate Based Elastomer as a Green Alternative to Styrene–Butadiene Rubber for Engineering Applications: Performance Comparison. *Processes* **2020**, *8*, 1527. [CrossRef]
30. Lee, S.H.; Park, S.Y.; Chung, K.H.; Jang, K.S. Phlogopite-Reinforced Natural Rubber (NR)/Ethylene-Propylene-Diene Monomer Rubber (EPDM) Composites with Aminosilane Compatibilizer. *Polymers* **2021**, *13*, 2318. [CrossRef]
31. Maciejewska, M.; Sowińska, A.; Grocholewicz, A. Zinc Complexes with 1,3-Diketones as Activators for Sulfur Vulcanization of Styrene-Butadiene Elastomer Filled with Carbon Black. *Materials* **2021**, *14*, 3804. [CrossRef] [PubMed]
32. Araujo-Morera, J.; Verdugo-Manzanares, R.; González, S.; Verdejo, R.; Lopez-Manchado, M.A.; Hernández Santana, M. On the Use of Mechano-Chemically Modified Ground Tire Rubber (GTR) as Recycled and Sustainable Filler in Styrene-Butadiene Rubber (SBR) Composites. *J. Compos. Sci.* **2021**, *5*, 68. [CrossRef]
33. Surya, I.; Waesateh, K.; Masa, A.; Hayeemasae, N. Selectively Etched Halloysite Nanotubes as Performance Booster of Epoxidized Natural Rubber Composites. *Polymers* **2021**, *13*, 3536. [CrossRef]
34. Farren, W.S.; Taylor, G.I. The Heat Developed during Plastic Extension of Metals. *Proc. R. Soc. Lond. Math. Phys. Eng. Sci.* **1925**, *107*, 422–451.
35. Taylor, G.I.; Quinney, H. The Latent Energy Remaining in a Metal after Cold Working. *Proc. R. Soc. Lond. Math. Phys. Eng. Sci.* **1934**, *143*, 307–326. [CrossRef]
36. Chrysochoos, A. Energy balance for elastic plastic deformation at finite strain (in French). *J. Méc. Appl.* **1985**, *5*, 589–614.
37. Chrysochoos, A.; Maisonneuve, O.; Martin, G.; Caumon, H.; Chezeau, J.O. Plastic and dissipated work and stored energy. *Nucl. Eng. Des.* **1989**, *114*, 323–333. [CrossRef]
38. Mason, J.; Rosakis, A.; Ravichandran, G. On the strain and strain rate dependence of the fraction of plastic work converted to heat: An experimental study using high speed infrared detectors and the Kolsky bar. *Mech. Mater.* **1994**, *17*, 135–145. [CrossRef]
39. Rittel, D. On the conversion of plastic work to heat during high strain rate deformation of glassy polymers. *Mech. Mater.* **1999**, *31*, 131–139. [CrossRef]
40. Oliferuk, W.; Maj, M.; Raniecki, B. Experimental analysis of energy storage rate components during tensile deformation of polycrystals. *Mater. Sci. Eng. A* **2004**, *374*, 77–81. [CrossRef]
41. Rittel, D. An investigation of the heat generated during cyclic loading of two glassy polymers. Part I: Experimental. *Mech. Mater.* **2000**, *32*, 131–147. [CrossRef]
42. Rittel, D.; Rabin, Y. An investigation of the heat generated during cyclic loading of two glassy polymers. Part II: Thermal analysis. *Mech. Mater.* **2000**, *32*, 149–159. [CrossRef]
43. Benaarbia, A.; Chrysochoos, A.; Robert, G. Kinetics of stored and dissipated energies associated with cyclic loadings of dry polyamide 6.6 specimens. *Polym. Test.* **2014**, *34*, 155–167. [CrossRef]
44. Benaarbia, A.; Chrysochoos, A.; Robert, G. Influence of relative humidity and loading frequency on the PA6.6 thermomechanical cyclic behavior: Part II. Energy aspects. *Polym. Test.* **2015**, *41*, 92–98. [CrossRef]

45. Lachhab, A.; Robin, E.; Le Cam, J.B.; Mortier, F.; Tirel, Y.; Canevet, F. Energy stored during deformation of crystallizing TPU foams. *Strain* **2018**, *54*, e12271. [CrossRef]
46. Loukil, M.; Corvec, G.; Robin, E.; Miroir, M.; Le Cam, J.B.; Garnier, P. Stored energy accompanying cyclic deformation of filled rubber. *Eur. Polym. J.* **2018**, *98*, 448–455. [CrossRef]
47. Khiem, V.; Le Cam, J.B.; Charlès, S.; Itskov, M. Thermodynamics of strain-induced crystallization in filled natural rubber under uni- and biaxial loadings. Part I: Complete energetic characterization and crystallinity evaluation. *J. Mech. Phys. Solids* **2022**, *159*, 104701. [CrossRef]
48. Marchal, J. Cristallisation des Caoutchoucs Chargés et Non Chargés Sous Contrainte: Effet sur les Chaînes Amorphes. Ph.D. Thesis, Université de Paris 11, Paris, France, 2006.
49. Chrysochoos, A.; Huon, V.; Jourdan, F.; Muracciole, J.; Peyroux, R.; Wattrisse, B. Use of full-Field digital image correlation and infrared thermography measurements for the thermomechanical analysis of material behaviour. *Strain* **2010**, *46*, 117–130. [CrossRef]
50. Caborgan, R. Contribution à L'analyse Expérimentale du Comportement Thermomécanique du Caoutchouc Naturel. Ph.D. Thesis, Université de Montpellier, Montpellier, France, 2011.
51. Chrysochoos, A. Analyse du comportement des matériaux par thermographie Infra Rouge. *Colloq. Photomécanique* **1995**, *95*, 201–211.
52. Chrysochoos, A.; Louche, H. An infrared image processing to analyse the calorific effects accompanying strain localisation. *Int. J. Eng. Sci.* **2000**, *38*, 1759–1788. [CrossRef]
53. Charlès, S.; Le Cam, J.B. Inverse identification from heat source fields: A local approach applied to hyperelasticity. *Strain* **2020**, *56*, e12334. [CrossRef]
54. Primel, A.; Ferec, J.; Ausias, G.; Tirel, Y.; Veille, J.M.; Grohens, Y. Solubility and interfacial tension of thermoplastic polyurethane melt in super-critical carbon dioxide and nitrogen. *J. Supercrit. Fluids* **2017**, *122*, 52–57. [CrossRef]
55. Toki, S.; Fujimaki, T.; Okuyama, M. Strain-induced crystallization of natural rubber as detected real-time by wide-angle X-ray diffraction technique. *Polymer* **2000**, *41*, 5423–5429. [CrossRef]
56. Trabelsi, S.; Albouy, P.A.; Rault, J. Effective local deformation in stretched filled rubber. *Macromolecules* **2003**, *36*, 9093–9099. [CrossRef]
57. Grady, B.; Cooper, S. Thermoplastic elastomers. In *The Science and Technology of Rubber*; Mark, J.E., Erman, B., Eirich, F.R., Eds.; Elsevier Inc.: San Diego, CA, USA, 2005; pp. 555–618.
58. Blundell, D.; Eeckhaut, G.; Fuller, W.; Mahendrasingam, A.; Martin, C. Real time SAXS/stress-strain studies of thermoplastic polyurethanes at large strains. *Polymer* **2002**, *43*, 5197–5207. [CrossRef]
59. Yeh, F.; Hsiao, B.; Sauer, B.; Michael, S.; Siesler, H. In-situ studies of structure development during deformation of a segmented poly(urethane-urea) elastomer. *Macromolecules* **2003**, *36*, 1940–1954. [CrossRef]
60. Unsal, E.; Yalcin, B.; Yilgor, I.; Yilgor, E.; Cakmak, M. Real time mechano-optical study on deformation behavior of PTMO/CHDI-based polyetherurethanes under uniaxial extension. *Polymer* **2009**, *50*, 4644–4655. [CrossRef]
61. Scetta, G.; Euchler, E.; Ju, J.; Selles, N.; Heuillet, P.; Ciccotti, M.; Creton, C. Self-Organization at the Crack Tip of Fatigue-Resistant Thermoplastic Polyurethane Elastomers. *Macromolecules* **2021**, *54*, 8726–8737. [CrossRef]

Article

The Influence of Colloidal Properties of Carbon Black on Static and Dynamic Mechanical Properties of Natural Rubber

William Amoako Kyei-Manu [1], Charles R. Herd [2], Mahatab Chowdhury [2], James J. C. Busfield [1,*] and Lewis B. Tunnicliffe [2]

[1] School of Engineering and Materials Science, Queen Mary University of London, London E1 4NS, UK; w.a.kyei-manu@qmul.ac.uk
[2] Birla Carbon, Marietta, GA 30062, USA; charles.herd@adityabirla.com (C.R.H.); mahatab.chowdhury@adityabirla.com (M.C.); lewis.tunnicliffe@adityabirla.com (L.B.T.)
* Correspondence: j.busfield@qmul.ac.uk; Tel.: +44-20-7882-8866

Abstract: The influence of carbon black (CB) structure and surface area on key rubber properties such as monotonic stress-strain, cyclic stress–strain, and dynamic mechanical behaviors are investigated in this paper. Natural rubber compounds containing eight different CBs were examined at equivalent particulate volume fractions. The CBs varied in their surface area and structure properties according to a wide experimental design space, allowing robust correlations to the experimental data sets to be extracted. Carbon black structure plays a dominant role in defining the monotonic stress–strain properties (e.g., secant moduli) of the compounds. In line with the previous literature, this is primarily due to strain amplification and occluded rubber mechanisms. For cyclic stress–strain properties, which include the Mullins effect and cyclic softening, the observed mechanical hysteresis is strongly correlated with carbon black structure, which implies that hysteretic energy dissipation at medium to large strain values is isolated in the rubber matrix and arises due to matrix overstrain effects. Under small to medium dynamic strain conditions, classical strain dependence of viscoelastic moduli is observed (the Payne effect), the magnitude of which varies dramatically and systematically depending on the colloidal properties of the CB. At low strain amplitudes, both CB structure and surface area are positively correlated to the complex moduli. Beyond ~2% strain amplitude the effect of surface area vanishes, while structure plays an increasing and eventually dominant role in defining the complex modulus. This transition in colloidal correlations reflects the transition in stiffening mechanisms from flexing of rigid percolated particle networks at low strains to strain amplification at medium to high strains. By rescaling the dynamic mechanical data sets to peak dynamic stress and peak strain energy density, the influence of CB colloidal properties on compound hysteresis under strain, stress, and strain energy density control can be estimated. This has considerable significance for materials selection in rubber product development.

Keywords: carbon black; elastomer; rubber; tensile; Mullins effect; Payne effect; dynamic strain; hysteresis; natural rubber

1. Introduction

The use of carbon black (CB) as a reinforcing agent for rubber allows for very precise tuning of rubber compound behavior via the appropriate selection of the colloidal properties and loading level of the CB in the compound. The incorporation of CB into rubber affects practically every aspect of rubber behavior. Properties of particular interest include the stress–strain and dynamic mechanical behavior of rubbers. These properties have major influences on final product performance, including the static and dynamic stiffness and deflection of the rubber under various loading conditions, material compliance and traction (in the case of rubber–surface contact, experienced for example in tires and dynamic/static seals), and the heat buildup and mechanical energy dissipation of the rubber product,

which is of particular concern for tire fuel efficiency. These properties play a critical role in defining the fatigue and lifetime performance of rubber products, although fatigue performance is beyond the scope of this particular work.

A substantial body of historical experimental work has been performed in order to understand the role of CB in rubber reinforcement, particularly in regard to selection of CB and the effects of CB on the functional and material properties of rubber compounds [1–9]. However, large gaps in the understanding of the exact mechanisms of reinforcement persist. Specifically, the mechanisms by which rubber is stiffened by incorporation of CB and other particulates as well as the strain history (Mullins Effect) and dynamic strain amplitude dependence (Payne Effect) of CB-reinforced rubbers remain incompletely understood, despite being of broad industrial significance. The basic stiffening effect of CB has to various degrees been attributed to strain amplification/overstrain of the rubber matrix [10–13], particulate networking/flocculation [14–16], and modifications to the local dynamics of the rubber matrix [17,18], all of which are dependent on the strain, temperature, and strain rate conditioning of the rubber compound. The strain history, or Mullins Effect, which is typically measured at moderate to large strains, has been extensively investigated [19–22] and attributed to microstructural damage or reorganizations originating from strain amplification/overstrain of the rubber matrix [23], strain dependent rupture or damage of flocculated particle clusters (which may [17] or may not [24,25] be percolated via surface immobilized polymer fractions), and slippage or rupture of physically or chemisorbed rubber chain–CB bonds [3,26]. The Payne Effect is observed at low to moderate strains, typically by application of dynamic strain ramps, and has been attributed to a breakdown and reformation of a particle network within the rubber which is percolated via direct particle contacts, van der Waals interactions [14,27,28], or surface-immobilized rubber [17]. Conversely, it has also been proposed that labile connections between the rubber matrix and the particle surface are responsible for the Payne Effect [29,30]. A complete and fully accepted microstructural explanation of these various manifestations of particulate reinforcement of rubber is yet to be realised. It is important to note that any such explanation should be capable of describing all observed phenomenology of rubber reinforcement. To that end, it is important to comprehensively map out the influence of CB colloidal properties on these manifestations of rubber reinforcement.

The fundamental particulate unit of carbon black is the aggregate, which is formed by a fused assembly of broadly spherical para-crystalline primary particles with diameters ranging from ~200 nm to ~5 nm. The number and spatial arrangement of primary particles comprising the aggregate define its "structure" level. Particle size and structure level are parameters which can be independently controlled during production of carbon black in furnace reactors. Particle size can be measured directly using transmission electron microscopy [31] or inferred from bulk measurements of surface area using gas adsorption techniques [32]. Structure is typically measured by oil adsorption tests [33,34]. These parameters play a key role in defining the levels of reinforcement imparted to a rubber compound by carbon black. For example, primary particle size is the key parameter defining both the contact area between CB and rubber and the number of aggregates per unit volume of rubber, and therefore governs the average inter-aggregate distance and aggregate–aggregate "networking". From simple geometrical considerations, the number density of primary CB particles per unit volume of rubber compound scales with the cube of the surface area. The aggregate structure is related to the volume of rubber occluded or screened from globally-applied strains by aggregate branches [30,35]. The effective volume fraction of solid in a rubber compound is therefore the sum of the CB volume and the volume of occluded rubber, with the latter directly related to CB structure level [35–37]. This has a direct impact on levels of strain amplification in the compound. Other important parameters of carbon black include surface chemistry/activity [3], porosity, thermal history [3,38], and the distributional nature of primary particle size and aggregate structure [37]. A detailed exploration of the effects of these parameters on rubber compound behavior is beyond the scope of this study.

In this work, we comprehensively examine the influence of CB surface area and structure on key properties of natural rubber compounds prepared with iso-loading (and volume fraction) of CB. A very wide colloidal space experimental design approach is taken, using furnace CBs varying only in their respective levels of structure and surface area. The wide colloidal space approach allows interpolation of our results and conclusions to cover the majority of industrial furnace CBs. From the resulting rubber compound data, new correlations and key insights can be drawn into the origins of CB reinforcement and design, and selection guides for CBs are consequently revisited.

2. Materials and Methods

2.1. Materials

Compounds of SMR CV60 natural rubber (NR) reinforced with eight different carbon black (CB) grades at 50 parts per hundred (phr) loading were prepared. An unreinforced NR counterpart was included in the tests for comparison. The CBs used in this study were selected to cover a broad range of surface area and structure, roughly correlating to a dual-factor central composite experimental design (as shown in Figure 1). Figure 1 shows several commonly used CB grades (N772, N660, N347, N330, N220, N115). These CBs are not evaluated in this work but are included in the figure to provide additional context. Table 1 shows the compound formulation used in this study. Table 2 provides the structure and surface area of the various CB grades. For the purposes of this paper, a naming convention is adopted which allows the reader to immediately identify the type of CB based on its CB structure and surface area; this is provided in Table 2 as well. The structure (measured by compressed oil absorption number, COAN) and surface area (measured by statistical thickness surface area, STSA) values of the CBs are listed as a superscript and a subscript, respectively. For example, N550, which has a structure value of 84 cc.100 g^{-1} and a surface area value of 37 $m^2 \cdot g^{-1}$, is referred to as CB^{84}_{37}. The corresponding rubber compound produced using N550 is referred to by the same naming convention. Table 2 shows the interferometric microscope (IFM) dispersion index (DI) values of the final compounds measured according to ASTM D2663 (method D) [39]. Transmission electron microscopy (TEM) images of the tested carbon black samples are provided in the Supplementary Materials.

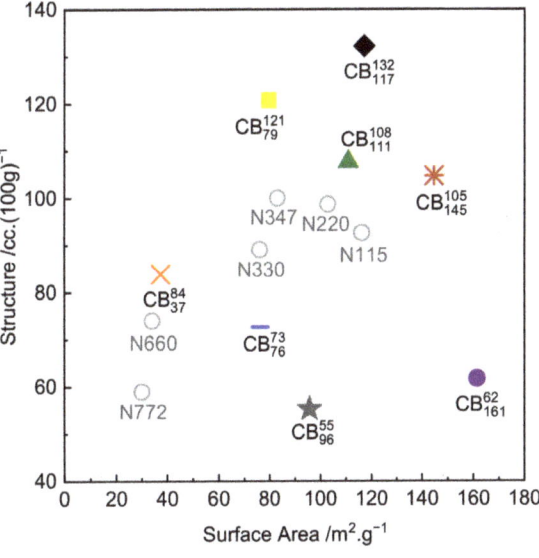

Figure 1. Colloidal plot of tested carbon blacks, with commonly used carbon black grades for reference (open circles).

Table 1. Compound formulation.

Component	Loading/Parts per Hundred Rubber (phr)	Manufacturer of Component
NR—SMR CV-60	100	Herman Weber & Co.
Carbon Black	50	Birla Carbon
Zinc Oxide	5	Akrochem
Stearic Acid	3	PMC Biogenix
Anti-ozonant/Antioxidant	3	Americas International
Micro-wax	2	Strahl & Pitsch
Sulphur	2.5	R.E. Carroll
TBBS *-75	0.8	Akrochem

* N-Tertiarybutyl-2-benzothiazole sulfonamide, 75% assay.

Table 2. CB structure and surface area; CB naming convention adopted in this paper; compounds and compound dispersion index.

Carbon Black/ Compound Code	Structure (COAN)/ cc.(100 g)$^{-1}$	Surface Area (STSA)/ m$^2\cdot$g^{-1}	Carbon Black Commercial Name	Corresponding Compound Dispersion Index
Unfilled NR	NA	NA	NA	NA
CB^{132}_{117}	132	117	BC2005	99.3
CB^{105}_{145}	105	145	BC2115	98.8
CB^{121}_{79}	121	79	BC2013	98.8
CB^{108}_{111}	108	111	N234	99.4
CB^{73}_{76}	73	76	N326	98.0
CB^{55}_{96}	55	96	Raven 1200	90.2
CB^{62}_{161}	62	161	Raven 2000	81.5
CB^{84}_{37}	84	37	N550	98.7

Compounds were prepared by Birla Carbon (Marietta, GA, USA) using a 1.6 L capacity Banbury mixer. Vulcanized sheets measuring 11 mm × 11 mm × ~2 mm were prepared via compression molding at 150 °C for a time of T_{90} + 5 min where T_{90} (time at 150 °C required for the specimen to reach 90% maximum torque) was measured using a moving die rheometer (MDR) from Alpha Technologies located in Hudson, OH, USA. The mixing procedure used to prepare the compounds is provided in the Supplementary Materials.

2.2. Shore A Hardness, Tensile to Break, and Cyclic Tensile Tests

Shore A hardness measurements were performed according to ASTM D2240 [40].

Dumbbells for uniaxial tensile testing were stamped from sheets of compound using a hydraulic die press. The dumbbells had approximate gauge length, width and thickness dimensions similar to ASTM D412 die C.

For uniaxial tensile testing to break, five dumbbell specimens were pulled until failure using a five station United tensile tester with a 1 kN load cell. Strain was defined using traveling contact extensometers; the extension rate was 500 mm/min (~strain rate of 0.19/s), following ASTM D412 [41].

For cyclic tensile tests, five dumbbells were extended at 500 mm/min to an initial target strain of 20% then retracted to 0% strain. The first cycle at this specified strain was followed by two cycles to the same target strain of 20%. The same specimen was then extended to a higher target strain of 50% for three cycles. This sequence was continued sequentially to higher target strains of 100%, 200%, and 300%.

All tensile and hardness testing was performed at 21 °C ± 1 °C, 55% relative humidity, and atmospheric pressure conditions.

2.3. Dynamic Strain Sweep Characterization

Dynamic strain sweeps between 0.1% and 62.5% single strain amplitude were performed at 10 Hz with zero mean strain using an ARES G2 torsional rheometer from TA

Instruments located in New Castle, DE, USA at 60 °C. Specimen geometries were cylinders measuring 8 mm in diameter and with ~2 mm thickness, which were stamped from sheets of vulcanized compound and bonded to the rheometer parallel plate geometry using Loctite 480 adhesive by Henkel, Hemel Hempstead, UK. A slight compressive normal force of 100 g was applied to the cylinders during the test procedure. We note that this particular specimen geometry has a non-uniform strain field, varying with the radius of the cylinder during deformation; the reported strain amplitude values are for the extremity of the cylinder radius. The strain sweep test was performed by pre-conditioning the specimen six times at the specified dynamic strain amplitude before collecting torque–time data. This process was repeated from low to high strain amplitudes.

3. Results and Discussion

3.1. Compound Dispersion Index

Compound dispersion indices as determined by IFM are presented in Table 2. All compounds except CB_{96}^{55} and CB_{161}^{62} had DI values > 98, indicating nearly full macro-incorporation of the CB into the rubber. The compounds CB_{96}^{55} and CB_{161}^{62} contain the lowest structure CBs in the colloidal experimental design and have comparatively high surface area values. At a fixed surface area, lower structure CBs are notoriously more difficult to disperse in rubber due to (i) the higher number of attractive contacts between aggregates in pelletized CB prior to mixing on a unit volume basis and (ii) the resulting lower mix viscosities versus medium-high structure CBs [42]. Despite the somewhat lower DI values for CB_{96}^{55} and CB_{161}^{62}, the compound dispersions are reasonable and well in line with dispersion indices observed for commercially prepared rubber compounds.

3.2. Tensile Stress Strain and Shore A Hardness Measurements

Figure 2 shows the five stress–strain to failure data sets for each rubber compound. From a visual examination of the data, it is clear that despite the CBs being at equivalent loading/volume fraction in the compounds, the colloidal differences between the CBs impart major differences to the observed stress–strain behavior. CBs with higher structure impart higher secant moduli values in comparison to CBs with lower structure. Table 3 summarizes the modulus at 300% (reported as stress value at 300% elongation), percent elongation at break, tensile strength, and Shore A hardness values of the CB-reinforced and gum NR specimens.

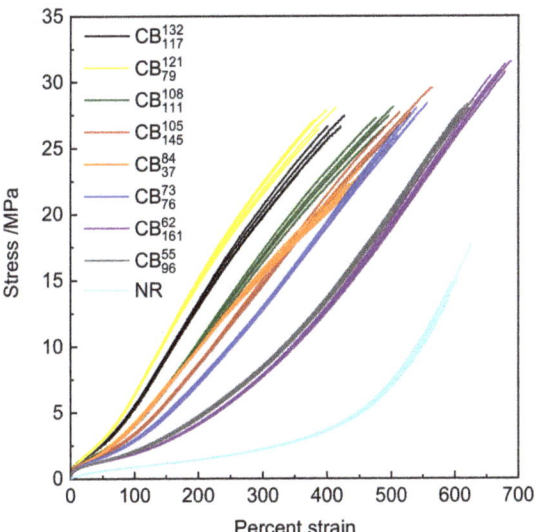

Figure 2. Stress–strain to failure data for the various rubber compounds.

Table 3. Summary of the hardness and tensile properties of the various rubber compounds.

Compound Code	300% Modulus /MPa	Percent Elongation at Break	Tensile Strength /MPa	Shore A Hardness /Shore A
Unfilled NR	2.23	598	14.6	41.6
CB^{132}_{117}	19.89	402	26.6	74.8
CB^{105}_{145}	14.58	530	27.7	70.8
CB^{121}_{79}	21.76	389	27.1	74.6
CB^{108}_{111}	16.69	491	27.3	71.5
CB^{73}_{76}	12.77	536	27.2	66.7
CB^{55}_{96}	8.33	620	28.0	63.7
CB^{62}_{161}	7.82	679	30.8	64.4
CB^{84}_{37}	15.38	461	23.1	66.1

For a quantitative analysis, multiple linear regression analyses of tensile properties to CB colloidal properties were conducted (the NR gum compound data were not included in the regressions). Multiple regressions were performed via error minimization in Origin 2019 per Equation (1), where Y is the dependent property being analyzed, C is an intercept constant, β_{St} is the coefficient of the structure of CB and β_{SA} is the coefficient of the surface area of CB, and ϵ is the error term. The full regression results (intercept, coefficients, p values, and regression R^2) are provided in Table 4. Note that coefficients with p values > 0.05 are highlighted in bold and are not statistically correlated to the observed parameter.

Table 4. Results of multiple linear regression analysis of tensile properties–CB colloidal properties.

Regression Parameter	300% Modulus /MPa	Tensile Strength /MPa	Percent Elongation at Break	Shore A Hardness/Shore A
C	3.52	24.79	671.32	54.33
β_{St}	0.1657	−0.0232	−3.1103	0.1545
COAN p	3.47×10^{-5}	0.1881	1.20×10^{-6}	7.28×10^{-5}
β_{SA}	−0.0404	0.0444	1.2572	0.0047
STSA p	0.0049	0.0091	1.91×10^{-5}	0.6330
Adjusted R^2	0.97	0.71	0.99	0.95

From the multiple regression results, it can be seen that the tensile modulus (here, the example is 300% modulus) is strongly correlated with the structure of CB. The structure coefficient is positive, indicating that increasing the structure of the CB increases the resulting tensile modulus. By contrast, the CB surface area is less strongly correlated (higher p value) and has only a slight negative correlation coefficient with tensile modulus, meaning that increasing the surface area actually very slightly decreases the resulting tensile modulus. The fact that CB structure is the key parameter controlling stress–strain moduli is quite well known and is typically attributed to strain amplification/overstraining of the rubber matrix through occlusion/screening of a certain volume of rubber by the CB aggregate structure [7,8,37]. The effect by which increased surface area slightly reduces modulus values has been tentatively attributed to increasing deactivation of the cure system, for example, by the adsorption of accelerators on the surface of CB and consequent net reduction in compound crosslink density [43].

In addition, tensile failure parameters are correlated to varying extents with CB colloidal properties. Tensile strength increases with increasing CB surface area, likely due to an increase in compound critical tear energy [44]. Structure and surface area have opposing correlations with elongation at break. Higher structure CBs (which produce higher modulus compounds) reach their work at fracture at lower strains, while increasing CB surface area increases tensile strength and therefore increases elongation at break.

Shore A hardness is only correlated with structure, reflecting the predominant trend for the tensile moduli.

$$Y = C + \beta_{St}(COAN) + \beta_{SA}(STSA) + \epsilon \tag{1}$$

3.3. Cyclic Tensile Tests

Figure 3 shows cyclic tensile data for the various rubber compounds. All compounds display evidence of the classical Mullins strain history effect to varying extents. Following initial strain cycles, the compounds display a pronounced softening effect upon subsequent cycling. When the specimen is stretched to higher peak strains than the previous cycles, the initial (virgin) stress–strain curve is recovered. Upon retraction, there are set effects (residual extension remaining) as well as evident mechanical hysteresis between the loading and unloading curves. As observed for the uniaxial tensile tests to break, CBs with higher structure impart a higher initial modulus to the compounds. Compounds containing higher structure CBs show a larger relative drop in stress–strain properties following the initial strain cycles than those containing lower structure CBs. From an initial visual examination, the role of CB particle size/surface area is not obvious.

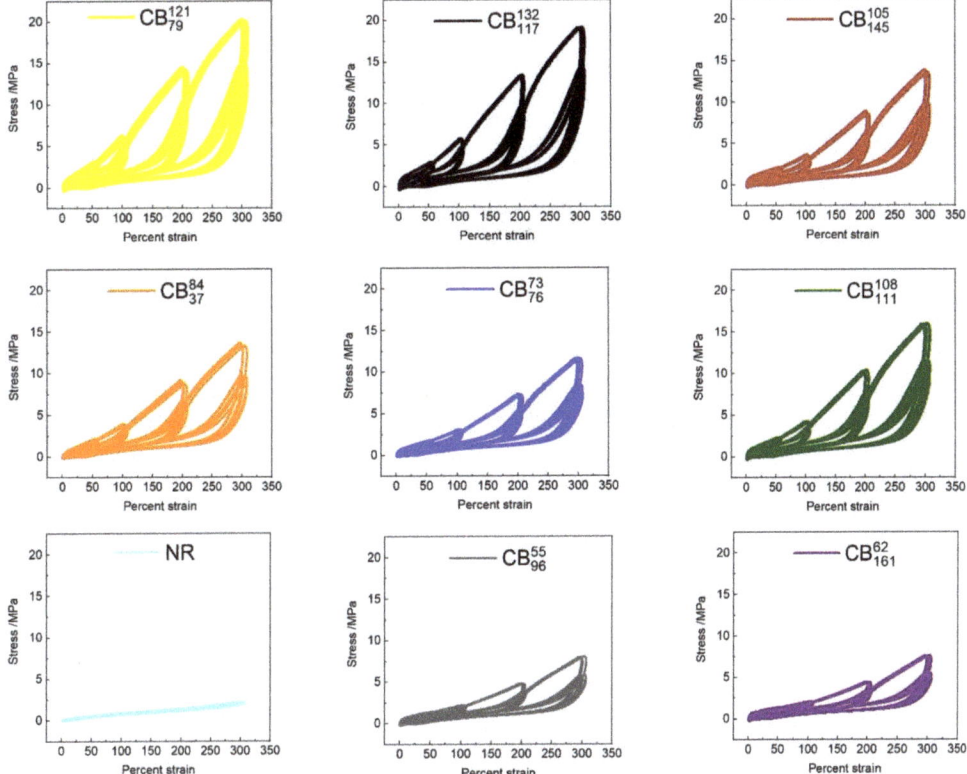

Figure 3. Stress–strain data, showing cyclic tensile test effects of the various rubber compounds.

In order to quantify the magnitude of these softening effects as a function of strain and cycle number, the mechanical hysteresis apparent between loading and unloading of the compounds was calculated as the difference between the respective integrals. Figure 4A–C show the mechanical hysteresis as a function of the first, second, and third strain cycles for each peak tensile strain. As expected, the magnitude of hysteresis increases with increasing tensile strain. The unfilled rubber shows the lowest levels of hysteresis, though nevertheless appreciable. The CB reinforced compounds vary substantially in the levels of hysteresis observed, with the ranking appearing to be predominantly dependent on CB structure. The magnitude of hysteresis observed at a given peak strain reduces upon sequential strain cycling, while the ranking of compounds remains consistent. The influence of the colloidal properties of CB on the magnitude and strain dependence of the mechanical hysteresis was evaluated by performing multiple linear regression analyses of hysteresis values at each peak strain level compared to the structure and surface area of the carbon blacks. The full results of these regression analyses are provided in the Supplementary Materials. Figure 4D shows the multiple regression coefficients of both the structure and surface area of CB at each peak strain. Data points are solid where their regression p values are <0.05 and dashed when p values are >0.05. CB surface area is only correlated with hysteresis at the lowest strain levels, whereas structure is strongly correlated across the entire strain history range. This finding has a number of implications:

- At a practical level, the degree of mechanical hysteresis (and therefore softening) of a rubber compound at a fixed strain level scales with the virgin modulus of the compound at that strain. All other parameters being equal (such as CB volume fraction, polymer type, and crosslink density), this modulus is determined by the structure of the CB in the formulation [45].
- At a microstructural level, the strong correlation between hysteresis and CB structure provides several hints as to the origin of the Mullins-type hysteresis and softening. It suggests that the hysteretic energy dissipation at these large strains is isolated in the rubber matrix and arises due to strain amplification/matrix overstrain, as opposed to hysteretic polymer–particle surface slippage and/or hysteretic breakup of flocculated aggregate clusters, which have been proposed in the literature. Note that strain amplification as described by hydrodynamic-type equations is independent of CB particle size/surface area, which is consistent with our observations [10–12]. In these experiments, specimens have been cycled to specified strain levels. Harwood, Mullins, and Payne [23] conducted highly relevant experiments where specimens were cycled to specified stress levels. Under these conditions, the resulting mechanical hysteresis values were found to be identical for a wide range of CB reinforced and gum NR compounds. These findings are consistent with our results in the sense that they can both be explained if we assume that energy dissipation occurring at these large strains is isolated predominantly in the overstrained rubber matrix.

3.4. Dynamic Strain Sweeps

Figure 5 shows example stress–strain-time raw data at selected strain amplitudes collected for the CB^{107}_{111} specimen. As can be seen, the stress–strain response of the material is elliptical and the stress–time response shows sinusoidal behavior, even up to relatively large strain amplitudes. This is both an unusual aspect of rubber rheology (as most complex materials show non-sinusoidal distortions in their dynamic mechanical behavior at moderate-large strains [46]) and a highly advantageous aspect, as it allows application of linear viscoelasticity for analysis of commercially relevant materials at commercially relevant deformation amplitudes [15,28,47,48].

Key linear viscoelastic parameters and interrelationships are provided by Equations (2)–(6), assuming a strain-controlled experiment with application of a sinusoidal shear strain amplitude γ_0 at fixed frequency ω, eliciting a time dependent stress response, $\sigma(t)$, which can be decomposed into elastic and loss moduli (G' and G'', respectively). The magnitude of the complex modulus $|G^*|$, loss tangent $\tan \delta$, and loss compliance J'' values can then be defined in terms of the elastic and loss moduli as follows:

$$\gamma(t) = \gamma_0 \sin(\omega t) \qquad (2)$$

$$\sigma(t) = \gamma_0 \left[G'(\omega) \sin \omega t + G''(\omega) \cos \omega t \right] \qquad (3)$$

$$|G^*| = \sqrt{G'^2 + G''^2} \qquad (4)$$

$$\tan \delta = \frac{G''}{G'} \qquad (5)$$

$$J'' = \frac{G''}{G'^2 + G''^2} = \frac{G''}{|G^*|^2} \tag{6}$$

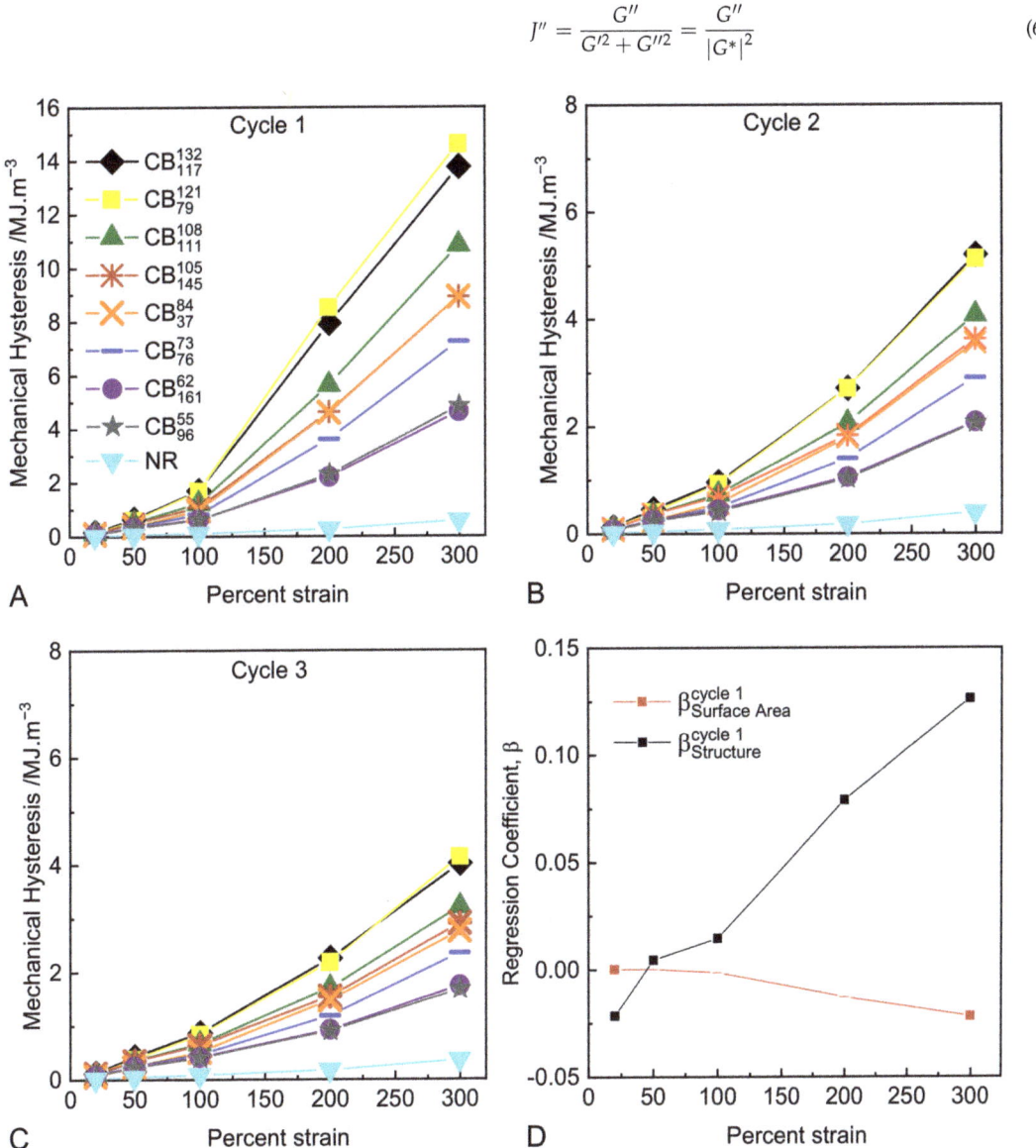

Figure 4. (**A**–**C**) Mechanical hysteresis after the first, second and third, strain cycles respectively (note the difference in ordinate scale between the first, second, and third cycle data); (**D**) regression coefficients of structure and surface area compared to mechanical hysteresis as measured on the first strain cycle (data points with p values > 0.05 are excluded).

Plots of $|G^*|$ versus strain amplitude for each compound in this study are presented in Figure 6A. The strain dependence of the viscoelastic parameters of CB reinforced rubbers versus unreinforced rubber is widely known as the Payne effect, and is ascribed to dynamic breakdown and reformation of a particle network within the rubber compound. The detailed physics and micro-mechanics of such networks of particles remain rather poorly understood. The complex moduli data in Figure 6A exhibit several interesting features. There are large variations in the magnitude and ranking of $|G^*|$ at the smallest strain amplitudes, dependent on the colloidal properties of the CB. However, the ranking of $|G^*|$ magnitude changes substantially as the strain amplitude is increased, and data set "crossover" effects are observed.

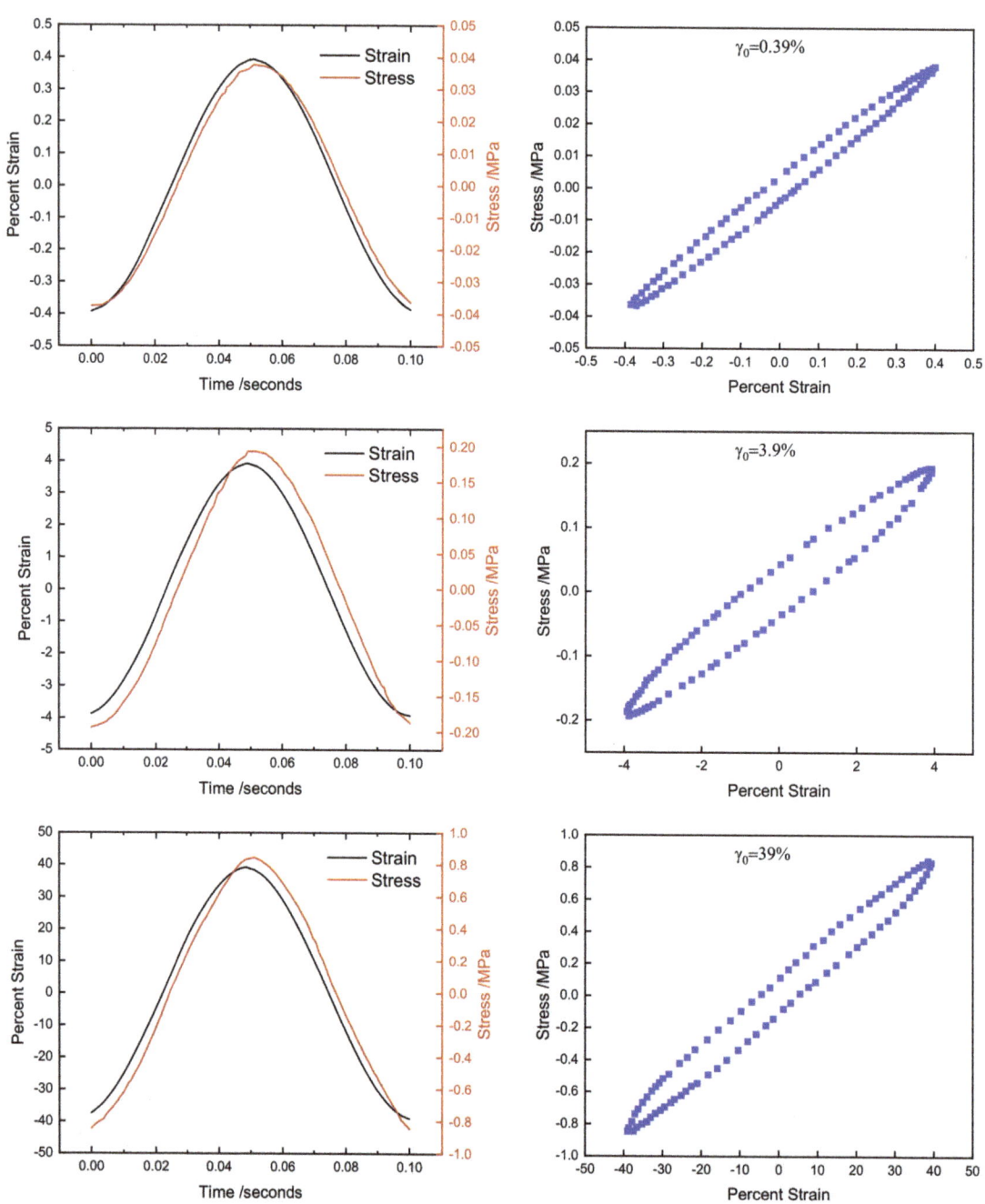

Figure 5. Stress–strain–time data for compound CB_{111}^{107} at selected strain amplitudes.

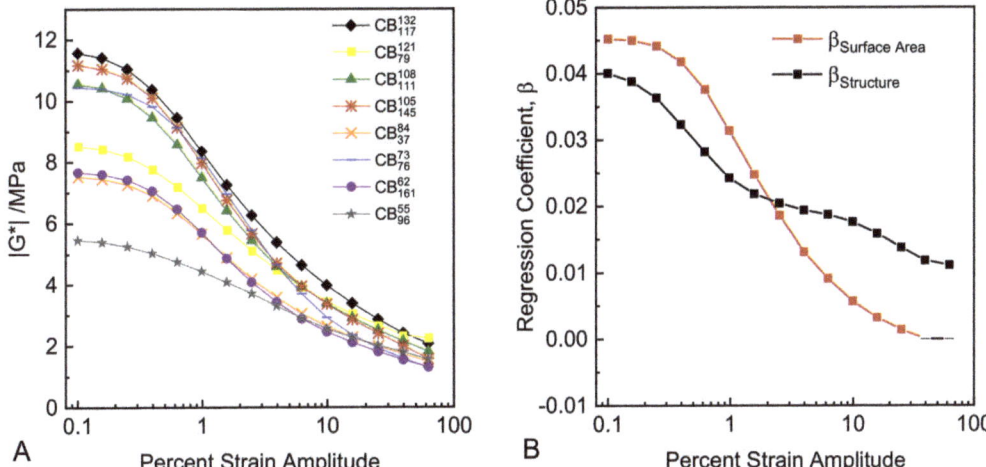

Figure 6. (**A**) $|G^*|$ versus strain amplitude; (**B**) multiple regression coefficients for structure and surface area versus strain amplitude (points with p values > 0.05 are shown as dashes).

The influence of the colloidal properties of CB on the magnitude and strain dependence of the complex moduli were evaluated by performing multiple linear regression analyses of $|G^*|$ values at each strain amplitude as compared to the structure and surface area of the carbon blacks. The full results of these regression analyses are provided in the Supplementary Materials. Figure 6B shows the multiple regression coefficients of both the structure and surface area of CB at each strain amplitude. At low strain amplitudes (0.1–2%), both the structure and surface area of CB are positively correlated with $|G^*|$ to similar extents. The magnitudes of the coefficients decrease with increasing strain amplitude, and structure starts to dominate over surface area at higher strain amplitudes on a relative basis. With increasing strain amplitude, the coefficient of surface area to $|G^*|$ reduces to zero, while structure plays an increasing and eventually dominating role in defining $|G^*|$. In Figure 6B, coefficient data points are solid where their regression p values are <0.05 and dashed when p values are >0.05; p values for both structure and surface area are <0.05 at strain amplitudes <2%, indicating strong statistical correlation of both colloidal properties to $|G^*|$. After ~2% strain amplitude, the p value of the surface area steadily increases with increasing strain amplitude until it exceeds 0.05 at higher strains, indicating no statistically significant correlation with $|G^*|$.

Therefore, with increasing strain amplitude there is a clear transition of the colloidal properties of CB controlling $|G^*|$, and by extension, other viscoelastic parameters. At the microstructural level this transition can be interpreted as the result of the strain-dependent breakdown of rigid particle networks, with the degree of particle networking being governed by the number of aggregates per unit volume of rubber, which at fixed volume fraction is roughly the cube power of the surface area. This continues until at higher strains the primary stiffening mechanism becomes that of strain amplification alone, as defined by solid and occluded rubber fractions, CB loading, and structure level. This observed transition is consistent with the earlier observation that CB structure is strongly correlated with the tensile stress–strain moduli, which are measured at larger strain ranges than in the dynamic experiments.

At a practical level, this separability of surface area and structure effects on $|G^*|$ depending on the applied strain level allows for precise engineering of rubber compounds with strain-dependent viscoelastic moduli via appropriate selection of CB surface area and structure.

In service, rubber components are deformed under conditions of either strain, stress, strain energy control, or more commonly in complex combinations of several of these conditions. This has enormous practical consequences for materials selection and component design, particularly as relates to energy dissipation and fatigue life performance [49,50]. Under constant cyclic strain amplitude γ_0, stress amplitude σ_0, and strain energy density amplitude W_0, the energy dissipation density, W_d, can be derived from linear viscoelastic parameters (see Supplementary Materials), yielding Equations (7)–(9), respectively:

$$W_d(\gamma_0) = \pi \gamma_0^2 G'' \tag{7}$$

$$W_d(\sigma_0) = \pi \sigma_0^2 J'' \tag{8}$$

$$W_d(W_0) \propto \tan \delta \tag{9}$$

The hysteresis performance of the rubber compounds under various deformation controls can therefore be predicted by comparison of the appropriate viscoelastic parameter at the equivalent deformation control condition (stress, strain, strain energy density).

For strain control, per Equation (7), compound hysteresis scales according to the loss modulus. Figure 7A shows the loss moduli derived from the raw data of the strain sweep experiments via Equations (2) and (3). The loss moduli data exhibit a pronounced strain dependence, with a peak in magnitude at around 2% strain amplitude. The magnitude of the G'' values vary substantially depending on the exact colloidal properties of the CB. Figure 7B shows the regression coefficients of surface area and structure for G'' values at each strain amplitude from multiple linear regression analyses; the full results of the regression analyses are provided in the Supplementary Materials. As can be seen, both the surface area and structure of CB are positively correlated with G'', with surface area having the larger contribution at lower strains and structure having the larger contribution at high strains, similar to the observed correlations for the complex moduli. Therefore, in order to minimize compound hysteresis in strain control, CB with low surface area and low structure should be selected, with the resulting compound dynamic stiffness being reduced. A particularly interesting data set in Figure 7A is CB_{161}^{62}, which contains a CB having very high surface area and very low structure. This compound has the highest G'' value at low to medium strains, owing to the significant stiffening effect of the surface area contribution, and a mid-ranked G'' value at high strains, where the more limited contribution of its structure to G'' dominates.

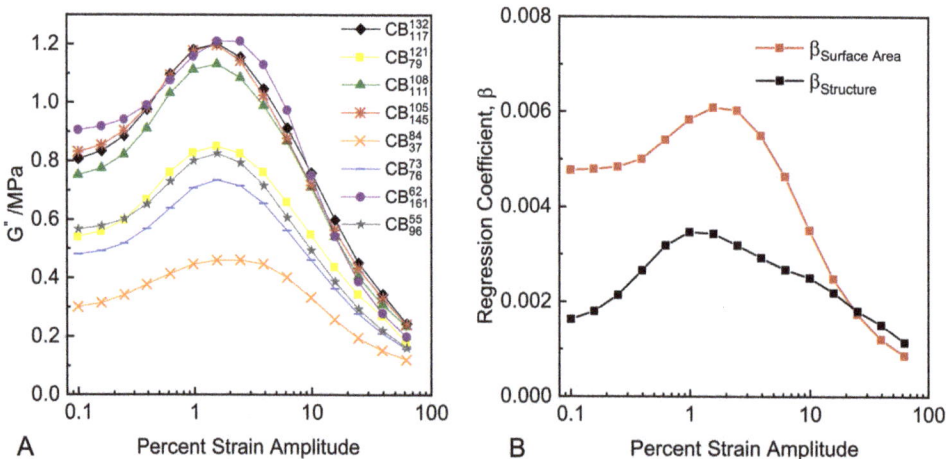

Figure 7. (**A**) G'' plotted versus strain amplitude; (**B**) multiple regression coefficients for structure and surface area plotted versus strain amplitude.

Characterizations of rubber material viscoelasticity are typically performed in strain control; however, it is possible to re-scale data collected in strain control to peak stress and peak strain energy density values collected during testing in order to obtain insight into the potential compound performance in other modes of deformation control. There are two key differences between this re-scaling approach and the direct collection of stress and strain energy density-controlled experimental data: (i) strain rate is controlled as opposed to stress rate or energy rate; and (ii) deformation history (the sequence of application of deformation cycles during the experiment) is strain controlled as opposed to stress or energy controlled. These limitations should be borne in mind for the proceeding analysis; the ideal situation is full experimental characterization of rubber compounds under each mode of deformation control.

Figure 8A shows the peak oscillatory stress values measured at each strain amplitude for each compound. Figure 8B shows the corresponding strain energy densities at each strain amplitude for each compound, which are calculated by integration of the data sets in Figure 8A. These data can be used to re-scale the relevant viscoelastic parameters in an attempt to predict hysteresis performance for strain energy control and stress control from data collected under strain control.

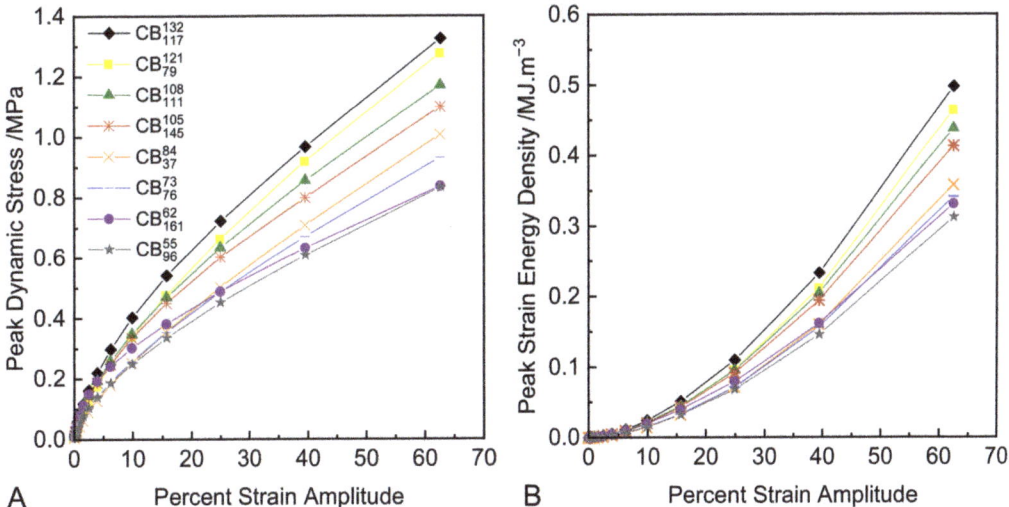

Figure 8. (**A**) Peak dynamic stress plotted versus strain amplitude; (**B**) peak dynamic strain energy density plotted versus strain amplitude.

For the case of strain energy density control, the loss tangent is predictive of hysteresis (Equation (9)). Figure 9A shows the loss tangent as a function of strain amplitude. The classical peak in loss tangent with increasing strain amplitude is observed for each compound. Figure 9B shows the loss tangent re-plotted as a function of peak strain energy density, and Figure 9C shows the coefficients for CB surface area and structure to $\tan \delta$ as determined by multiple linear regressions carried out on data sets interpolated from Figure 9B. The full results of the regression analyses and examples of the interpolated $\tan \delta$–stress data sets are provided in the Supplementary Materials. As can be seen from Figure 9C, only the surface area of CB is statistically correlated with $\tan \delta$ over the majority of the strain energy density range. Therefore, in order to minimize hysteresis in strain energy density control, the CB surface area should be minimized.

For the case of stress control, loss compliance is predictive of hysteresis (Equation (8)). Figure 10A shows loss compliance as a function of strain amplitude. Figure 10B shows loss compliance re-plotted as a function of peak stress amplitude. The loss compliance of the various compounds increases with increasing strain amplitude as the compound softens, and shows complex and stress-dependent material rankings over the range of peak stress values observed. Figure 10C shows the coefficients for CB surface area and structure as determined by multiple linear regressions carried out on data sets interpolated from Figure 10B. The full results of the regression analyses and examples of the interpolated J''-stress data sets are provided in the Supplementary Materials. The regression results show a complex relationship between CB colloidal properties and J''.

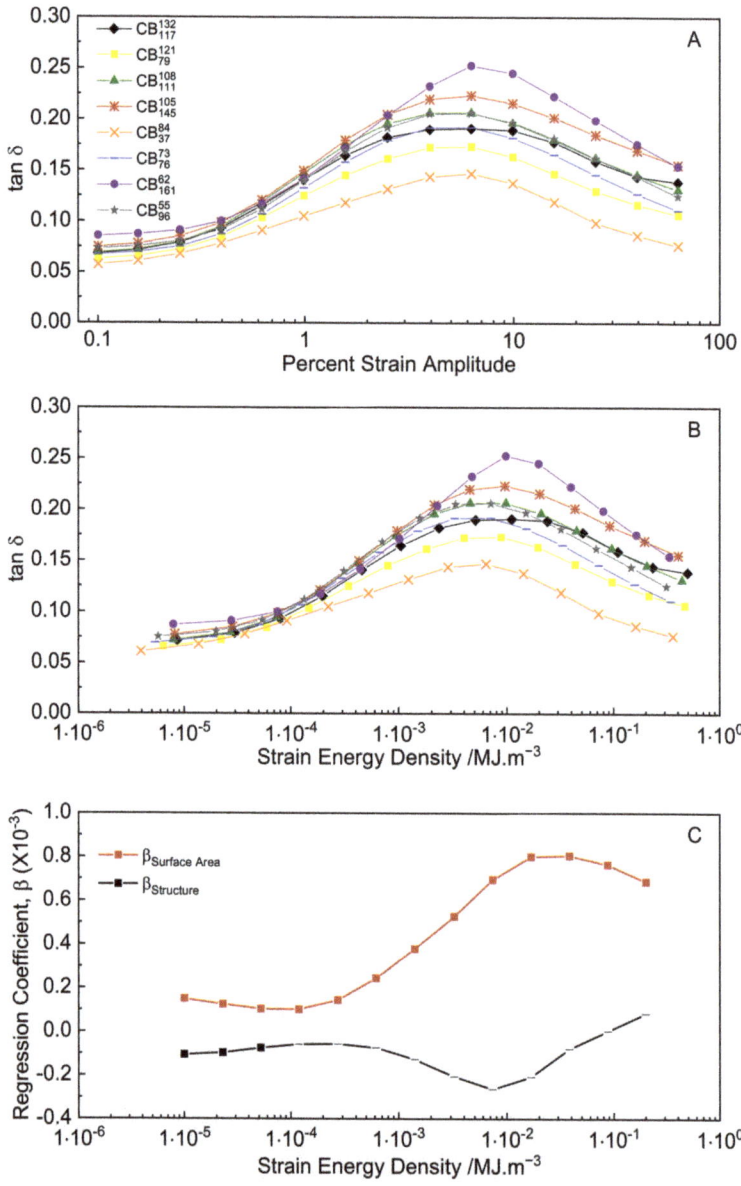

Figure 9. (**A**) tan δ versus strain amplitude; (**B**) tan δ versus peak dynamic strain energy density; (**C**) multiple regression coefficients for structure and surface area versus peak dynamic strain energy density (points with p values > 0.05 are shown as dashes).

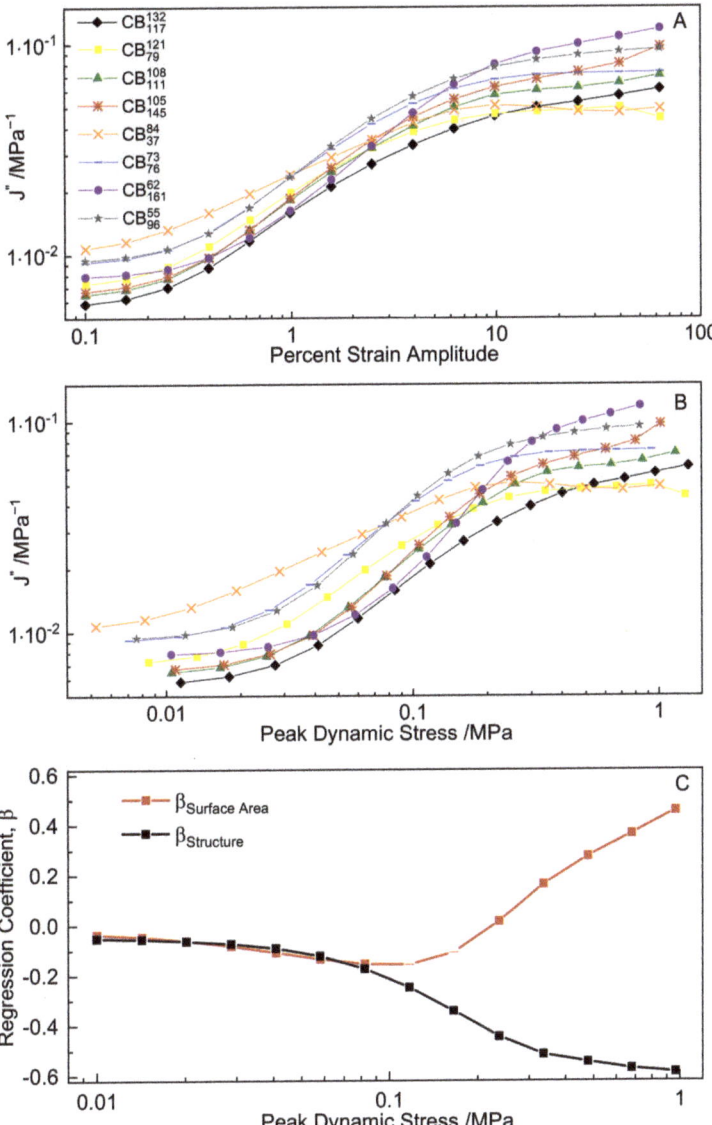

Figure 10. (**A**) J'' plotted versus strain amplitude; (**B**) J'' plotted versus peak dynamic stress; (**C**) multiple regression coefficients for structure and surface area plotted versus peak dynamic stress (points with p values > 0.05 are shown as dashes).

At low stress levels (<0.1 MPa), both surface area and structure are negatively correlated with J'', implying that an increase in CB surface area and structure reduces J'', and therefore hysteresis as well. In this stress region, stiffer compounds experience less deflection, and the predicted hysteresis is therefore minimized. At larger stresses, the correlations diverge. Structure remains negatively correlated with J'' up to high strains, while surface area reverses sign and becomes positively correlated with J''. This complex dependence of J'' on CB colloidal properties can be rationalized by considering the definition of loss compliance (see Equations (4) and (6)). In order to reduce compound hysteresis, J'' must be minimized. Therefore, G'' must be minimized and $|G^*|$ maximized, which is a conflicting requirement. In the low stress region, maximizing $|G^*|$ requires selection of CB with high structure and high surface area (Figure 6B). Although this would increase G'', the $|G^*|$ term in Equation (6) is squared, and thus dominates J'' values in this stress range. At higher stresses, the contribution of surface area to $|G^*|$ is diminished (see Figure 6B), while for G'' it remains appreciable (Figure 7B), leading to the change in sign of the surface area coefficient. In this medium to high stress region, maximizing CB structure in order to increase stiffness and minimize compound deflection while minimizing surface area to suppress G'' yields lower J'', and therefore, hysteresis.

This nuanced result has implications for material selection tradeoffs when designing rubber components. Should the component be operated entirely in stress control, knowledge of the levels/distributions of applied stresses in the component is critical in order to enable optimum selection of CB properties. Should the component operate in a mix of stress, strain, and strain energy control, the contradictory requirements for optimizing hysteresis performance should be taken into account; finite element simulations of in-service component deformations can guide this process.

4. Conclusions

The effects of CB structure and surface area on the static and dynamic mechanical properties of rubber were studied; CB structure increases the stress–strain moduli due to overstraining effects in the rubber matrix, while increased CB surface area slightly reduces the observed moduli. The latter effect is proposed to be due to adsorption of accelerators on the surface of CB, resulting in a net reduction in matrix crosslink density. Mechanical hysteresis from cyclic tensile testing to fixed peak strain is predominantly controlled by CB structure. At a fixed strain level, the mechanical hysteresis, and therefore the softening of the rubber compound, scales with the compound's virgin modulus at that strain. This suggests that the hysteretic energy dissipation at these large strains occurs in the rubber matrix and arises due to matrix overstrain effects rather than through rupture of flocculated particle clusters or interfacial polymer slippage. This is broadly in line with the conclusions drawn from cyclic experiments conducted to specified stress levels by Harwood, Mullins, and Payne. Further cyclic tensile testing of the compounds in the current study for comparison to controlled stress and controlled strain energy density levels would be highly informative in this regard. Dynamic mechanical properties show classical strain amplitude dependence (the Payne effect), which varies in magnitude according to the colloidal properties of CB. At low to medium strain amplitudes (0.1–2%), both the structure and the surface area of CB are positively correlated to $|G^*|$ to similar extents. Beyond 2%, however, the effect of surface area is reduced and structure plays an increasing and eventually dominant role in defining $|G^*|$. This transition in colloidal correlations is further evidence that a transition in stiffening mechanisms occurs with increasing strain amplitude. At low strains, the viscoelastic moduli are controlled by the flexing of percolated particle–particle networks, the formation of which is promoted by higher CB surface area. At higher strains, stiffening appears to be described best by matrix overstraining/strain amplification effects, which at iso-volume fractions are controlled by CB structure, and which map to correlations observed in the monotonic tensile stress–strain data. Replotting of the dynamic data versus peak stress and energy density allows an approximate assessment of the hysteresis response of the rubber compounds under various modes of deformation control. These results are summarized in Table 5, which presents proposed CB selection criteria for both dynamic and static conditions under iso-strain, iso-stress, and iso-strain energy density control. The table assumes a desire to minimize mechanical hysteresis; in practice, mechanical hysteresis is typically not the sole performance parameter guiding material selection. The need to achieve a specified compound tensile modulus (or hardness), tensile and tear strength, abrasion resistance, or friction performance level introduces material selection conflicts which run contrary to the trends shown in Table 5. Further forthcoming work on these materials will highlight these tradeoffs for this set of compounds, for example, in terms of their respective failure properties.

Table 5. Summary of CB selection recommendations to minimize mechanical hysteresis based on strain sweep and cyclic tensile tests.

Deformation Mode	Dynamic Deformation Conditions	Static Deformation Conditions
Strain control	• Reduce CB surface area • Reduce CB structure	• Reduce CB structure
Stress control	• Low Stress Levels • Increase CB surface area • Increase CB structure • Medium-Large Stress Levels • Reduce CB surface area • Increase CB structure	• Per Harwood, Mullins, and Payne, mechanical hysteresis at large strains under static stress-controlled conditions is independent of CB surface area and structure, although achieved elongations will depend on CB structure.
Strain energy control	• Reduce CB surface area	• Not determined in this study.

Supplementary Materials: The following are available online at https://www.mdpi.com/article/10.3390/polym14061194/s1: Figure S1: Transmission electron microscopy (TEM) images of tested carbon blacks, Figure S2: Interpolated $\tan\delta(W_0)$ data versus experimental $\tan\delta(W_0)$ data for three compounds, Figure S3: Interpolated $J''(\sigma_0)$ data versus experimental $J''(\sigma_0)$ data for three compounds, Table S1: Stage 1 compounding of natural rubber (NR) compounds, Table S2: Stage 2 compounding of natural rubber (NR) compounds, Table S3: Stage 3 compounding of natural rubber (NR) compounds, Table S4: Multiple regression results of mechanical hysteresis, Table S5: Multiple regression results of $|G^*|(\gamma_0)$, Table S6: Multiple regression results of $G''(\gamma_0)$, Table S7: Multiple regression results of $\tan\delta(W_0)$, Table S8: Multiple regression results of $J''(\sigma_0)$.

Author Contributions: Conceptualization, L.B.T., W.A.K.-M. and J.J.C.B.; Methodology, L.B.T. and M.C.; validation, L.B.T., W.A.K.-M., J.J.C.B. and C.R.H.; formal analysis, L.B.T. and W.A.K.-M.; investigation, M.C.; resources, L.B.T., M.C. and C.R.H.; data curation, L.B.T. and W.A.K.-M.; writing—original draft preparation, L.B.T., W.A.K.-M. and J.J.C.B.; writing—review and editing, L.B.T., W.A.K.-M., J.J.C.B. and C.R.H.; visualization, L.B.T. and W.A.K.-M.; supervision, L.B.T., J.J.C.B. and C.R.H.; project administration, L.B.T. and C.R.H.; funding acquisition, C.R.H. All authors have read and agreed to the published version of the manuscript.

Funding: This research was funded by Birla Carbon USA Inc., Marietta, GA, USA.

Institutional Review Board Statement: Not applicable.

Informed Consent Statement: Not applicable.

Data Availability Statement: The data presented in this study are available on request from the corresponding author.

Acknowledgments: The authors would like to thank Birla Carbon for funding and for providing the materials studied in this article.

Conflicts of Interest: The authors declare that the research was conducted in the absence of any commercial or financial relationships that could be construed as a potential conflict of interest.

References

1. Dannenberg, E.M. Bound Rubber and Carbon Black Reinforcement. *Rubber Chem. Technol.* **1986**, *59*, 512–524. [CrossRef]
2. Dannenberg, E.M. Carbon Black Dispersion and Reinforcement. *Rubber Chem. Technol.* **1952**, *25*, 843–857. [CrossRef]
3. Dannenberg, E.M. The Effects of Surface Chemical Interactions on the Properties of Filler-Reinforced Rubbers. *Rubber Chem. Technol.* **1975**, *48*, 410–444. [CrossRef]
4. Donnet, J.-B. Black and White Fillers and Tire Compound. *Rubber Chem. Technol.* **1998**, *71*, 323–341. [CrossRef]
5. Kraus, G. Interactions of Elastomers and Reinforcing Fillers. *Rubber Chem. Technol.* **1965**, *38*, 1070–1114. [CrossRef]
6. Kraus, G. Mechanical losses in carbon-black-filled rubbers. In *Chemistry and Technology of Rubber, Proceeding of the International Rubber Conference, Paris, France, 2–4 June 1982*; Vidal, A., Donnet, J.-B., Eds.; John Wiley & Sons: New York, NY, USA, 1984; pp. 75–93.
7. Medalia, A.I. Heat Generation in Elastomer Compounds: Causes and Effects. *Rubber Chem. Technol.* **1991**, *64*, 481–492. [CrossRef]
8. Medalia, A.I. Effect of Carbon Black on Dynamic Properties of Rubber Vulcanizates. *Rubber Chem. Technol.* **1978**, *51*, 437–523. [CrossRef]
9. Wang, M.-J. Effect of polymer-filler and filler-filler interactions on dynamic properties of filled vulcanizates. *Rubber Chem. Technol.* **1998**, *71*, 520–589. [CrossRef]
10. Smallwood, H.M. Limiting Law of the Reinforcement of Rubber. *J. Appl. Phys.* **1944**, *15*, 758. [CrossRef]
11. Guth, E. Theory of Filler Reinforcement. *J. Appl. Phys.* **1945**, *16*, 20–25. [CrossRef]
12. Allegra, G.; Raos, G.; Vacatello, M. Theories and simulations of polymer-based nanocomposites: From chain statistics to reinforcement. *Prog. Polym. Sci.* **2008**, *33*, 683–731. [CrossRef]
13. Domurath, J.; Saphiannikova, M.; Heinrich, G. *The Concept of Hydrodynamic Amplification in Filled Elastomers*; KGK Rubberpoint: Heidelberg, Germany, 2017; pp. 40–43.
14. Payne, A.R.; Whittaker, R.E. Low Strain Dynamic Properties of Filled Rubbers. *Rubber Chem. Technol.* **1971**, *44*, 440–478. [CrossRef]
15. Randall, A.M.; Robertson, C.G. Linear-nonlinear dichotomy of the rheological response of particle-filled polymers. *J. Appl. Polym. Sci.* **2014**, *131*, 40818. [CrossRef]
16. Tunnicliffe, L.B. Thixotropic flocculation effects in carbon black–reinforced rubber: KINETICS and thermal activation. *Rubber Chem. Technol.* **2021**, *94*, 298–323. [CrossRef]
17. Merabia, S.; Sotta, P.; Long, D.R. A Microscopic Model for the Reinforcement and the Nonlinear Behavior of Filled Elastomers and Thermoplastic Elastomers (Payne and Mullins Effects). *Macromolecules* **2008**, *41*, 8252–8266. [CrossRef]
18. Mujtaba, A.; Keller, M.; Ilisch, S.; Radusch, H.-J.; Beiner, M.; Thurn-Albrecht, T.; Saalwächter, K. Detection of Surface-Immobilized Components and Their Role in Viscoelastic Reinforcement of Rubber–Silica Nanocomposites. *ACS Macro Lett.* **2014**, *3*, 481–485. [CrossRef]
19. Mullins, L. Effect of Stretching on the Properties of Rubber. *Rubber Chem. Technol.* **1948**, *21*, 281–300. [CrossRef]
20. Mullins, L. Softening of Rubber by Deformation. *Rubber Chem. Technol.* **1969**, *42*, 339–362. [CrossRef]
21. Diani, J.; Fayolle, B.; Gilormini, P. A review on the Mullins effect. *Eur. Polym. J.* **2009**, *45*, 601–612. [CrossRef]
22. Plagge, J.; Lang, A. Filler-polymer interaction investigated using graphitized carbon blacks: Another attempt to explain reinforcement. *Polymer* **2021**, *218*, 123513. [CrossRef]
23. Harwood, J.; Mullins, L.; Payne, A. Stress softening in natural rubber vulcanizates. Part II. Stress softening effects in pure gum and filler loaded rubbers. *J. Appl. Polym. Sci.* **1965**, *9*, 3011–3021. [CrossRef]
24. Huang, M.; Tunnicliffe, L.B.; Thomas, A.G.; Busfield, J.J. The glass transition, segmental relaxations and viscoelastic behaviour of particulate-reinforced natural rubber. *Eur. Polym. J.* **2015**, *67*, 232–241. [CrossRef]
25. Robertson, C.; Roland, C.M. Glass Transition and Interfacial Segmental Dynamics in Polymer-Particle Composites. *Rubber Chem. Technol.* **2008**, *81*, 506–522. [CrossRef]
26. Bueche, F. Mullins effect and rubber-filler interaction. *J. Appl. Polym. Sci.* **1961**, *5*, 271–281. [CrossRef]
27. Kraus, G. Mechanical losses in carbon-black filled rubbers. *J. Appl. Polym. Sci. Appl. Polym. Symp.* **1984**, *39*, 75.
28. Heinrich, G.; Klüppel, M. Recent advancces in the theory of filler networking in elastomers, in Filled elastomers drug delivery systems. In *Advances in Polymer Science*; Springer: Berlin/Heidelberg, Germany, 2002.
29. Goöritz, D.; Raab, H.; Froöhlich, J.; Maier, P.G. Surface Structure of Carbon Black and Reinforcement. *Rubber Chem. Technol.* **1999**, *72*, 929–945. [CrossRef]
30. Goeritz, D.; Hofmann, R.W.P.; Bissem, H.H. Mobility of adsorbed polymer chains. *Rubber Chem. Technol.* **2010**, *83*, 323–330. [CrossRef]
31. *ASTM D3849*; Standard Test method for Carbon Black-Morphological Characterization of Carbon Black Using electron Microscopy. ASTM International: West Conschohocken, PA, USA, 2016.
32. *ASTM D6556*; Standard Test Method for Carbon Black—Total and External Surface area by Nitrogen Adsorption. ASTM International: West Conschohocken, PA, USA, 2019.
33. *ASTM D2414*; Standard Test Method for Carbon Black Oil Absorption Number (OAN). ASTM International: West Conschohocken, PA, USA, 2019.
34. *ASTM D3493*; Standard Test Method for Carbon Black—Oil Absoprtion Number of Compressed Sample (COAN). ASTM International: West Conschohocken, PA, USA, 2019.

35. Medalia, A.I. Morphology of aggregates. VI. Effective volume of aggregates of carbon black from electron microscopy;application to vehicle absorption and to die swell of filled rubber. *J. Colloid Interface Sci.* **1970**, *32*, 115–131. [CrossRef]
36. Medalia, A.I. Effective Degree of Immobilization of Rubber Occluded within Carbon Black Aggregates. *Rubber Chem. Technol.* **1972**, *45*, 1171–1194. [CrossRef]
37. Wang, M.-J.; Wolff, S.; Tan, E.-H. Filler-elastomer interactions. Part VIII. The role of the distance between filler aggregates in the dynamic properties of filled vulcanizates. *Rubber Chem. Technol.* **1993**, *66*, 178–195. [CrossRef]
38. Robertson, C.G.; Hardman, N.J. Nature of Carbon Black Reinforcement of Rubber: Perspective on the Original Polymer Nanocomposite. *Polymers* **2021**, *13*, 538. [CrossRef]
39. ASTM D2663; Standard Test Methods for Carbon Black—Dispersion in Rubber. ASTM International: West Conschohocken, PA, USA, 2019.
40. ASTM D2240; Standard Test Method for Rubber Property—Durometer Hardness. ASTM International: West Conschohocken, PA, USA, 2017.
41. ASTM D412; Standard Test Methods for Vulcanized Rubber and Thermoplastic Elastomers—Tension. ASTM International: West Conschohocken, PA, USA, 2002.
42. Klüppel, M. The Role of Disorder in Filler Reinforcement of Elastomers on Various Length Scales. In *Filler-Reinforced Elastomers Scanning Force Microscopy*; Advances in Polymer Science; Springer: Berlin/Heidelberg, Germany, 2003; pp. 1–86. [CrossRef]
43. Kraus, G. A Carbon Black Structure-Concentration Equivalence Principle. Application to Stress-Strain Relationships of Filled Rubbers. *Rubber Chem. Technol.* **1971**, *44*, 199–213. [CrossRef]
44. Robertson, C.G.; Tunnicliffe, L.B.; Maciag, L.; Bauman, M.A.; Miller, K.; Herd, C.R.; Mars, W.V. Characterizing Distributions of Tensile Strength and Crack Precursor Size to Evaluate Filler Dispersion Effects and Reliability of Rubber. *Polymers* **2020**, *12*, 203. [CrossRef]
45. Kyei-Manu, W.A.; Tunnicliffe, L.B.; Plagge, J.; Herd, C.R.; Akutagawa, K.; Pugno, N.M.; Busfield, J.J.C. Thermomechanical Characterization of Carbon Black Reinforced Rubbers During Rapid Adiabatic Straining. *Front. Mater.* **2021**, *8*, 421. [CrossRef]
46. Hyun, K.; Wilhelm, M.; Klein, C.O.; Cho, K.S.; Nam, J.G.; Ahn, K.H.; Lee, S.J.; Ewoldt, R.; McKinley, G. A review of nonlinear oscillatory shear tests: Analysis and application of large amplitude oscillatory shear (LAOS). *Prog. Polym. Sci.* **2011**, *36*, 1697–1753. [CrossRef]
47. Xiong, W.; Wang, X. Nonlinear responses of carbon black-filled polymer solutions to forced oscillatory shear. *J. Non-Newtonian Fluid Mech.* **2020**, *282*, 104319. [CrossRef]
48. Xiong, W.; Wang, X. Linear-nonlinear dichotomy of rheological responses in particle-filled polymer melts. *J. Rheol.* **2018**, *62*, 171–181. [CrossRef]
49. Futamura, S. Deformation Index—Concept for Hysteretic Energy-Loss Process. *Rubber Chem. Technol.* **1991**, *64*, 57–64. [CrossRef]
50. Futamura, S. Analysis of Ice- and Snow Traction of Tread Material. *Rubber Chem. Technol.* **1996**, *69*, 648–653. [CrossRef]

Article

Testing of Rubber Composites Reinforced with Carbon Nanotubes

Dana Bakošová [1,*] and Alžbeta Bakošová [2]

[1] Department of Material Engineering, Faculty of Industrial Technologies in Púchov, Alexander Dubček University of Trenčín, Ivana Krasku 491/30, 020 01 Púchov, Slovakia
[2] Department of Numerical Methods and Computational Modeling, Faculty of Industrial Technologies in Púchov, Alexander Dubček University of Trenčín, Ivana Krasku 491/30, 020 01 Púchov, Slovakia; alzbeta.bakosova@student.tnuni.sk
* Correspondence: dana.bakosova@tnuni.sk

Abstract: Carbon nanotubes (CNTs) have attracted growing interest as a filler in rubber nanocomposites due to their mechanical and electrical properties. In this study, the mechanical properties of a NR/BR/IR/SBR compound reinforced with single-wall carbon nanotubes (SWCNTs) were investigated using atomic force microscopy (AFM), tensile tests, hardness tests, and a dynamical mechanical analysis (DMA). The tested materials differed in SWCNT content (1.00–2.00 phr) and were compared with a reference compound without the nanofiller. AFM was used to obtain the topography and spectroscopic curves based on which local elasticity was characterized. The results of the tensile and hardness tests showed a reinforcing effect of the SWCNTs. It was observed that an addition of 2.00 phr of the SWCNTs resulted in increases in tensile strength by 9.5%, Young's modulus by 15.44%, and hardness by 11.18%, while the elongation at break decreased by 8.39% compared with the reference compound. The results of the temperature and frequency sweep DMA showed higher values of storage and loss moduli, as well as lower values of tangent of phase angle, with increasing SWCNT content.

Keywords: nanocomposites; carbon nanotubes; mechanical properties; atomic force microscopy; dynamical mechanical analysis

1. Introduction

Polymer nanocomposites have attracted research interest and have found a broad field of application in recent years because of their enhanced material properties compared with original polymers. The extent of improvement depends, generally, on a number of parameters, including the type of nanofiller, particle size, aspect ratio, filler dispersion status, and surface properties, which determine the interaction between the filler and the polymer chain. Polymer carbon nanotube composites are objects of particular interest due to the structural characteristics of carbon nanotubes (CNTs) and their large surface area available for stress transmission, as well as their exceptionally high modulus of elasticity and excellent electrical and thermal properties [1].

Carbon nanotubes are seamless cylinders formed by rolling sheets of graphene atoms with open or closed ends. They can be divided into two main categories: single-wall carbon nanotubes (SWCNTs), with a diameter in the nanometer scale, and multi-wall carbon nanotubes (MWCNTs), consisting of several concentrically connected nanotubes or nanotubes rolled in a spiral. The diameter of a MWCNT can reach more than 100 nm. The length of CNTs can reach several micrometers or even millimeters. The structure of rolled CNTs is given by a chiral vector. Based on their structure, CNTs can be divided into the zigzag, chiral, and armchair types. The quality of CNTs relates to the fact that all carbons are bound in a hexagonal lattice except their ends. Similar to graphene, CNTs are chemically bound by sp2 bonds, which are extremely strong forms of molecular interaction.

This property, combined with the natural tendency of carbon nanotubes to bond with van der Waals forces, provides an opportunity to develop ultra-high-strength, low-weight materials with high electrical and thermal conductivity. This makes them very attractive for many applications [2–5]. In addition to their electrical properties, which they inherit from graphene, CNTs also have unique thermal and mechanical properties. They have high Young's modulus (up to 1 TPa) and tensile strength (11–63 GPa). They are very light with good thermal conductivity. As graphite, they have high chemical stability and are extremely resistant to corrosion, unless they are simultaneously exposed to high temperatures and oxygen [6].

Due to their properties, CNTs have been used as functional fillers in rubber compounds. Rubber nanocomposites reinforced with CNTs have been investigated in order to achieve required dynamic mechanical properties, gas resistance, flammability resistance, and thermal and electrical conductivity [7–13]. Several studies [14–18] have reported increases in tensile strength, elastic modulus, and hardness and decreases in the elongation at break of natural rubber composites reinforced with CNTs. In addition to improving the mechanical properties, the incorporation of conductive fillers into rubbers that are thermal and electrical insulators can produce composites with certain electrical conductivities. The potential applications of CNT-filled rubber composites vary from industrial applications such as rubber hoses, tire components, and sensing devices to electrically conductive systems and biomedical applications [19,20].

The final properties of CNT-reinforced elastomer nanocomposites mainly depend on the CNT type, the rate of CNT dispersion and their orientation in the matrix, the physical and chemical interactions of polymer chains with the CNTs, and the crosslinking chemistry of the rubbers. Orientation effects are mainly due to their high aspect ratio. The degree of alignment of nanotubes can be determined with X-ray diffraction and polarized Raman spectroscopy. As presented in [21], rolling direction affects the final alignment of the CNTs. If the composites are manufactured using extrusion or injection molding, tuning of the alignment degree can be achieved by regulating the shear rate, as well as the pressure applied. Other methods of alignment also include mechanical stretching, filtration, plasma-enhanced chemical vapor deposition, electrospinning, force-field-induced alignment, magnetic-field-induced alignment, liquid-crystalline-phase-induced alignment, etc. [22].

Carbon nanotubes are difficult to process due to their low dispersibility and their tendency to form aggregates [23], and in order to utilize CNT properties, several strategies have been proposed to improve the compatibility between polymer matrices and carbon nanotubes [24–27]. Acids and organic solvents are used to functionalize the carbon nanotubes [28–30], and another alternative is ionic liquids [31–33].

This study focuses on testing and evaluating selected material properties of rubber nanocomposites that differ in their contents of single-wall nanotubes and compares them with a rubber compound with the same base material but without nanofillers. The rubber compounds are examined using atomic force microscopy (AFM), hardness tests, tensile tests, and a dynamical mechanical analysis (DMA) in order to determine the influence of the CNT content on the mechanical properties. Microscopy is an essential tool for understanding the morphology of rubber compounds, including the size, shape and distribution of filler phases and particles, as well as for determining the effects of fillers and processing additives on the properties of rubber compounds. Atomic force microscopy (AFM) is a versatile and powerful analytical tool for the development and research of rubber materials. In [34–38], AFM was used to determine morphology, as well as for the observation of the microdispersion of the fillers in rubber compounds. AFM can also be used to characterize the local elasticity of rubber materials [39], to study the homogeneity of rubber compounds [40,41], to study the aging of rubber composites [42], and to determine the effects of fillers on the properties and qualities of compounds [42–44]. AFM force spectroscopy presents much information regarding the surface and mechanical properties of tested materials. Information on the elasticity and stiffness of individual macromolecules of the tested material can also be obtained from the measured curves. In [45,46], force spectroscopy was used to test rubber compounds. A more detailed, state-of-art review of advances in the use of AFM in polymer investigation was presented by Wang et al. in [47]. In the present study, AFM is used to observe the topography of the examined materials and to evaluate their elastic properties based on the measured spectroscopic curves.

2. Materials and Methods

2.1. Materials

The rubber compounds examined in this study had the same polymer matrices, and they differed in proportions of single-wall carbon nanotubes. These compounds were labelled as CNT 1–CNT 5, and they were compared with a reference compound without carbon nanofillers labelled as CNT 0. The composition of the base material is listed in Table 1, and the contents of single-wall carbon nanotubes of the individual compounds can be seen in Table 2. The used carbon nanotubes had a diameter of 2 nm, a length of 5–20 µm, and a purity of 95%.

Table 1. Composition of the examined compounds.

Component	Amount (phr)
natural rubber (NR)	40
butadiene rubber (CIS BR)	20
isoprene rubber (CIS IR)	10
styrene-butadiene rubber (SBR 1500)	30
carbon black filler (N 339)	30
reclaimed rubber	10
silica filler ULTRASIL	8
resin	2
antidegradant	3
antioxidant	2
vulcanization activator—STEARIN	2
distillate aromatic extract DAE	2
oxidized polyethylene wax (OPW)	2
vulcanization activator—ZnO	2.5
vulcanizing agent—insoluble sulphur 67%	3.3
sulfenamide vulcanization accelerator—CBS	1.1

Table 2. The contents of the single-wall carbon nanotubes in the compounds.

Compound label:	CNT 0	CNT 1	CNT 2	CNT 3	CNT 4	CNT 5
Content of carbon nanotubes (phr):	0.00	1.00	1.25	1.50	1.75	2.00

The production process of the CNT 1–CNT 5 compounds was divided into several phases. The first phase involved the dispersion of the carbon nanotubes in a dispersant (ethanol) and a distillate aromatic extract (DAE), as nanotubes have tendency to form aggregates, which are difficult to process. This solution was heated to a temperature just above the boiling point of the solvent (80 °C) and then sonified for 120 min, using a mechanical ultrasonic sonifier with a probe capable of vibrating at appropriate ultrasonic frequencies (30 kHz) in order to induce the efficient dispersion of the nanotubes. As a part of the sonication, the dispersing agent also evaporated. The second phase involved dissolving the individual components of the rubber compound in an organic solvent (ethanol) with the addition of oil (DAE), followed by mixing at the temperature of 80 °C for 120 min until the components were mixed and part of the dispersant evaporated. Then the preparation process was followed with the phase of mixing the rubber compound with the carbon nanotube solution, which included a two-stage mixing process to mix the prepared solutions thoroughly and to ensure the evaporation of the excess solvent. Mixing was performed with a Farrel Technolab BR 1600 Banbury mixer. In the first stage, rubber components, including oil, carbon black, silica, resin, antioxidants, antidegradants, regenerates, stearin, wax, and nanotubes, were added in the oil solution. The temperature of the chamber, rotors, and cap was 70 °C. The total mixing time was 360 s. The maximum number of revolutions of the mixer was 55 rpm, and the maximum pressure was 196 kPa. The highest temperature of the compound was 150 °C. Before the second mixing stage, the compound was rolled for 30 min with a Servitec double roller in order to achieve better dispersion of the fillers in the compound. In the second stage of mixing, in order to vulcanize the compound, the remaining components were added: insoluble sulphur, zinc oxide (ZnO), and CBS. The compound was mixed for 150 s at a maximum temperature of 105 °C. Subsequently, the compound was rolled again with a double roller to achieve the required thickness of the compound for the preparation of test samples. The compound was then allowed to cool and stabilize in an oven at ambient temperature for 5 days. Test samples were prepared in accordance with the standards for individual tests of vulcanized, semi-finished products by cutting.

2.2. Methods

2.2.1. Tensile Test

Tensile tests give an orientation view about rubber material properties. Tensile curves are characteristic for rubber compounds. The following material characteristics were evaluated using the tensile test results:

- Tensile strength: the maximum tensile stress recorded during the elongation of the testing sample until the breaking moment;
- Elongation at break: tensile deformation of the sample working length in the breaking moment;
- Young's modulus: defined as the initial slope of the stress–strain response.

Tensile tests were performed for the CNT 0–CNT 5 samples in accordance with the ISO 37 standard, which specifies a method for determining the tensile deformation characteristics of vulcanized and thermoplastic rubbers. Ten dumbbell-shaped samples of each compound made in accordance with the ISO 23529 standard were tested. The working length of the samples was 20 mm, and the loading speed was 100 mm/min.

2.2.2. Hardness Test

Hardness is the ability of a material to withstand compressive forces. It depends on several factors, namely the Young's modulus, the viscoelastic properties of the elastomer, the thickness of the test sample, the geometry of the indenter, the applied pressure, the rate of pressure increase, and the interval in which the hardness is recorded. Choosing the appropriate hardness test method is important in order to obtain accurate and reliable results. The most commonly used method for rubber compounds is the Shore A method, which follows the ISO 7619-1 standard, according to which a test material with a thickness of 6 mm is required.

Hardness measurements can be used to roughly estimate the Young's modulus. The best-known correlation of hardness values to Young's modulus was introduced by A.N. Gent [48] in 1958 and is given by the following equation:

$$E = \frac{0.0981(56 + 7.62336S)}{0.137505(254 - 2.54S)} \text{ (MPa)}, \tag{1}$$

where E is Young's modulus, and S is Shore A hardness in the range of 20–80 Shore A. There are several other correlations, such as the equation postulated by Ruess [49,50]:

$$log_{10}E = 0.0235S - 0.6403, \qquad (2)$$

where E (MPa) is Young's modulus, and S is Shore A hardness. The correlation by Lindeman [51] is another example:

$$E = 11.427S - 0.4445S^2 + 0.0071S^3 \ (psi), \qquad (3)$$

where E (psi) is Young's modulus, and S is Shore A hardness.

In this study, the correlations mentioned above were used to estimate Young's modulus, and the estimated values were compered to results from the tensile tests.

2.2.3. Dynamical Mechanical Analysis

Rubber compounds are viscoelastic materials, and a dynamic mechanical analysis (DMA) is often used to characterize their viscoelastic properties. In this study, a Pyris Diamond DMA analyzer was used for the DMA of the CNT 0–CNT 5 rubber compounds. Samples were prepared in the form of a strip with measurements of 20 mm × 10 mm × 2.1 mm. They were subjected to tensile loading in the temperature range of −80–100 °C using a heating rate of 5 °C/min at a frequency of 1 Hz. A cryogenic nitrogen vessel was used to achieve low temperatures. The samples were also subjected to frequencies of 0.01 Hz, 0.05 Hz, 0.2 Hz, 0.5 Hz, 1 Hz, 5 Hz, 10 Hz, 20 Hz, and 50 Hz at 20 °C. The frequency and temperature dependencies of the storage modulus E', the loss modulus E'', and the tangent of the phase angle $tan\ \delta$ were evaluated.

2.2.4. Atomic Force Microscopy

Atomic force microscopy (AFM) is a nonoptical imaging technique that allows accurate and nondestructive measurements of topographic, mechanical, electrical, magnetic, chemical, and optical properties of a sample surface at very high resolutions. An AFM microscope works on the principle of measuring the intermolecular force. It uses a cantilever with an attached, sharp, several-micrometers-long tip, which scans the surface of the sample. Atomic forces between the tip and the sample surface result in the bending of the cantilever. To detect any changes in the cantilever deflection, a laser beam is used. It is directed at the end of the cantilever, from which it is reflected into a photodetector. Change in the distance of the tip from the surface causes a change in the force and bending of the cantilever and, thus, the direction of the reflection of the laser beam into the photodetector changes. The deflection of the laser beam is recorded, and the resulting surface topography is generated with further software processing [52].

In spectroscopic measurements, the deflection of the cantilever tip is recorded as a function of the force and distance between the AFM probe and the sample. Due to the different stiffnesses of the tested systems, the stiffness constant of the AFM probe must be higher than the stiffness of the examined sample [37,53,54]. The spectroscopic curves were sufficiently specific for each material; however, they could be divided into general characteristic sections, as shown in Figure 1. The solid line shows the curves measured in a vacuum. The dashed line is the variation of the curves measured in air with the presence of layers of moisture and microscopic impurities.

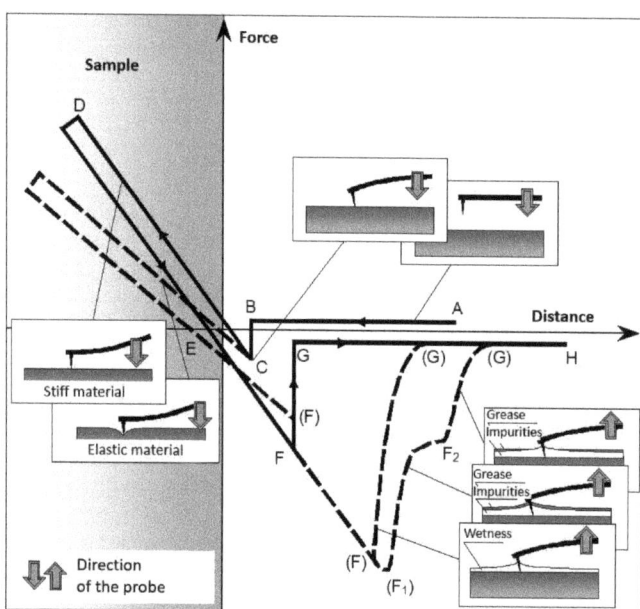

Figure 1. Schematic of force spectroscopic curve [55].

Between the A and B points, the tip and the sample are far apart, and there is no deflection of the cantilever with the tip. At the B point, long-distance interactions, mainly of Van der Waals and electrostatic origins, occur upon approach. At the C point, the tip touches the surface of the sample. The shape of the curve is also influenced by the surface moisture and impurities. The C–D section is characterized by further approach of the tip to the sample while they are physically in contact, and it results in pressing the tip against the sample surface and the deflection of the probe. At the D point, the sample surface is punctured due to the maximum force that the sample surface is able to withstand. The D point characterizes the end of the approach and the beginning of the departure of the tip from the surface. According to the slope of the C–D section, the Young's modulus of the probe–surface system can be evaluated. If the probe is softer than the sample surface, the slope of the curve mostly presents the modulus of the probe; otherwise, if the stiffness of the probe is higher, the slope of the section allows the Young's modulus of the sample to be examined. If the C–D and D–E sections are not parallel, the time-reversible elastic or plastic deformation of the sample can be evaluated. The probe deflection is neutral at the E point. The probe moves away from the surface between the E and F points, and it begins to tilt towards the sample due to attractive or adhesive forces. In a vacuum, Van der Waals and electrostatic forces act on the tip, and in an air environment, the tip is also subjected to capillary force from the moisture on the surface. The F point is a separation point at which the maximum adhesive force acts between the tip and the surface of the sample, and it gives information for adhesion evaluation. The number of the separation points depends on the viscosity and thickness of the surface layers (moisture, impurities, and grease). The probe separates from the surface at the G point after overcoming the adhesive force [53,55].

General approximation and Snedonn's model [56,57] were used to evaluate the measured data and to calculate the ratios of the Young's moduli. Snedonn's model formulates the dependence between Young's modulus E and the load gradient dP/dh and is given by the following equation:

$$\frac{dP}{dh} = \frac{2A^{1/2}}{\pi^{1/2}} E \ \left[Nm^{-1}\right], \qquad (4)$$

where A (m^2) is the contact area, and E (Pa) is a combined modulus of the elasticity of the probe and the examined surface, which is given by the following equation:

$$E = \left\{ \left[\left(1 - v_m^2\right)/E_m \right] + \left[\left(1 - v_c^2\right)/E_c \right] \right\}^{-1} [Pa], \tag{5}$$

where E_m, E_c (Pa) are the Young's moduli of the examined material and the cantilever, respectively; v_m, v_c (-) are the Poisson ratios of the examined material and the cantilever, respectively, h (m) is the indentation depth, and P (N) is the normal load. It can be assumed that the Young's modulus of the cantilever is much higher compared with that of the examined material ($E_c \gg E_m$). Therefore, Equation (5) can be simplified to $E_m = E$, and the following equations representing the moduli of two different samples E1 and E2 in (6) and (7), respectively, and their ratio (8) can be derived:

$$\frac{dP_1}{dh_1} = \frac{2A^{1/2}}{\pi^{1/2}} E_1 \Rightarrow E_1 = \frac{dP_1}{dh_1} \frac{\pi^{1/2}}{2A^{1/2}}, \tag{6}$$

$$\frac{dP_2}{dh_2} = \frac{2A^{1/2}}{\pi^{1/2}} E_2 \Rightarrow E_2 = \frac{dP_2}{dh_2} \frac{\pi^{1/2}}{2A^{1/2}}, \tag{7}$$

$$\frac{E_1}{E_2} = \frac{\frac{dP_1}{dh_1} \frac{\pi^{1/2}}{2A^{1/2}}}{\frac{dP_2}{dh_2} \frac{\pi^{1/2}}{2A^{1/2}}} \Rightarrow \frac{E_1}{E_2} = \frac{\frac{dP_1}{dh_1}}{\frac{dP_2}{dh_2}}. \tag{8}$$

After a linear approximation (9) of the C–D section of the spectroscopic curve from which the Young's modulus can be determined in order to find the slope k, Equation (11), which expresses the ratio of the Young's moduli of the two samples, can be formulated:

$$y = kx + q, \text{ where } k = \frac{dP}{dh}, \tag{9}$$

$$\frac{dP_1}{dh_1} = k_1 \text{ and } \frac{dP_2}{dh_2} = k_2, \tag{10}$$

$$\Rightarrow \frac{E_1}{E_2} = \frac{k_1}{k_2} \rightarrow E_2 = \frac{k_2 E_1}{k_1}. \tag{11}$$

3. Results

3.1. Tensile Test Results

In relation to the tensile test, each CNT 0–CNT5 compound was subjected to ten measurements, and subsequently, the tensile strength, elongation at break, and Young's modulus were determined. The average values of these material properties are listed in Table 3, and a graphical comparison of the examined rubber compound properties can be seen in Figure 2.

Table 3. Tensile test results.

Compound	Tensile Strength (MPa)	Elongation at Break (%)	Young's Modulus (MPa)
CNT 0	16.29 ± 0.35	643.21 ± 13.21	3.03 ± 0.03
CNT 1	16.55 ± 0.27	625.33 ± 12.55	3.15 ± 0.02
CNT 2	16.71 ± 0.31	619.65 ± 14.25	3.27 ± 0.03
CNT 3	16.95 ± 0.39	618.24 ± 10.22	3.33 ± 0.04
CNT 4	17.43 ± 0.32	611.45 ± 11.45	3.42 ± 0.04
CNT 5	17.79 ± 0.41	589.25 ± 13.99	3.60 ± 0.04

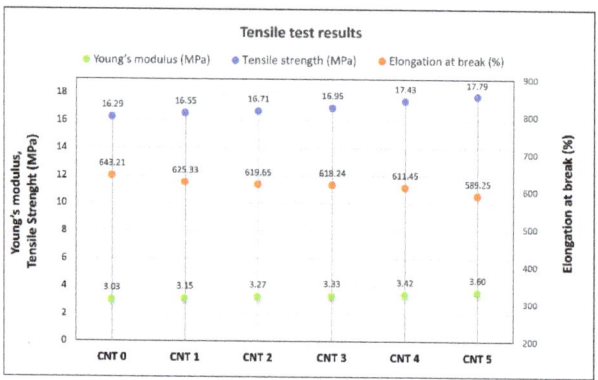

Figure 2. Tensile test results.

The tensile strength and Young´s modulus of the compounds with nanofillers in the form of the carbon nanotubes (CNT 1–CNT 5) increased with increasing content in the nanofiller. The tensile strength of the CNT 5 compound with the highest nanofiller content (2.00 phr) was higher by 9.5% compared with CNT 0 (without nanofillers). The Young´s modulus of CNT 5 was higher by 15.44% compared with CNT 0. The presence of SWCNTs in the tested compounds caused reductions in the values of the elongation at break. The elongation at break of CNT 5 decreased by 8.39% compared with CNT 0.

3.2. Hardenss Test Results

Approximately ten Shore A hardness measurements for each compound were performed, and Young´s moduli were calculated using Equations (1)–(3). The results are listed in Table 4, and dependence of Young's modulus on hardness given by Equations (1)–(3) is shown in Figure 3.

Table 4. Hardness test results and estimation of Young´s modulus using Equations (1)–(3).

Compound	Shore A Hardness	Young's Modulus (MPa)		
		Gent's Equation	Ruess's Equation	Lindeman's Equation
CNT 0	50.44 ± 0.21	2.50 ± 0.35	3.51 ± 0.43	2.46 ± 0.36
CNT 1	52.21 ± 0.35	2.67 ± 0.43	3.86 ± 0.54	2.73 ± 0.45
CNT 2	53.45 ± 0.39	2.80 ± 0.49	4.13 ± 0.57	2.93 ± 0.50
CNT 3	54.35 ± 0.24	2.89 ± 0.37	4.33 ± 0.45	3.09 ± 0.39
CNT 4	55.12 ± 0.29	2.98 ± 0.38	4.52 ± 0.47	3.23 ± 0.40
CNT 5	56.08 ± 0.32	3.09 ± 0.41	4.76 ± 0.52	3.41 ± 0.44

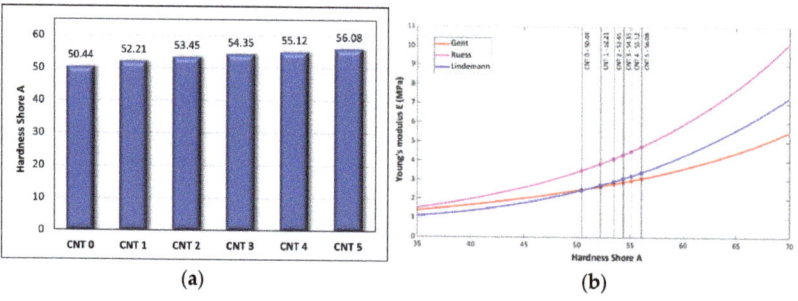

Figure 3. (**a**) Hardness test results and (**b**) estimation of Young's modulus from Shore A hardness.

The hardness of the compounds increased with increasing content in nanotubes. The hardness of CNT 5 was higher by 11.18% compared with the CNT 0 compound without the nanotubes.

3.3. Evaluation of Rubber Compounds Using Atomic Force Microscopy

An NT-206 atomic force microscope was used to evaluate the rubber compounds, and along with the appropriate hardware and software, it allowed the analysis of the topography and micromechanical properties of the solid surfaces up to a nanometer-level resolution. The topography examples of the examined compounds are shown in Figure 4. The spectroscopic curves were measured for ten different locations of each compound. The C–D section of the spectroscopic curve (Figure 1) was approximated using a linear function in order to evaluate the Young´s moduli of the individual compounds. The ratios of the Young's moduli of the compounds with nanotubes (CNT 1–CNT 5) to the compound without carbon nanotubes (CNT 0) were determined. The examples of spectroscopic curves for the CNT 0 and CNT 5 compounds are shown in Figure 5, and the slopes of the linear functions approximating the C–D section of the spectroscopic curves are listed in Table 5.

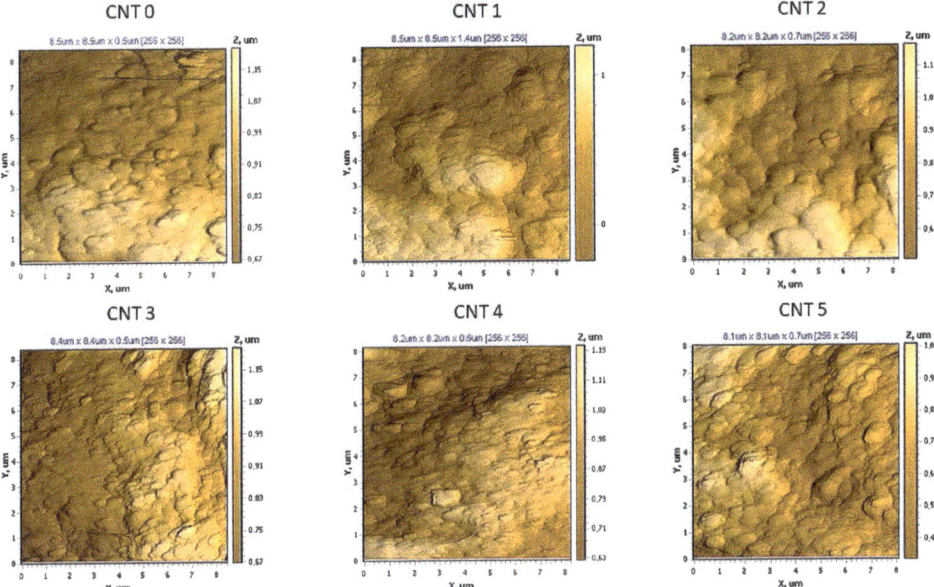

Figure 4. Examples of the topography of the individual compounds obtained by AFM.

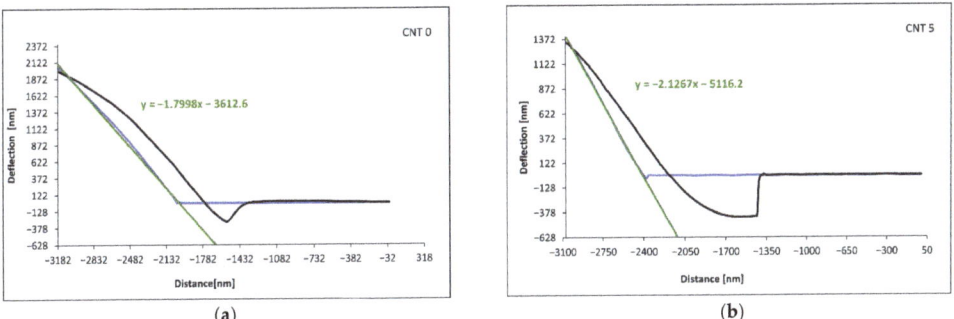

Figure 5. Examples of the spectroscopic curves of the compounds: (**a**) CNT 0 and (**b**) CNT 5.

Table 5. The slopes of the linear functions approximating the spectroscopic curves.

Measurement	CNT 0	CNT 1	CNT 2	CNT 3	CNT 4	CNT 5
1	−1.7998	−1.8642	−1.9354	−1.9824	−2.0322	−2.1267
2	−1.7954	−1.8524	−1.9521	−1.9724	−2.0452	−2.1454
3	−1.7969	−1.8852	−1.9754	−1.9853	−2.0563	−2.1541
4	−1.7912	−1.8921	−1.9551	−1.9621	−2.0725	−2.1145
5	−1.7954	−1.8245	−1.9254	−1.9994	−2.0168	−2.1354
6	−1.7945	−1.8526	−1.9278	−1.9685	−2.0698	−2.1078
7	−1.7997	−1.8354	−1.9245	−1.9824	−2.0597	−2.1298
8	−1.7921	−1.8759	−1.9154	−1.9974	−2.0125	−2.1758
9	−1.7991	−1.8875	−1.9285	−1.9899	−2.0137	−2.1267
10	−1.7991	−1.8522	−1.9354	−1.9678	−2.0045	−2.1045
Avarage value of slope $k_{CNT\,i}$	−1.7963 ± 0.0010	−1.8622 ± 0.0072	−1.9375 ± 0.0057	−1.9808 ± 0.0040	−2.0383 ± 0.0081	−2.1321 ± 0.0069

Based on Equation (11), the values of the Young's moduli of the CNT 1–CNT 5 nanocomposites were determined with respect to the CNT 0 reference compound without carbon nanotubes, and they are listed in Table 6. These ratios were calculated using the average slope values (Table 5).

Table 6. Young's modulus ratios of compounds CNT 1–CNT 5 with respect to CNT 0.

Rubber Compound	Young's Modulus Ratios $E_{CNT\,i} = \frac{k_{CNT\,i}\, E_{CNT0}}{k_{CNT\,0}}$
CNT 1	$E_{CNT1} = 1.037 E_{CNT0}$
CNT 2	$E_{CNT2} = 1.079 E_{CNT0}$
CNT 3	$E_{CNT3} = 1.103 E_{CNT0}$
CNT 4	$E_{CNT4} = 1.135 E_{CNT0}$
CNT 5	$E_{CNT5} = 1.187 E_{CNT0}$

The slope values of the spectroscopic curves of the individual compounds obtained from the linear approximation of the C–D section differed slightly, which indicated a slight inhomogeneity. By comparing the average values of the slopes, it can be stated that the CNT 5 compound had the highest value for the Young's modulus, and the CNT 0 compound had the lowest value of the modulus.

3.4. Dynamical Mechanical Analysis Results

The DMA was performed, during which the samples were subjected to a tensile loading in the temperature range of −80–100 °C and to the frequencies of 0.01 Hz, 0.05 Hz, 0.2 Hz, 0.5 Hz, 1 Hz, 5 Hz, 10 Hz, 20 Hz, and 50 Hz. The temperature dependencies of E', E'', and $\tan \delta$ at a frequency of 1 Hz for the CNT 0–CNT 5 rubber compounds can be seen in Figures 6 and 7. The frequency dependencies of E', E'', and $\tan \delta$ at the temperature of 20 °C are shown in Figure 8.

The dependence of the elastic portion of the complex elasticity modulus on temperature (−80–100 °C) at a frequency of 1 Hz is shown in Figure 6. The storage modulus E' of the tested rubber compounds increased with increasing CNT proportion. The storage modulus reflects the elastic properties of the tested materials and the renewable energy in the deformed samples. At a low temperature, the modulus E' had a relatively high value that was attributed to the inert semicrystalline structure, and as the mobility of the polymer chains increased with temperature, the elastic modulus decreased.

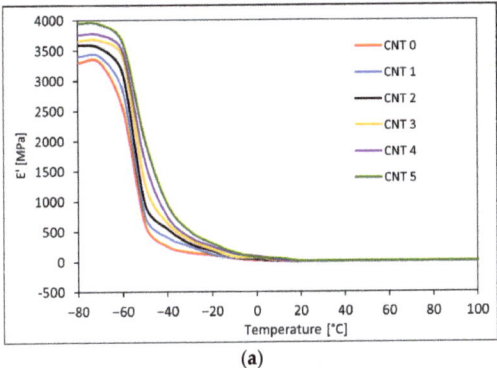 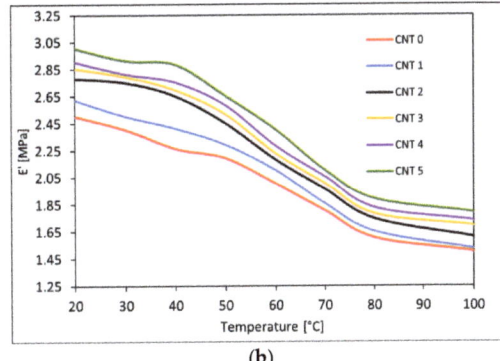

(a) (b)

Figure 6. (a) Temperature dependence of storage modulus E' at frequency of 1 Hz and temperature range of −80–100 °C; (b) detail of temperature dependence of storage modulus E' at frequency of 1 Hz and temperature range of 20–100 °C.

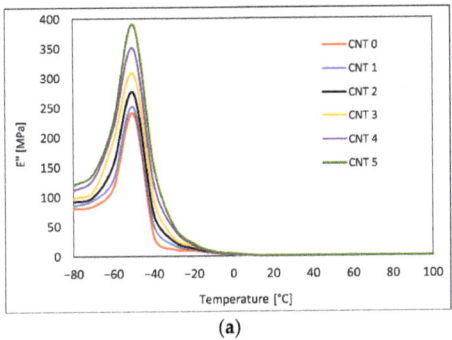 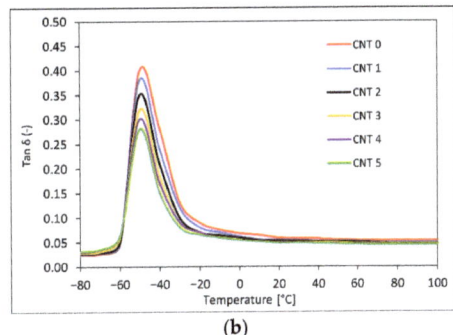

(a) (b)

Figure 7. (a) Temperature dependence of loss modulus E'' at frequency of 1 Hz and temperature range of −80–100 °C; (b) temperature dependence of the tangent of phase angle $tan\ \delta$ at frequency of 1 Hz and temperature range of −80–100 °C.

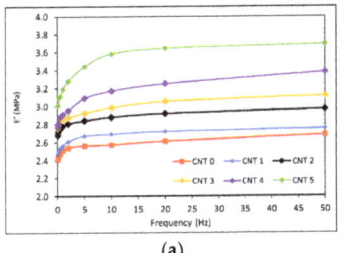 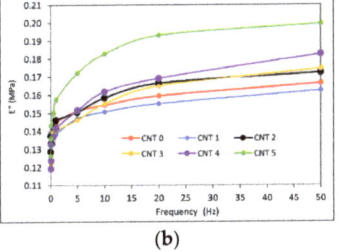 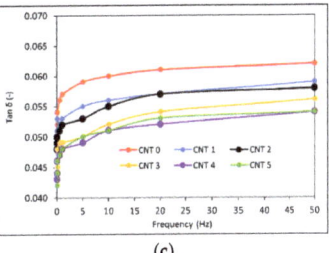

(a) (b) (c)

Figure 8. Frequency dependencies (frequency range: 0.1–50 Hz) at a temperature of 20 °C: (a) the storage modulus E', (b) the loss modulus E'', and (c) the tangent of phase angle $tan\ \delta$.

The dependence of the viscous portion of the complex elasticity modulus on temperature (−80–100 °C) at a frequency of 1 Hz can be seen in Figure 7a. The loss modulus E'' corresponds to the viscous properties of a viscoelastic material and is a measure of a material's ability to dissipate energy in the form of heat due to viscous movements in the material. The values of the loss modulus were significantly lower than the values of the storage modulus, with elastic properties predominant in the compounds. The loss modulus increased slightly with the increasing CNT content in the tested compounds.

Figure 7b shows the dependence of the tangent of phase angle *tan δ* on temperature at a frequency of 1 Hz and a temperature range of −80–100 °C. The *tan δ* is determined from the ratio of the loss modulus to the storage modulus and represents the ratio of dissipated, lost energy to the energy stored during the deformation cycle. The *tan δ* characterized the damping material properties, which decreased with the increased CNT content and could be attributed to CNT stiffness. The glass transition temperature T_g determined from the peak of the temperature dependency of *tan δ* was around −50 °C, and there was not noticeable change in the T_g after addition of nanotubes.

The dependencies of the storage modulus E', the loss modulus E'' and the *tan δ* on the frequency at 20 °C can be seen in Figure 8. The storage modulus E', the loss modulus E'', and the *tan δ* showed increasing tendency for all the compounds within the investigated frequency interval. With further increase in the frequency, increases in the storage moduli and decreases in the loss moduli and *tan δ* past their peaks were expected due to the viscoelastic nature of rubber compounds. The CNT 5 nanocomposite with the highest SWCNT content showed higher E' and E'' values and lower values of *tan δ* compared with CNT 0.

4. Discussion

The tensile test results showed a reinforcing effect of the single-wall carbon nanotube filler. As can be seen in Table 3 and Figure 2, the tensile strength and Young's modulus values increased with the increasing SWCNT content, and the elongation at break values decreased. The same trends were observed in [16–20]. The increase in the tensile strength and the Young's modulus could be attributed to good dispersion and interatomic interaction between the rubber matrix and the nanofiller, as the CNTs (with their high aspect ratio) could improve the crosslinking of the compound [58]. The strengthening effect of CNTs was also reflected in the hardness of the tested compounds, and resistance to the penetration of foreign objects into the material increased. Hardness can also be used to roughly estimate Young's modulus with a suitable correlation model. The estimation of Young's modulus based on Shore A hardness measurements (calculated using Equations (1)–(3)) and its comparison to the tensile test results can be seen in Figure 9. In comparison to the tensile tests results, the closest estimation of Young's modulus was calculated using Lindeman's Equation (3). However, such calculation of Young's modulus provides just approximate values and more measurements are needed to determine the most appropriate correlation model for this type of nanocomposite. The advantages of Young's modulus estimation from hardness measurements are that it is quick and inexpensive, and it can find application, for example, in the testing of material properties during a production process [59].

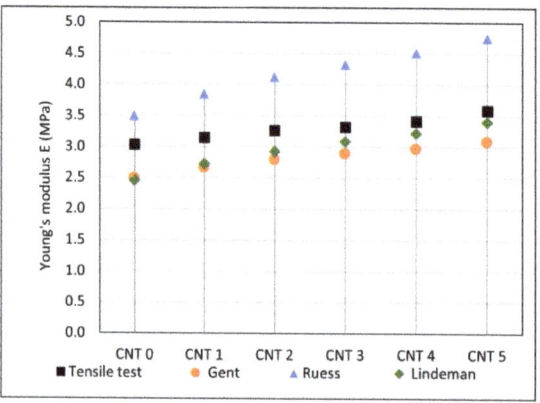

Figure 9. Comparison of Young's modulus from tensile tests with values estimated from hardness tests.

AFM spectroscopic curves can be a good tool for monitoring the Young's moduli of various materials, as they provide opportunities to compare materials in terms of their stiffness and elastic behavior. If the value of the Young's modulus of the reference material is known, it allows the calculation of values of the moduli of other materials. Based on the results of force spectroscopy, the Young's modulus values of the CNT 1–CNT 5 nanocomposites were calculated using Equation (11), as well as the reference value of the Young's modulus ($E_{CNT\,0}$ = 3.032 MPa) from the tensile test. The results obtained from the AFM and spectroscopic curves were comparable to the results obtained from the static tensile test and are summarized in Table 7 and graphically represented in Figure 10. Force spectroscopy also allows the comparison of the slope of a spectroscopic curve and the local moduli within one sample in order to determine the properties of selected material phases and to evaluate its homogeneity. With the results summarized in the table, it is possible to observe a larger variance of local values compared with the reference sample. This could be caused by the presence of CNTs in the polymer matrix, as number of studies have suggested that the interaction of a polymer matrix with CNTs results in an interfacial region with properties and a morphology different than the bulk [60,61]. Using AFM, the significant agglomerates of the CNTs were not detected. However, for a thorough characterization of the dispersion, further research is needed using an AFM microscope with better resolution or with different methods, such as scanning electron microscopy [62] or transmission electron microscopy [63]. These might be useful for determining the efficiency of the mixing process, as well as for potentially further improving the material characteristics of the tested materials.

Table 7. Comparison of Young's modulus obtained from tensile tests and AFM force spectroscopy.

Compound	Young's Modulus (MPa)	
	AFM	Tensile Test
CNT 1	3.144	3.151
CNT 2	3.272	3.265
CNT 3	3.344	3.325
CNT 4	3.441	3.421
CNT 5	3.599	3.595

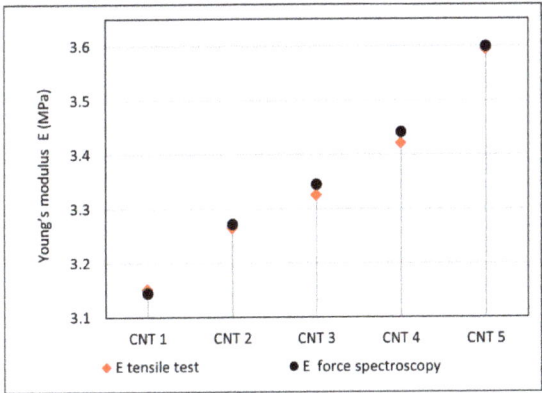

Figure 10. Comparison of Young's modulus obtained from tensile test and force spectroscopy.

The viscoelastic properties of a rubber composite depend on the interactions of its components, the crystalline behavior, and the extent of crosslinking between the polymer chains and the filler. These properties improve with the addition of suitable fillers [58]. A temperature and frequency sweep DMA was performed to investigate the viscoelastic properties of the tested nanocomposites. The values of the storage modulus and the loss modulus increased, and the tangent of phase angle decreased with the increasing content in

SWCNTs. Similar trends have been observed in [30,64–66], where different types of rubber nanocomposites filled with SWCNTs have been investigated. Dynamic stiffness improved with the addition of SWCNT nanofiller.

5. Conclusions

In recent years, rubber nanocomposites reinforced with a low-volume fraction of carbon-based nanofillers have attracted research interest due to their properties. The incorporation of nanofillers into various elastomers has been found to improve their overall mechanical properties. The properties of rubber nanocomposites depend significantly on the structure of the polymer matrices, the nature of the nanofillers, and the technological processes used for their preparation. A uniform dispersion of a nanofiller in a rubber matrix is a general prerequisite for achieving the desired material characteristics.

In the present study, the influence of single-wall CNTs on the mechanical properties of a NR/BR/IR/SBR compound were investigated. Five compounds that differed in SWCNT content (1.00–2.00 phr) and one without CNTs were tested and compared mutually. It was observed that the tensile strength, the Young's modulus, and the hardness increased with increasing SWCNT content, while the elongation at break decreased. A comparison of the DMA results for the compounds showed an increase in the loss modulus and the storage modulus and a decrease in the tangent of phase angle values with increasing CNT content. The addition of a small amount of well-dispersed nanotubes in the rubber compound improved its mechanical properties. AFM force spectroscopy combined with Snedonn's model was used to evaluate the local elasticity of the samples, and the results showed good agreement with the tensile test results. AFM force spectroscopy was a useful tool for obtaining information about the elasticity and stiffness of individual phases of the tested material.

Author Contributions: Conceptualization, D.B. and A.B.; methodology, D.B.; validation, D.B.; formal analysis, D.B. and A.B.; investigation, D.B. and A.B.; resources, D.B.; data curation, D.B.; writing—original draft preparation, A.B.; writing—review and editing, A.B. and D.B.; visualization, A.B.; supervision, D.B.; project administration, D.B.; funding acquisition, D.B. All authors have read and agreed to the published version of the manuscript.

Funding: This research work was supported by the Operational Programme Integrated Infrastructure and cofinanced by the European Regional Development Fund through the Advancement and Support of Research and Development project for the "Centre for diagnostics and quality testing of materials" in the RIS3 SK specialization domain (acronym: CEDITEK II., ITMS2014+ code 313011W442 and KEGA 011TnUAD-4/2021) for the implementation of progressive methods of analysis and synthesis of mechanical systems in the educational process.

Institutional Review Board Statement: Not applicable.

Informed Consent Statement: Not applicable.

Data Availability Statement: Data are contained within the article.

Conflicts of Interest: The authors declare no conflict of interest.

References

1. Bokobza, L. Multiwall carbon nanotube-filled natural rubber: Electrical and mechanical properties. *Express Polym. Lett.* **2012**, *6*, 213–223. [CrossRef]
2. Ma, P.C.; Kim, J.K. *Carbon Nanotubes for Polymer Reinforcement*; Taylor & Francis: Boca Raton, FL, USA, 2011.
3. Meyyappan, M. *Carbon Nanotubes: Science and Applications*; CRC Press: Boca Raton, FL, USA, 2004.
4. Harris, P.J.F. *Carbon Nanotube Science: Synthesis, Properties and Applications*; Cambridge University Press: Cambridge, UK, 2009.
5. Ibrahim, K.S. Carbon nanotubes-properties and applications: A review. *Carbon Lett.* **2013**, *14*, 131–144. [CrossRef]
6. Yu, M.F.; Lourie, O.; Dyer, M.J.; Moloni, K.; Kelly, T.F.; Ruoff, R.S. Strength and breaking mechanism of multiwalled carbon nanotubes under tensile load. *Science* **2000**, *287*, 637–640. [CrossRef] [PubMed]
7. Jančíková, Z.; Koštial, P.; Bakošová, D.; Ružiak, I.; Frydrýšek, K.; Valíček, J.; Farkašová, M.; Puchký, R. The study of electrical transport in rubber blends filled by single wall carbon nanotubes. *J. Nano Res.* **2013**, *21*, 1–6. [CrossRef]

8. Bokobza, L.; Rahmani, M.; Belin, C.; Bruneel, J.-L.; El Bounia, N.-E. Blends of carbon blacks and multiwall carbon nanotubes as reinforcing fillers for hydrocarbon rubbers. *J. Polym. Sci.* **2008**, *46*, 1939–1951. [CrossRef]
9. Park, S.M.; Lim, Y.W.; Kim, C.H.; Kim, D.J.; Moon, W.-J.; Kim, J.-H.; Lee, J.-S.; Hong, C.H.; Seo, G. Effect of carbon nanotubes with different lengths on mechanical and electrical properties of silica-filled styrene butadiene rubber compounds. *J. Ind. Eng. Chem.* **2013**, *19*, 712–719. [CrossRef]
10. Liu, Y.; Li, L.; Wang, Q.; Zhang, X. Fracture properties of natural rubber filled with hybrid carbon black/nanoclay. *J. Polym. Res.* **2011**, *18*, 859–867. [CrossRef]
11. Lorenz, H.; Fritzsche, J.; Das, A.; Stöckelhuber, K.W.; Jurk, R.; Heinrich, G.; Klüppel, M. Advanced elastomer nano-composites based on CNT-hybrid filler systems. *Compos. Sci. Technol.* **2009**, *69*, 2135–2143. [CrossRef]
12. Le, H.H.; Pham, T.; Henning, S.; Klehm, J.; Wießner, S.; Stöckelhuber, K.-W.; Das, A.; Hoang, X.T.; Do, Q.K.; Wu, M.; et al. Formation and stability of carbon nanotube network in natural rubber: Effect of non-rubber components. *Polymer* **2015**, *73*, 111–121. [CrossRef]
13. Bakošová, D. The study of the distribution of carbon black filler in rubber compounds by measuring the electrical conductivity. *Manuf. Technol.* **2019**, *19*, 366–370. [CrossRef]
14. Gumede, J.I.; Carson, J.; Hlangothi, S.P.; Bolo, L.L. Effect of single-walled carbon nanotubes on the cure and mechanical properties of reclaimed rubber/natural rubber blends. *Mater. Today Commun.* **2020**, *23*, 100852. [CrossRef]
15. Boonmahitthisud, A.; Chuayjuljit, S. NR/XSBR nanocomposites with carbon black and carbon nanotube prepared by latex compounding. *J. Met. Mater. Miner.* **2012**, *22*, 1.
16. Darestani Farahani, T.; Bakhshandeh, G.R.; Abtahi, M. Mechanical and viscoelastic properties of natural rubber/reclaimed rubber blends. *Polym. Bull.* **2006**, *56*, 495–505. [CrossRef]
17. Srivastava, S.K.; Mishra, Y.K. Nanocarbon Reinforced Rubber Nanocomposites: Detailed Insights about Mechanical, Dynamical Mechanical Properties, Payne, and Mullin Effects. *Nanomaterials* **2018**, *8*, 945. [CrossRef]
18. Azam, M.A.; Talib, E.; Mohamad, N.; Kasim, M.S.; Rashid, M.W.A. Mechanical and thermal properties of single-walled carbon nanotube filled epoxidized natural rubber nanocomposite. *J. Appl. Sci.* **2014**, *14*, 2183–2188. [CrossRef]
19. Bokobza, L. Natural Rubber Nanocomposites: A Review. *Nanomaterials* **2019**, *9*, 12. [CrossRef]
20. Kaushik, B.K.; Majumder, M.K. Carbon nanotube: Properties and applications. In *Carbon Nanotube Based VLSI Interconnects*; Springer: New Delhi, India, 2015; pp. 17–37. [CrossRef]
21. Kueseng, P.; Sae-oui, P.; Sirisinha, C.; Jacob, K.I.; Rattanasom, N. Anisotropic studies of multi-wall carbon nanotube (MWCNT)-filled natural rubber (NR) and nitrile rubber (NBR) blends. *Polym. Test.* **2013**, *32*, 1229–1236. [CrossRef]
22. Ponnamma, D.; Sadasivuni, K.K.; Grohens, Y.; Guo, Q.; Thomas, S. Carbon nanotube based elastomer composites–an approach towards multifunctional materials. *J. Mater. Chem. C* **2014**, *2*, 8446–8485. [CrossRef]
23. Polo-luque, M.L.; Simonet, B.M.; Valcárcel, M. Functionalization and dispersion of carbon nanotubes in ionic liquids. *TrAC Trends Anal. Chem.* **2013**, *47*, 99–110. [CrossRef]
24. Sui, G.; Zhong, W.H.; Yang, X.P.; Yu, Y.H.; Zhao, S.H. Preparation and properties of natural rubber composites reinforced with pretreated carbon nanotubes. *Polym. Adv. Technol.* **2008**, *19*, 1543–1549. [CrossRef]
25. Bhattacharyya, S.; Sinturel, C.; Bahloul, O.; Saboungi, M.-L.; Thomas, S.; Salvetat, J.-P. Improving reinforcement of natural rubber by networking of activated carbon nanotubes. *Carbon* **2008**, *46*, 1037–1045. [CrossRef]
26. Jiang, H.X.; Ni, Q.Q.; Natsuki, T. Design and evaluation of the interface between carbon nanotubes and natural rubber. *Polym. Compos.* **2011**, *32*, 236–242. [CrossRef]
27. Likozar, B. The effect of ionic liquid type on the properties of hydrogenated nitrile elastomer/hydroxy-functionalized multi-walled carbon nanotube/ionic liquid composites. *Soft Matter* **2001**, *7*, 970–977. [CrossRef]
28. Zhan, Y.H.; Liu, G.Q.; Xia, H.S.; Yan, N. Natural rubber/carbon black/carbon nanotubes composites prepared through ultrasonic assisted latex mixing process. *Plast. Rubber Compos.* **2011**, *40*, 32–39. [CrossRef]
29. Danafar, F.; Kalantari, M. A review of natural rubber nanocomposites based on carbon nanotubes. *J. Rubber Res.* **2018**, *21*, 293–310. [CrossRef]
30. Abdullateef, A.A.; Thomas, S.P.; Al-Harthi, M.A.; De, S.K.; Bandyopadhyay, S.; Basfar, A.A.; Atieh, M.A. Natural rubber nanocomposites with functionalized carbon nanotubes: Mechanical, dynamic mechanical, and morphology studies. *J. Appl. Polym. Sci.* **2012**, *125* (Suppl. S1), E76–E84. [CrossRef]
31. Carrión, F.J.; Sanes, J.; Bermúdez, M.-D.; Arribas, A. New single-walled carbon nanotubes–ionic liquid lubricant. Application to polycarbonate–stainless steel sliding contact. *Tribol. Lett.* **2011**, *41*, 199–207. [CrossRef]
32. Das, A.; Stöckelhuber, K.W.; Jurk, R.; Fritzsche, J.; Klüppel, M.; Heinrich, G. Coupling activity of ionic liquids between diene elastomers and multi-walled carbon nanotubes. *Carbon* **2009**, *47*, 3313–3321. [CrossRef]
33. Le, H.H.; Hoang, X.T.; Das, A.; Gohs, U.; Stoeckelhuber, K.-W.; Boldt, R.; Heinrich, G.; Adhikari, R.; Radusch, H.-J. Kinetics of filler wetting and dispersion in carbon nanotube/rubber composites. *Carbon* **2012**, *50*, 4543–4556. [CrossRef]
34. Wang, C.C.; Donnet, J.B.; Wang, T.K.; Pontier-Johnson, M.; Welsh, F. AFM study of rubber compounds. *Rubber Chem. Technol.* **2005**, *78*, 17–27. [CrossRef]
35. Jeon, I.H.; Kim, H.; Kim, S.G. Characterization of rubber micro-morphology by atomic force microscopy (AFM). *Rubber Chem. Technol.* **2003**, *76*, 1–11. [CrossRef]

36. Innes, J.R.; Young, R.J.; Papageorgiou, D.G. Graphene Nanoplatelets as a Replacement for Carbon Black in Rubber Compounds. *Polymers* **2022**, *14*, 1204. [CrossRef] [PubMed]
37. Lin, D.C.; Dimitriadis, E.K.; Horkay, F. Elasticity of rubber-like materials measured by AFM nanoindentation. *Express Polym. Lett.* **2007**, *1*, 576–584. [CrossRef]
38. Yerina, N.; Magonov, S. Atomic force microscopy in analysis of rubber materials. *Rubber Chem. Technol.* **2003**, *76*, 846–859. [CrossRef]
39. Miyaji, K.; Sugiyama, T.; Ohashi, T.; Saalwächter, K.; Ikeda, Y. Study on Homogeneity in Sulfur Cross-Linked Network Structures of Isoprene Rubber by TD-NMR and AFM–Zinc Stearate System. *Macromol.* **2020**, *53*, 8438–8449. [CrossRef]
40. Leng, L.; Han, Q.Y.; Wu, Y.P. The aging properties and phase morphology of silica filled silicone rubber/butadiene rubber composites. *RSC Adv.* **2020**, *10*, 20272–20278. [CrossRef]
41. Bakošová, D.; Bakošová, A. Analysis of Homogeneity and Young's Moduli of Rubber Compounds by Atomic Force Microscopy. *Manuf. Technol.* **2021**, *21*, 749–756. [CrossRef]
42. Ning, N.; Mi, T.; Chu, G.; Zhang, L.-Q.; Liu, L.; Tian, M.; Yu, H.-T.; Lu, Y.-L. A quantitative approach to study the interface of carbon nanotubes/elastomer nanocomposites. *Eur. Polym. J.* **2018**, *102*, 10–18. [CrossRef]
43. Ito, M.; Liu, H.; Kumagai, A.; Liang, X.; Nakajima, K.; Jinnai, H. Direct Visualization of Interfacial Regions between Fillers and Matrix in Rubber Composites Observed by Atomic Force Microscopy-Based Nanomechanics Assisted by Electron Tomography. *Langmuir* **2022**, *38*, 777–785. [CrossRef]
44. Morozov, I.A. A novel method of quantitative characterization of filled rubber structures by AFM. *KGK Rubberpoint* **2011**, *64*, 24–27.
45. Muravyeva, T.I.; Gainutdinov, R.V.; Morozov, A.V.; Shcherbakova, O.O.; Zagorskiy, D.L.; Petrova, N.N. Influence of antifriction fillers on the surface properties of elastomers based on propylenoxide rubbers. *J. Frict. Wear* **2017**, *38*, 339–348. [CrossRef]
46. Longo, M.; De Santo, M.P.; Esposito, E.; Fuoco, A.; Monteleone, M.; Giorno, L.; Jansen, J.C. Force spectroscopy determination of Young's modulus in mixed matrix membranes. *Polymer* **2018**, *156*, 22–29. [CrossRef]
47. Wang, D.; Russell, T.P. Advances in atomic force microscopy for probing polymer structure and properties. *Macromolecules* **2018**, *51*, 3–24. [CrossRef]
48. Gent, A.N. On the relation between indentation hardness and Young's modulus. *Rubber Chem. Technol.* **1958**, *31*, 896–906. [CrossRef]
49. Larson, K. *Can You Estimate Modulus from Durometer Hardness for Silicones*; Dow Corning Corporation: Midland, MI, USA, 2016; pp. 1–6.
50. Manohar, D.M.; Chakraborty, B.C.; Begum, S.S. Hardness–Elastic Modulus Relationship for Nitrile Rubber and Nitrile Rubber–Polyvinyl Chloride Blends. In *Advances in Design and Thermal Systems*; Springer: Singapore, 2021; pp. 301–314. [CrossRef]
51. Altidis, P.A.; Warner, B.V. *Analyzing Hyperelastic Materials w/Some Practical Considerations*; Midwest ANSYS Users Group: Canonsburg, PA, USA, 2005.
52. Kubínek, R.; Mašláň, M.; Vůjtek, M. *Mikroskopie Skenující Sondou*; Vydavatelství Univerzity Palackého: Olomouc, Czech Republic, 2003.
53. Garcia-Manyes, S.; Sanz, F. Nanomechanics of lipid bilayers by force spectroscopy with AFM: A perspective. *Biochim. Biophys. Acta (BBA)-Biomembr.* **2010**, *1798*, 741–749. [CrossRef]
54. Seo, Y.; Jhe, W. Atomic force microscopy and spectroscopy. *Rep. Prog. Phys.* **2007**, *71*, 23. [CrossRef]
55. C o, A.A.; ; C.A. Ca oo oo (oo). *aa oo* **1997**, *2*, 78–89.
56. Sneddon, I.N. The relation between load and penetration in the axisymmetric Boussinesq problem for a punch of arbitrary profile. *Int. J. Eng. Sci.* **1965**, *3*, 47–57. [CrossRef]
57. Tsukruk, V.V.; Huang, Z.; Chizhik, S.A.; Gorbunov, V.V. Probing of micromechanical properties of compliant polymeric materials. *J. Mater. Sci.* **1998**, *33*, 4905–4909. [CrossRef]
58. Tang, L.C.; Zhao, L.; Qiang, F.; Wu, Q.; Gong, L.-X.; Peng, J.-P. Mechanical properties of rubber nanocomposites containing carbon nanofillers. In *Carbon-Based Nanofillers and Their Rubber Nanocomposites*; Elsevier: Amsterdam, The Netherlands, 2019; pp. 367–423. [CrossRef]
59. Zhao, H.; Allanson, D.; Ren, X.J. Use of shore hardness tests for in-process properties estimation/monitoring of silicone rubbers. *J. Mater. Sci. Chem. Eng.* **2015**, *3*, 142–147. [CrossRef]
60. Wang, D.; Fujinami, S.; Nakajima, K.; Niihara, K.-I.; Inukai, S.; Ueki, H.; Magario, A.; Noguchi, T.; Endo, M.; Nishi, T. Production of a cellular structure in carbon nanotube/natural rubber composites revealed by nanomechanical mapping. *Carbon* **2010**, *48*, 3708–3714. [CrossRef]
61. Coleman, J.N.; Khan, U.; Blau, W.J.; Gun'ko, Y.K. Small but strong: A review of the mechanical properties of carbon nanotube–polymer composites. *Carbon* **2006**, *44*, 1624–1652. [CrossRef]
62. Kovacs, J.Z.; Andresen, K.; Pauls, J.R.; Garcia, C.P.; Schossig, M.; Schulte, K.; Bauhofer, W. Analyzing the quality of carbon nanotube dispersions in polymers using scanning electron microscopy. *Carbon* **2007**, *45*, 1279–1288. [CrossRef]
63. Park, C.; Ounaies, Z.; Watson, K.A.; Crooks, R.E.; Smith, J., Jr.; Lowther, S.E.; Connell, J.W.; Siochi, E.J.; Harrison, J.S.; Clair, T.L. Dispersion of single wall carbon nanotubes by in situ polymerization under sonication. *Chem. Phys. Lett.* **2002**, *364*, 303–308. [CrossRef]
64. Yan, N.; Wu, J.K.; Zhan, Y.H.; Xia, H.S. Carbon nanotubes/carbon black synergistic reinforced natural rubber composites. *Plast. Rubber Compos.* **2009**, *38*, 290–296. [CrossRef]

65. López-Manchado, M.A.; Biagiotti, J.; Valentini, L.; Kenny, J.M. Dynamic mechanical and Raman spectroscopy studies on interaction between single-walled carbon nanotubes and natural rubber. *J. Appl. Polym. Sci.* **2004**, *92*, 3394–3400. [CrossRef]
66. Bakošová, D. Dynamic mechanical analysis of rubber mixtures filled by carbon nanotubes. *Manuf. Technol.* **2018**, *18*, 345–351. [CrossRef]

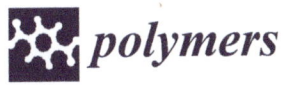

Article

Rheometer Evidences for the Co-Curing Effect of a Bismaleimide in Conjunction with the Accelerated Sulfur on Natural Rubber/Chloroprene Rubber Blends

Marek Pöschl, Shibulal Gopi Sathi *, Radek Stoček and Ondřej Kratina

Centre of Polymer Systems, Tomas Bata University in Zlín, Třída Tomáše Bati 5678, 760 01 Zlín, Czech Republic; poschl@utb.cz (M.P.); stocek@utb.cz (R.S.); okratina@utb.cz (O.K.)
* Correspondence: sathi@utb.cz

Abstract: The rheometer curing curves of neat natural rubber (NR) and neat chloroprene rubber (CR) with maleide F (MF) exhibit considerable crosslinking torque at 180 °C. This indicates that MF can crosslink both these rubbers via Alder-ene reactions. Based on this knowledge, MF has been introduced as a co-crosslinking agent for a 50/50 blend of NR and CR in conjunction with accelerated sulfur. The delta (Δ) torque obtained from the curing curves of a blend with the addition of 1 phr MF was around 62% higher than those without MF. As the content of MF increased to 3 phr, the Δ torque was further raised to 236%. Moreover, the mechanical properties, particularly the tensile strength of the blend with the addition of 1 phr MF in conjunction with the accelerated sulfur, was around 201% higher than the blend without MF. The overall tensile properties of the blends cured with MF were almost retained even after ageing the samples at 70 °C for 72 h. This significant improvement in the curing torque and the tensile properties of the blends indicates that MF can co-crosslink between NR and CR via the Diels–Alder reaction.

Keywords: rubber; curing; bismaleimide; tensile strength; Diels–Alder reaction

Citation: Pöschl, M.; Gopi Sathi, S.; Stoček, R.; Kratina, O. Rheometer Evidences for the Co-Curing Effect of a Bismaleimide in Conjunction with the Accelerated Sulfur on Natural Rubber/Chloroprene Rubber Blends. *Polymers* **2021**, *13*, 1510. https://doi.org/10.3390/polym13091510

Academic Editor: Francesca Lionetto

Received: 19 April 2021
Accepted: 4 May 2021
Published: 7 May 2021

Publisher's Note: MDPI stays neutral with regard to jurisdictional claims in published maps and institutional affiliations.

Copyright: © 2021 by the authors. Licensee MDPI, Basel, Switzerland. This article is an open access article distributed under the terms and conditions of the Creative Commons Attribution (CC BY) license (https://creativecommons.org/licenses/by/4.0/).

1. Introduction

It is well-known that natural rubber (NR) is a polymer of isoprene (2-methyl-1, 3-butadiene) and chloroprene rubber (CR) is a polymer of 2-chloro-1, 3-butadiene. Therefore, the main structural difference between NR and CR is that a methyl group in NR is substituted by a chlorine atom in CR. Because of the presence of this electronegative chlorine atom, CR has many unique properties such as improved heat, oil, ozone, and chemical resistance. Moreover, it has better resilience and weather resistance compared to NR [1,2]. It has been reported in the literature that commercial-grade polychloroprene comprises four isomeric forms such as trans-1, 4-polychloroprene (80–90%), cis-1, 4-polychloroprene (5–15%), 1, 2-polychloroprene (1–2%), and 3, 4-polychloroprene (3–4%) [3–8]. Out of these four isomeric forms, the 1, 2-isomer has been identified as the major isomer responsible for the curing process because of its ability to undergo the allylic rearrangement of the tertiary chlorine atom [9–11]. The rearrangement of the 1, 2-isomer can occur on the heating of the neat polychloroprene. However, the rearrangement occurs much faster in the presence of zinc oxide (ZnO) [12]. It is well known that ZnO or ethylene thiourea (ETU) are used as the main crosslinking agents for CR either separately or in combination [13,14]. Several mechanisms have been reported in the literature concerning the curing of CR with ZnO or ETU [13–15]. The mechanism proposed by Vukov based on the theory of Kuntz et al. using model compounds is considered as the most appropriate mechanism for the crosslinking of CR with ZnO [16,17]. Apart from ZnO or ETU, other chemicals such as, tribasic lead sulphate; thiophosphoryl disulfides; dimethyl L-cystine; and cetyltrimethylammonium maleate (CTMAM) have also been used as curing agents for CR [18–21]. Recently, Dziemidkiewicz et al. have used certain metal acetylacetonate as pro-ecological crosslinking agents for CR based on the Heck coupling reaction [22].

Unlike CR, the electropositive methyl group adjacent to the double-bonded carbon atom in NR enhances the activity of the unsaturated double bonds in the backbone of the polymer chains. As a result, NR has inferior weather, ozone, and oil resistance. Moreover, the high-temperature performance of NR is also limited because of its degradation above 70 °C. However, due to the active double bond, NR can be cured with accelerated sulfur systems. Based on the accelerator to sulfur ratio, accelerated sulfur curing systems are classified as conventional vulcanization (CV), efficient vulcanization (EV), and semi-efficient vulcanization (SEV). Generally, the CV system is characterized by a high dosage of sulfur (2–3.5 phr) and a low dosage of accelerators (0.4–1.2 phr) [1,2]. Even gum NR cured with a CV system can exhibit a tensile strength (TS) of around 20 to 25 MPa due to its unique strain-induced crystallization behaviour [23–25]. Blending between NR and CR can exploit certain unique properties of these individual rubbers. However, the structural disparity due to the electropositive methyl group in NR and the electronegative chlorine atom in CR makes the blending of NR and CR and the subsequent co-curing of the resultant blends very difficult. Therefore, a chemical that can co-cure both the NR and CR chains is essential for developing a compatible blend of NR/CR.

The rheometer is one of the key pieces of characterization equipment in the rubber industry and is used to check the feasibility of a new chemical as a curing agent in rubber compounds. Generally, the rheometer analysis gives a clear spectrum concerning the processing behaviors, such as the viscosity, scorch time, and optimum cure time of rubber compounds. Based on this knowledge, the compounder can select specific ingredients and determine the dosage of each ingredient needed to meet the required target. Moreover, with rheometer cure data, rubber scientists can quickly arrive at certain predictions concerning the crosslinking mechanisms based on the available theory. This may help the scientist establish the actual chemical reaction mechanisms involved in the curing process at the molecular level using advanced characterization techniques such as differential scanning calorimetry (DSC), nuclear magnetic resonance (HNMR), and infrared (IR) spectrometry, etc. From our previous experimental investigation, it has been observed that bismaleimide can react with halogenated rubbers during curing [26–29], and can also interact with NR/CV and butadiene rubber (BR)/CV systems via Diels–Alder and Alder-ene reactions [30–32]. Sadao Inoue filed a patent based on a chloroprene rubber and bismaleimide/ZnO composition for the development of vulcanizate with a high degree of crosslink density [33].

In the present investigation, we explored the curing behavior of a 50/50 blend of NR/CR with different contents of MF using a rheometer. The swelling behavior and mechanical properties of the vulcanizate derived from these blends were also evaluated. To the best of our knowledge, no reports are available in the literature concerning the curing behavior of an NR/CR blend with MF.

2. Materials

Natural rubber (standard Vietnamese rubber with a Mooney viscosity ML (1 + 4) at 100 °C: 60 ± 5) supplied by Binh Phuoc, Vietnam under the trade name SVR CV60 and chloroprene rubber (Neoprene 9243P, DuPont elastomer with a Mooney viscosity ML (1 + 4) at 100 °C: 87) were used as the base elastomers. Maleide F (MF) is a combination of 75% N, N'-meta phenylene dimaleimide and a 25% blending agent was procured from Krata Pigment, Tambov, Mentazhnikov, Russia. The chemical structure of MF is shown in Figure 1. Other ingredients such as sulfur; n-cyclohexyl-2-benzothiazole sulfenamide (CBS); stearic acid; zinc oxide (ZnO); and magnesium oxide (MgO) were purchased from Sigma-Aldrich, Czech Republic.

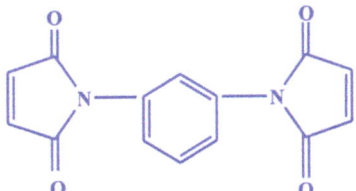

Figure 1. Chemical structure of N, N'-meta phenylene dimaleimide (Maleide F).

2.1. Preparation of Rubber Compounds

The formulation of the mixes with designations are displayed in Table 1. All the compounds were prepared using an internal mixer (Brabender Plastograph, GmbH & Co, KG, Duisburg, Germany) with a chamber volume of 50 cc. A fill factor of 0.8 was taken for the efficient mixing of the ingredients. To prepare the CR-based compound, the neat CR was masticated at 50 °C under 50 rpm for 2 min. To this, the ZnO, MgO, stearic acid, and MF were added, and the mixing was continued under the same rotor speed and temperature for another 2 min. After the mixing, the compound was discharged and homogenized using a two-roll mill. To prepare the blend-based compounds, the individual rubbers were masticated separately for 2 min under the same processing conditions. The pre-masticated rubbers were mixed for 1 min. To this, the ZnO, MgO, stearic acid, and MF were added, and the mixing was continued for 2 more minutes. Finally, the sulfur and CBS were added and mixed for an additional minute. After the mixing, the compound was discharged and homogenized using a two-roll mill. It was then molded into sheets with a thickness of 2 mm by applying a constant force of 200 N using a compression molding heat press LaBEcon 300 (Fontijne Presses, Delft, Netherlands) for the respective cure time obtained from the rheometer cure data at 180 °C.

Table 1. Formulation of the mixes.

Ingredients → Mix Code ↓	NR	CR	ZnO	MgO	Stearic Acid	Sulfur	CBS	Maleide F
M-1	-	100	5	4	0.5	-	-	-
M-2	-	100	-	-	-	-	-	3
M-3	-	100	5	4	0.5	-	-	3
M-4	50	50	5	2	1.25	1.25	0.25	-
M-5	50	50	5	2	1.25	1.25	0.25	1
M-6	50	50	5	2	1.25	1.25	0.25	3
M-7	50	50	5	2	1.25	1.25	0.25	5

2.2. Characterization

2.2.1. Cure Characteristics

Maximum torque: M_H, minimum torque: M_L, the difference between maximum and minimum torque: ΔM, scorch time: T_{S2}, optimum cure time: T_{90} (the time required for the torque to reach 90% of the maximum torque) of the rubber compounds were determined from the cure curves from a moving die rheometer (MDR-3000, MonTech, Buchen, Germany) at 180 °C as per ASTM D 5289. The cure rate index (CRI), a measure of the rate of curing, was calculated using Equation (1).

$$CRI = 100/(T_{90-S2}). \tag{1}$$

2.2.2. Swelling Behavior

Samples with a diameter of 20 mm and a thickness of 2 mm with an initial weight (W_i) were swelled in toluene at room temperature until they reached an equilibrium state of swelling. The swelled samples were then taken out and wiped off the adhered toluene from the surface using a filter paper, and the weights (W_s) were immediately recorded. The swollen samples were dried at room temperature (20 °C) for 24 h, and ensured that the absorbed toluene is completely expelled out. Then, we measured the dried weight of the samples (W_d). From the values of W_i, W_s, and W_d the percentage swelling and the swell ratio were calculated using Equations (2) and (3), respectively [34,35].

$$Swelling\ (\%) = \frac{W_s - W_i}{W_S} \times 100 \tag{2}$$

$$Swell\ ratio(Q) = \frac{W_s}{W_d} - 1 \tag{3}$$

2.2.3. Mechanical (Tensile) Properties

The stress–strain behavior and the corresponding tensile properties of the vulcanizates were measured using a universal testing machine (Testometric M350, Testometric Company, Ltd. Rochdale, UK). The testing was performed under ambient conditions at a crosshead speed of 500 mm/min as per ISO 37 using S2 type specimen with a thickness of 2 mm. The results were reported at an average of six tested specimens. The properties of the cured samples were also measured after ageing at 70 °C and 100 °C for 72 h using a forced air circulating oven.

2.2.4. Hardness Testing

Cured samples having smooth surfaces were used to measure the indentation hardness using a Shore-A hardness tester (Bareiss Durometer, Oberdischingen, Germany) as per ASTM D 2240. Indentations were made on different areas of the samples by applying constant pressure for 15 s. Six readings were taken from different areas of the sample and we reported the average value.

3. Results and Discussion

3.1. Curing Behavior of Neat CR with ZnO and MF

Represented in Figure 2 are the curing curves of neat CR with ZnO, MF, and a combination of ZnO/MF at 180 °C for 1 h. Their cure characteristics are depicted in Table 2. The cure curve of CR/ZnO (M-1) exhibits a fast curing reaction, as evident from the short ts_2 value (0.65 min). However, after a rapid initial curing reaction, the cure curve turned into a marching modulus behavior and ended up with a maximum torque of 5.80 dNm at the given curing time. As a result, the t_{90} (40.2 min) value was higher than expected. Several mechanisms have been proposed in the literature to explain the curing behavior of CR with ZnO. Out those, a cationic mechanism proposed by Vukov is widely accepted [21]. As per this mechanism, the 1, 2-isomer of CR undergoes a rearrangement (isomerization) upon heating above 160 °C. The rearranged 1, 2-isomer then produces a conjugated diene in the presence of ZnO. The rearranged 1, 2-isomer and the in situ-formed diene catalyzed by $ZnCl_2$ produce the crosslinks.

The cure curve of M-1 shows a relatively low ΔM (4.42 dNm) value, indicating that the extent of curing of CR with ZnO is not high enough. On the other hand, the curing of CR with MF alone (M-2) progressed at a slow pace and exhibited a marching modulus curing right from the beginning until the end of the given curing time. As a result, the ts_1 (2.43 min) and ts_2 (4.42 min) values were higher compared to M-1. Though the curing curves of M-1 and M-2 exhibited a marching modulus curing behavior, the extent of vulcanization was greatly increased in M-2 as the cure time progressed. For instance, the ΔM (extent of vulcanization) after 30 min of curing in M-1 was 3.70 dNm. At the same time, the ΔM in M-2 after 30 min of curing was around 114% higher than M-1. This dramatic improvement in ΔM indicates that MF alone can substantially crosslink the CR chains. Since there are no other ingredients in the system other than CR and MF, one the plausible crosslinking reactions might be the Alder-ene reaction between the rearranged 1,2-isomer of CR and MF, as depicted in Figure 3.

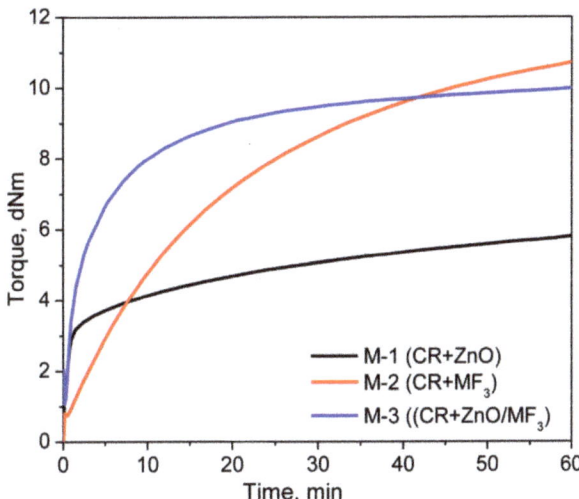

Figure 2. Cure curves of the mixes M-1, M-2, and M-3.

Table 2. Rheometer cure data of the mixes at 180 °C, 1 h.

Mix Code	M_L (dNm)	M_H (dNm)	ΔM (dNm)	T_{S1} (min)	T_{S2} (min)	T_{90} (min)	Cure Rate Index (min^{-1})
M-1	1.38	5.80	4.42	0.65	0.65	40.23	2.52
M-2	0.74	10.69	9.95	2.43	4.52	41.58	2.69
M-3	1.07	9.97	8.90	0.53	0.70	20.63	4.99
M-4	0.70	3.09	2.39	1.65	3.86	4.62	131.57
M-5	0.55	4.43	3.88	1.55	2.54	5.55	33.22
M-6	0.66	8.70	8.04	1.34	2.30	12.44	9.86
M-7	0.49	9.80	9.31	1.51	2.38	18.36	6.25

It was interesting to note that the curing of CR with a combination of ZnO and MF (M-3) exhibits an initial rapid reaction (ts_2: 0.60 min) with a plateau-type curing behavior. As a result, the t_{90} value of M-3 (20.63 min) was much lower compared to M-1 (t_{90}: 40.23 min) and M-2 (t_{90}: 41.58 min). Moreover, the extent of curing was also higher compared to M-1 and M-2. For instance, the ΔM generated after 10 min of curing in M-3 was around 152% higher than M-1 and 70% higher than M-2 under the same curing conditions. It has been reported that the 1, 2-isomer of CR undergoes a rearrangement and subsequently produces a conjugated diene in the presence of ZnO [12]. Therefore, it is reasonable to believe that one of the reasons behind the synergistic curing behavior in M-3 might be the Diels–Alder reaction between the in situ-formed diene and the maleimide moieties of MF, as shown in Figure 4a.

3.2. Curing Behavior of NR/CR Blend in the Presence of MF in Conjunction with Accelerated Sulfur

Represented in Figure 5 are the cure curves of NR/CR blends corresponding to the mixes 4–7 at 180 °C for 1 hr. The cure curve of the blend without MF (M-4) exhibits a scorch time of 3.86 min with a ΔM value of 2.39 dNm. The addition of 1 phr MF reduces the scorch time of M-4 from 3.86 min to 2.54 min and improves the ΔM value by 62%. Similarly, the addition of 3 phr MF improves the ΔM value of M-4 by 236%, which further rose to 289% with the addition of 5 phr MF. Here, it is interesting to note that although there are not many differences in the scorch time between M-4 and the rest of the mixes, the addition of MF gradually increases the optimum cure time as the content of MF increases. For instance, the t_{90} of M-4 was 4.62 min, which became 12.44 min after the addition of 3 phr

MF and 18.36 min with the addition of 5 phr MF. This might be due to the slightly marching modulus curing behavior of the mixes containing MF. From this study, it has been observed that CR undergoes a synergistic curing behavior in the presence of a combination of ZnO and MF. One of the reasons for this was suspected to be the Diels–Alder reaction between the in situ-formed diene from the isomerized CR and the malemide moieties of MF as shown in Figure 4a. From our previous experimental investigation, it has been identified that MF can act as an anti-reversion agent during the curing of NR and BR with a CV system [30–32]. One of the plausible mechanisms proposed to explain the anti-reversion ability of MF was also the Diels–Alder reaction between the in situ formed diene from the NR/CV system and the maleimide moieties of MF. Based on these experimental inferences, it is reasonable to believe that the three types of reactions given in Figure 4a–c might be possible during the curing of mixes 4–7. The occurrence of any of these three reactions can improve the overall crosslink density of the blend system. However, the co-curing reaction given in Figure 4c is essential for enhancing the compatibility between NR and CR.

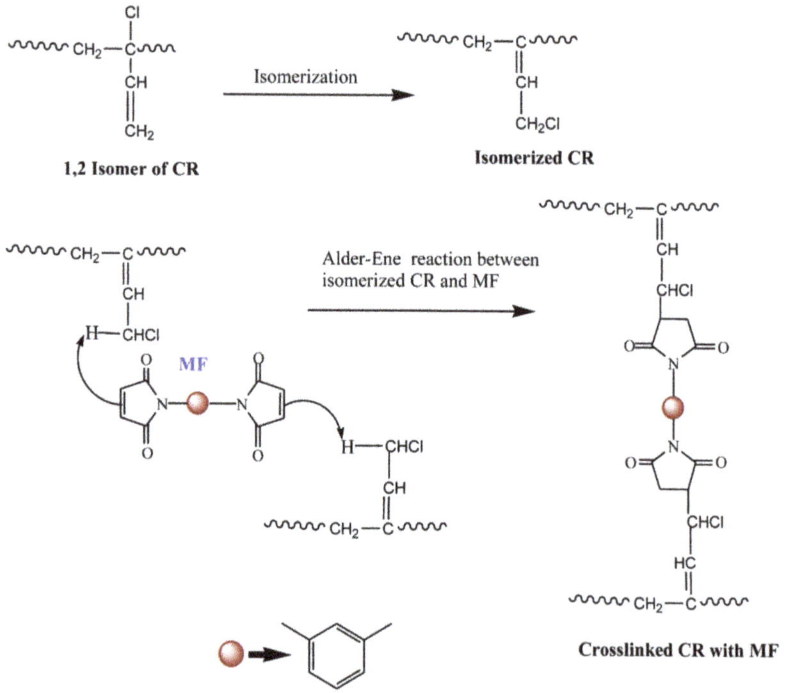

Figure 3. Plausible reaction mechanism for the curing behavior of CR with MF.

3.3. Swelling Behavior

It is well known that the crosslinked rubbers with a tight network structure generally show high swelling resistance. The percentage swelling and the swell ratio as per Equations (2) and (3) were calculated for the blends (M-4 to M-7), and the results are given in Table 3.

For comparative purposes, the swell ratio of M-1, M-2, and M-3 were also given. The blend with no MF (M-4) exhibits a solvent uptake of around 551%, corresponding to a swell ratio of 6.5. As the content of MF increased, the percentage swelling gradually decreased. This means that a higher concentration of MF produces more crosslinked points between the polymer chains as per the reactions proposed in Figure 4, thereby enhancing the extent of crosslinking and the network density. The ΔM values obtained from the rheometer cure data were also in line with this swelling behavior. However, the rheometer data and the swelling behavior do not give a clear indication of which reaction represented in Figure 4 is predominant during the curing of blends containing MF.

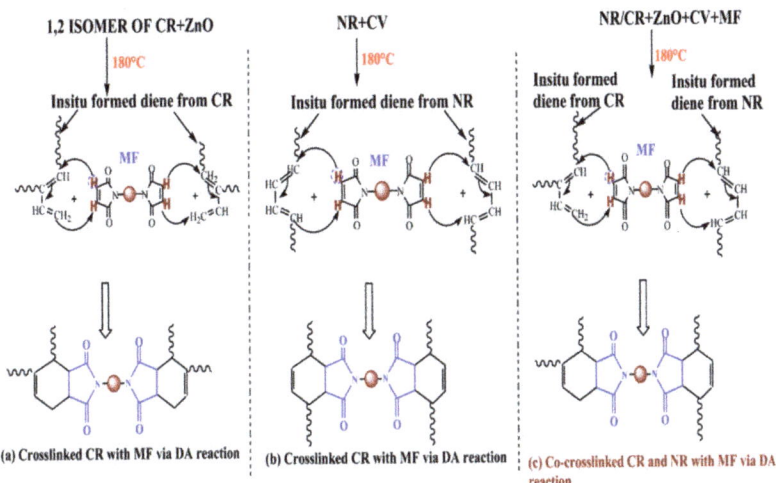

Figure 4. Plausible reactions in NR/CR blend during curing with MF in conjunction with the accelerated sulfur system.

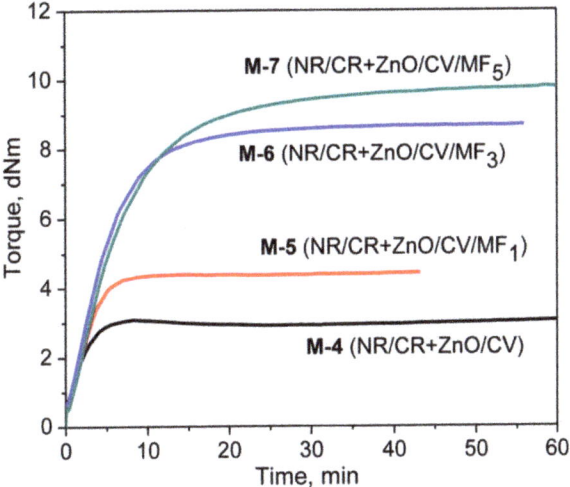

Figure 5. Cure curves of the mixes M-4, M-5, M-6, and M-7.

3.4. Mechanical Properties of the Blends

To understand the crosslinking effects of MF on the mechanical properties, the tensile strength (TS), the elongation at break (EB), the modulus at a different percentages of elongation, and the hardness of the cured blends were evaluated with different contents of MF, and the results are represented in Table 4.

Table 3. Percentage swelling and swell ratio.

Mix Code	Swelling (%)	Swell Ratio (Q)
M-1	281.7 ± 0.1	2.94 ± 0.1
M-2	237.2 ± 0.02	2.35 ± 0.003
M-3	177.0 ± 0.07	1.78 ± 0.08
M-4	562.9 ± 0.08	6.55 ± 0.13
M-5	428.2 ± 0.03	4.57 ± 0.12
M-6	314.9 ± 0.13	3.29 ± 0.1
M-7	267.9 ± 0.06	2.79 ± 0.06

Table 4. Tensile properties and hardness of the mixes.

Mie	Tensile Strength (MPa)	Elongation at Break (%)	Modulus at 50% (MPa)	Modulus at 100% (MPa)	Modulus at 300% (MPa)	Hardness (Shore A)
M-1	5.32 ± 0.98	292 ± 63	0.91 ± 0.13	1.56 ± 0.26	4.03 ± 1.87	39
M-3	2.84 ± 0.20	155 ± 9.0	1.08 ± 0.08	1.83 ± 0.19	-	52
M-4	3.07 ± 0.74	490 ± 98	0.37 ± 0.03	0.54 ± 0.05	1.38 ± 0.26	24
M-5	9.25 ± 1.40	566 ± 89	0.62 ± 0.13	0.90 ± 0.10	2.55 ± 0.67	34
M-6	9.34 ± 2.93	482 ± 77	0.70 ± 0.11	1.07 ± 0.12	3.08 ± 0.48	42
M-7	9.89 ± 2.81	425 ± 45	0.85 ± 0.11	1.33 ± 0.18	4.32 ± 1.34	47

For a comparative evaluation, the tensile properties of the vulcanizates of M-1 and M-3 are also given in Table 3. The vulcanizate of M-1 gives a TS of 5.32 MPa with an EB of 292%. Both the TS and the EB were reduced to 2.84 MPa and 154%, respectively, when CR was cured with a combination of ZnO and MF (M-3). The cured network of M-1 might not be strong enough because the network mainly composed of carbon–carbon crosslinks as per the reaction mechanism proposed by Vukov. Therefore, the cured network of M-1 might have certain flexibility to transfer the tensile load, and thereby exhibit a relatively high TS and EB compared to the vulcanizate of M-3. The additional crosslinks formed in the vulcanizate of M-3 owing to the proposed Diels–Alder reaction shown in Figure 4a, the cured network of M-3 will be rigid and strong. As a result, the transfer of tensile load becomes difficult and hence shows a low TS and strain at break. However, the strong network structure in M-3 significantly enhances its hardness and modulus. It was interesting to note that the vulcanizate of the blend with no MF (M-4) gives a TS of 3.07 MPa and an EB of 490%. However, after the incorporation of 1 phr MF, the TS of the blend vulcanizate (M-5) suddenly improved and was three times higher than M-4. Both the modulus and hardness of M-5 were also significantly improved with the addition of even 1 phr MF. The improved mechanical properties of the blends cured with MF in conjunction with the accelerated sulfur give a strong indication concerning the co-crosslinking reaction between NR and CR, as proposed in Figure 4c. It is worth noting that the TS of the blends did not improve further beyond 1 phr MF. That being said, the modulus and hardness were significantly improved up to the addition of 5 phr MF. To understand the strength of the blend vulcanizate, the abovementioned tensile properties were also evaluated after ageing them at 70 °C and 100 °C for 72 h. The results are depicted in Figure 6a–c.

From the results, it is clear that the properties of the unaged and aged samples at 70 °C for 72 h were comparable. However, the properties, particularly the TS were almost 3 times lower when the blends were aged at 100 °C for 72 h. This might be due to the degradation of the NR phase in the blend during the ageing process at 100 °C.

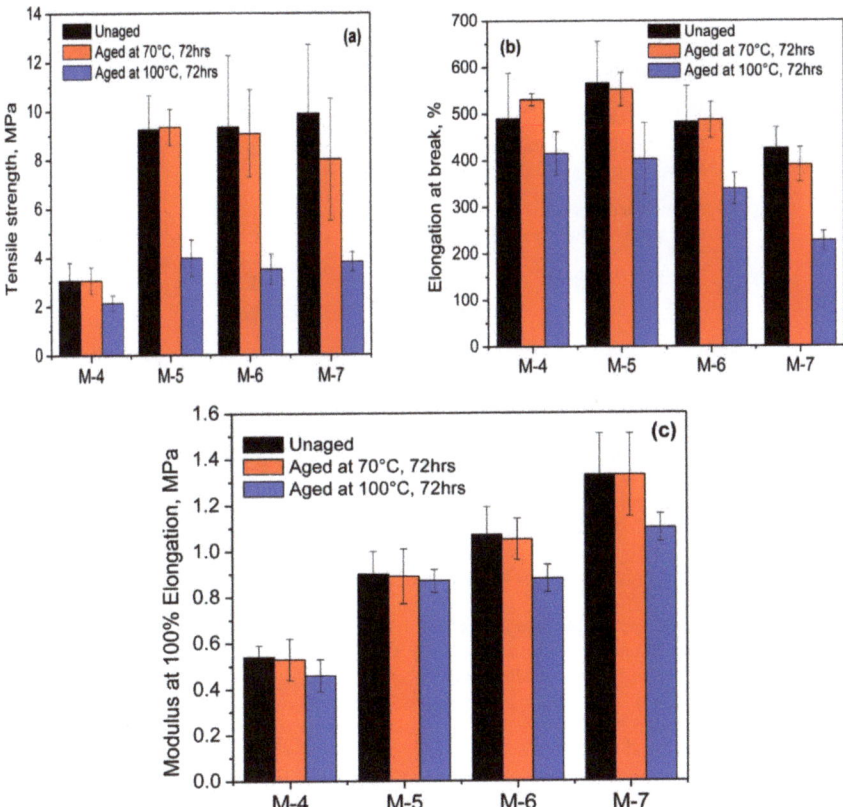

Figure 6. Tensile properties of blend vulcanizates (**a**) tensile strength, (**b**) elongation at break, and (**c**) modulus at 100% elongation before and after ageing.

4. Conclusions

The rheometer investigation of the curing of neat CR with a combination of ZnO and MF at 180 °C exhibits a synergistic curing curve with a plateau type cure behavior. In light of the mechanism proposed by Vukov, herein we propose that the Diels–Alder reaction between the in situ-formed diene from CR/ZnO and the maleimide moieties of MF were responsible for this synergistic curing behavior. Similarly, a synergistic curing behavior was also observed in the rheometer study when a 50/50 blend of NR/CR was cured with MF in conjunction with a conventional type accelerated sulfur (CV) system. The tensile properties and the hardness of the blends were significantly improved after curing with MF. A co-vulcanization reaction between the dienes generated from the CR and the NR phases of the blends with either ends of the maleimide moieties via Diels–Alder reaction was proposed to substantiate the observed synergistic curing behavior and the improved tensile properties of the blends. The rheometer evidence concerning the curing of CR with ZnO/MF demands an advanced investigation for further confirming the proposed Diels–Alder reaction.

Author Contributions: Conceptualization, S.G.S.; methodology, S.G.S.; software, M.P.; validation, S.G.S. and R.S.; formal analysis, M.P.; investigation, M.P. and O.K.; resources, R.S.; data curation, M.P. and O.K.; writing—original draft preparation, S.G.S.; writing—review and editing, S.G.S. and R.S.; visualization, S.G.S.; supervision, S.G.S.; project administration, R.S.; funding acquisition, R.S. All authors have read and agreed to the published version of the manuscript.

Funding: This work was supported by the Ministry of Education, Youth and Sports of the Czech Republic—DKRVO (RP/CPS/2020/004).

Institutional Review Board Statement: Not applicable.

Informed Consent Statement: Not applicable.

Data Availability Statement: The data presented in this study are available on request from the corresponding author.

Acknowledgments: The authors would like to thank the Ministry of Education, Youth and Sports of the Czech Republic for providing funding to support this research.

Conflicts of Interest: The authors declare no conflict of interest.

References

1. Morton, M. *Rubber Technology*; Van Nostrand Reinhold Company: New York, NY, USA, 1987; pp. 1–20.
2. Hofmann, W. *Rubber Technology Hand Book*; Hanser Publishers: New York, NY, USA, 1989.
3. Ferguson, R.C. Infrared and nuclear magnetic resonance studies of the microstructures of polychlo-roprenes. *J. Polym. Sci. Part A Gen. Pap.* **1964**, *2*, 4735–4741. [CrossRef]
4. Tabb, D.L.; Koenig, J.L.; Coleman, M.M. Infrared spectroscopic evidence of structural defects in the crystalline regions of trans-1,4-polychloroprene. *J. Polym. Sci. Polym. Phys. Ed.* **1975**, *13*, 1145–1158. [CrossRef]
5. Aufdermarsh, C.A.; Pariser, R. Cis-polychloroprene. *J. Polym. Sci. Part A Gen. Pap.* **1964**, *2*, 4727–4733. [CrossRef]
6. Sathasivam, K.; Haris, M.R.H.M.; Mohan, S. Vibrational spectroscopic studies on cis-1, 4-polychloroprene. *Int. J. Chemtech Res.* **2010**, *2*, 1780–1785.
7. Ferguson, R.C. Determination of Polychloroprene Isomers by High Resolution Infrared Spectrometry. *Anal. Chem.* **1964**, *36*, 2204–2205. [CrossRef]
8. Petcavich, R.J.; Painter, P.C.; Coleman, M.M. Application of infra-red digital subtraction techniques to the microstructure of polychloroprenes: 2. Mechanism of oxidative degradation at 60 °C. *Polymer* **1978**, *19*, 1249–1252. [CrossRef]
9. Alliger, G.; Sjothun, I.J. *Vulcanization of Elastomers*; Robert E. Krieger Publishers: New York, NY, USA, 1978.
10. Hofmann, W. *Vulcanization and Vulcanizing Agents*; Maclaren and Sons: London, UK, 1967.
11. Desai, H.; Hendrikse, K.G.; Woolard, C.D. Vulcanization of polychloroprene rubber. I. A revised cationic mechanism for ZnO crosslinking. *J. Appl. Polym. Sci.* **2007**, *105*, 865–876. [CrossRef]
12. Berry, K.; Liu, M.; Chakraborty, K.; Pullan, N.; West, A.; Sammon, C.; Topham, P.D. Mechanism for Cross-Linking Polychloroprene with Ethylene Thiourea and Zinc Oxide. *Rubber Chem. Technol.* **2015**, *88*, 80–97. [CrossRef]
13. Miyata, Y.; Atsumi, M. Zinc Oxide Crosslinking Reaction of Polychloroprene Rubber. *Rubber Chem. Technol.* **1989**, *62*, 1–12. [CrossRef]
14. Kovacic, P. Bisalkylation Theory of Neoprene Vulcanization. *Ind. Eng. Chem.* **1955**, *47*, 1090–1094. [CrossRef]
15. Mallon, P.E.; McGill, W.J.; Shillington, D.P. A DSC study of the crosslinking of polychloroprene with ZnO and MgO. *J. Appl. Polym. Sci.* **1995**, *55*, 705–721. [CrossRef]
16. Vukov, R. Zinc oxide cross-linking chemistry of halobutyl elastomers—A model compound approach. *Rubber Chem. Technol.* **1984**, *57*, 284–290. [CrossRef]
17. Kuntz, I.; Zapp, R.L.; Pancirov, R.J. The Chemistry of the Zinc Oxide Cure of Halobutyl. *Rubber Chem. Technol.* **1984**, *57*, 813–825. [CrossRef]
18. Joseph, R.; George, K.E.; Francis, D.J. Tribasic lead sulphate as efficient curing agent for Polychlo-roprene. *Angew. Makromol. Chem.* **1987**, *148*, 19–26. [CrossRef]
19. Das, A.; Naskar, N.; Datta, R.N.; Bose, P.P.; Debnath, S.C. Naturally occurring amino acid: Novel curatives for chloroprene rubber. *J. Appl. Polym. Sci.* **2006**, *100*, 3981–3986. [CrossRef]
20. Das, A.; Naskar, N.; Basu, D.K. Thiophosphoryl disulfides as crosslinking agents for chloroprene rubber. *J. Appl. Polym. Sci.* **2003**, *91*, 1913–1919. [CrossRef]
21. Ismail, H.; Ahmad, Z.; Ishak, Z.M. Effects of cetyltrimethylammonium maleate on curing characteristics and mechanical properties of polychloroprene rubber. *Polym. Test.* **2003**, *22*, 179–183. [CrossRef]
22. Dziemidkiewicz, A.; Pingot, M.; Maciejewska, M. Metal complexes as a new pro-ecological cross-linking agents for chloroprene rubber based on Heck coupling reaction. *Rubber Chem. Technol.* **2019**, *92*, 589–597. [CrossRef]
23. Gros, A.; Tosaka, M.; Huneau, B.; Verron, E.; Poompradub, S.; Senoo, K. Dominating factor of strain-induced crystallization in natural rubber. *Polymer* **2015**, *76*, 230–236. [CrossRef]
24. Sotta, P.; Albouy, P.-A. Strain-Induced Crystallization in Natural Rubber: Flory's Theory Revisited. *Macromolecules* **2020**, *53*, 3097–3109. [CrossRef]
25. Albouy, P.A.; Sotta, P. Draw ratio at the onset of strain-induced crystallization in cross-linked natural rubber. *Macromolecules* **2020**, *53*, 992–1000. [CrossRef]

26. Sathi, S.G.; Jang, J.Y.; Jeong, K.U.; Nah, C. Thermally stable bromobutyl rubber with a high cross-linking density based on a 4,4′ bismaleimidodiphenylmethane curing agent. *J. Appl. Polym. Sci.* **2016**, *133*, 44092. [CrossRef]
27. Sathi, S.G.; Jeon, J.; Won, J.; Nah, C. Enhancing the efficiency of zinc oxide vulcanization in brominated poly (isobutylene-co-isoprene) rubber using structurally different bismaleimides. *J. Polym. Res.* **2018**, *25*, 108–121. [CrossRef]
28. Sathi, S.G.; Jang, J.Y.; Jeong, K.U.; Nah, C. Synergistic effect of 4,4′-bis(maleimido) diphenylme-thane and zinc oxide on the vulcanization behavior and thermo-mechanical properties of chlorinated isobutylene–isoprene rubber. *Polym. Adv. Technol.* **2017**, *28*, 742–753.
29. Sathi, S.G.; Park, C.; Huh, Y.I.; Jeon, J.; Yun, C.H.; Won, J.; Jeong, K.U.; Nah, C. Enhancing the reversion resistance, crosslinking density and thermo-mechanical properties of accelerated sulfur cured chlorobutyl rubber using 4,4′-bis (maleimido) diphenyl methane. *Rubber Chem. Technol.* **2019**, *92*, 110–128.
30. Shibulal, G.S.; Jang, J.; Yu, H.C.; Huh, Y.I.; Nah, C. Cure characteristics and physico-mechanical properties of a conventional sulphur-cured natural rubber with a novel anti-reversion agent. *J. Polym. Res.* **2016**, *23*, 1–12. [CrossRef]
31. Sathi, S.G.; Harea, E.; Machů, A.; Stoček, R. Facilitating high-temperature curing of natural rubber with a conventional accelerated-sulfur system using a synergistic combination of bismaleimides. *EXPRESS Polym. Lett.* **2021**, *15*, 16–27. [CrossRef]
32. Sathi, S.G.; Stocek, R.; Kratina, O. Reversion free high-temperature vulcanization of cis-polybutadiene rubber with the accelerated-sulfur system. *Express Polym. Lett.* **2020**, *14*, 823–837. [CrossRef]
33. Inoue, S. Chloroprene Rubber Composition. JPS59109541A, 1984.
34. Jain, S.R.; Sekkar, V.; Krishnamurthy, V.N. Mechanical and swelling properties of HTPB-based copolyurethane networks. *J. Appl. Polym. Sci.* **1993**, *48*, 1515–1523. [CrossRef]
35. Hayeemasae, N.; Salleh, S.Z.; Ismail, H. Utilization of chloroprene rubber waste as blending compo-nents with natural rubber: Aspect on metal oxide contents. *J. Mater. Cycles Waste Manag.* **2019**, *21*, 1095–1105. [CrossRef]

Article

Identifying the Co-Curing Effect of an Accelerated-Sulfur/Bismaleimide Combination on Natural Rubber/Halogenated Rubber Blends Using a Rubber Process Analyzer

Marek Pöschl, Shibulal Gopi Sathi * and Radek Stoček

Centre of Polymer Systems, Tomas Bata University in Zlín, Třída Tomáše Bati 5678, 760 01 Zlín, Czech Republic; poschl@utb.cz (M.P.); stocek@utb.cz (R.S.)
* Correspondence: sathi@utb.cz

Abstract: The rheometer curing curves of 50/50 blends of natural rubber (NR) and two different halogenated rubbers with a combination of conventional accelerated sulfur (CV) and 3 phr of a bismaleimide (MF_3) at 170 °C indicates that a co-curing reaction has been taken place between NR and the halogenated rubbers via Diels–Alder reaction. To further confirm whether the co-curing reaction has taken place in the early stage of curing, a complex test methodology was applied with the help of a rubber process analyzer. In this test, the blends with CV and with $CVMF_3$ were subjected to cure at 170 °C for a predetermined time so that both the CV and $CVMF_3$ cured blends will have the same magnitude of curing torque. It is then cooled down to 40 °C and the storage modulus (G′) was evaluated as a function of strain from 0.5% to 100% at a constant frequency of 1 Hz. The results reveal that the blends cured with $CVMF_3$ exhibit a higher G′ due to the enhanced network strength because of the formation of bismaleimide crosslinks than the same cured with only the CV system. The swelling resistance and the mechanical properties of the blends cured with $CVMF_3$ were significantly higher than those cured with only the CV system.

Keywords: rubber; curing; strain sweep; rheometer; rubber process analyzer

1. Introduction

Manufactures of rubber products and suppliers of polymers and raw materials are forced to apply predictive and advanced laboratory test methods and experiments when seeking for high-performance elastomers for future rubber products, as well as for a better, overall, understanding of the properties of the materials. Moreover, they are strictly following the environmental requirements for reducing energy consumption for production while keeping constant or increasing the performance of rubber products. The curing of rubber compounds is one of the most important processes and it is almost the final step of the rubber product development technology. Moreover, the process of curing and curing systems will significantly influence the final performance of rubber products as well as the total time required for rubber production. Through curing, the entangled rubber chains turn into a network structure due to the formation of chemical crosslinks between the rubber chains.

Scientists and technologists have mechanically connected the progressive enhancement of the stiffness of the rubber stocks during curing and developed cure meters to precisely monitor the curing process. In the rubber industry, cure meters are commonly called rheometers. Oscillating disc rheometer (ODR), oscillating die (moving die) rheometer (MDR) and rubber process analyzer (RPA) are the commonly used equipment to characterize the curing behavior of compounded rubber stocks. This equipment can directly describe the kinetics of the crosslinking reaction due to the combination of mechanical representations of chemical processes. The ability of these cure meters to detect the minor changes in different batches of rubber compounds due to improper mixing, insufficient quantity of the compounding ingredients makes it a widely accepted production control instrument in the rubber industry [1,2]. Nowadays, the rheometers are modified in such a way that they can be utilized to test the curing behavior as well as the viscoelastic properties on the same sample to be tested. RPA is one such version of a modified rheometer that can be used to test the curing behavior as well as the viscoelastic properties of the sample after curing. It uses a rotorless biconical die design. The lower die of RPA can oscillate from 0.05° of arc to 90° of arc at an oscillation frequency of 0.1 to 2000 cycles per minute (0.33 Hz to 33 Hz). The temperature can be programmed to change upward or downward between 40 °C and 230 °C. Because of the possibilities of applying a wide range of strains and frequencies to the test sample, RPA can be employed to evaluate the viscoelastic properties of the rubber compounds after the curing has been completed [2–4]

It is well-known that natural rubber (NR) is a non-polar, highly unsaturated elastomer. Because of its non-polar nature, NR exhibits poor resistance to hydrocarbon solvents, oils and greases [5]. On contact with these substances, the NR-based compounds undergo failure due to swelling. Therefore, NR is not considered a material of choice for the development of oil seals or gaskets. Moreover, the unsaturated chemical structure of NR makes it vulnerable to weather elements such as oxidative ageing and ozone attack [6]. To overcome the above-mentioned limitations, NR is frequently blended with polar elastomers such as chloroprene rubber (CR) [7–10] or relatively less polar and less unsaturated elastomers such as bromobutyl rubber (BIIR), etc., [11] and curing the same with proper curing agent. It is well-known that the most appropriate curing system for the NR is the accelerated-sulfur system. The accelerated sulfur system is a package which comprises sulfur, accelerator, activator (Znic oxide, ZnO) and a fatty acid (stearic acid). The accelerated sulfur system produces sulfidic crosslinks in the cured network. Generally, three types of sulfidic crosslinks such as monosulfidic (C–S–C), disulfidic (C–S–S–C), and polysulfidic (C–Sx–C) have been identified in the cured network of NR after curing with the accelerator sulfur system. By adjusting the accelerator to sulfur (A/S) ratio, the level of mono, di, and the polysulfidic crosslinks in the cured network can be manipulated. The conventional accelerated-sulfur vulcanization (CV) system generally produces a cured network with 95% poly and di sulfidic crosslinks and 5% monosulfidic. Therefore, for a CV system, the accelerator to sulfur ratio should maintain in the range 0.1–0.6. This means in the CV system, the sulfur dose is around 2–3.5 phr and the accelerator dose is around 0.4–1.2 phr. [12–14]. Unlike NR, both CR and BIIR are generally cured with ZnO. However, some amount of magnesium oxide (MgO) is also used along with ZnO to cure CR. The cured network of CR or BIIR mainly consists of C–C crosslinks [15–17]. It is very important to note that being a polar elastomer, CR has many unique properties such as oil, ozone and weather resistance. Similarly, the bromobutyl rubber also possesses resistance to ageing and weathering from atmospheric exposure due to its predominantly saturated polyisobutylene backbone of the butyl rubber. Moreover, bromobutyl rubber possesses low gas and moisture permeability. Therefore, blending of NR with CR or BIIR can exploit certain unique properties of these individual elastomers. However, because of the microstructural differences and the cure rate incompatibility, the blending and curing of NR with either CR or BIIR is very difficult. Therefore, a chemical that can act as a reactive compatibilizing or co-curing agent between NR and CR (or BIIR) is essential for enhancing the compatibility between NR and CR or BIIR.

In our previous article, the co-curing effect of a combination of conventional accelerated sulfur (CV) and a bismaleimid on a 50/50 blend of NR/CR was reported based on a moving die rheometer (MDR-3000, Mon Tech, Buchen, Germany). Mainly based on this rheometer cure data, a mechanism responsible for the co-curing effect of the CV/bismalimide on the NR/CR blend has been proposed [18]. In the present work, the curing behavior of a 50/50 blend of NR with another grade of CR and a 50/50 blend of NR with another halogenated rubber was investigated; BIIR was again investigated in the presence of a combination of CV/bismaleimide using the same MDR. For a comparative analysis, the curing behavior of neat NR with CV system and the curing behaviors of neat

CR and BIIR with metal oxides were also reported. To check the validity of the curing behavior and the mechanism proposed in reference [18], a specially designed testing protocol was employed in this study with the help of a rubber process analyzer (RPA).

2. Materials

The natural rubber (standard Vietnamese rubber with a Mooney viscosity ML (1 + 4 at 100 °C: 60 ± 5) was obtained from Binh Phuoc, Vietnam under the trade name SVR CV60, Chloropre rubber (Chloroprene Denka M.40), DuPont elastomer with a Mooney viscosity ML (1 + 4, 100 °C: 48 ± 5) and Bromobutyl rubber (Exxon bromobutyl 224) with a Mooney viscosity ML (1 + 8, 125 °C: 46 ± 5) were used as the base elastomers. Maleide F (MF) is a combination of 75% N,N'-meta phenylene dimaleimide and a 25% blending agent was procured from Krata Pigment, Tambov, Mentazhnikov, Russia. The chemical structure of MF is shown in Figure 1. Other ingredients such as sulfur; n-cyclohexyl-2-benzothiazole sulfenamide (CBS); stearic acid; and zinc oxide; and magnesium oxide were purchased from Sigma-Aldrich, Prague, Czech Republic.

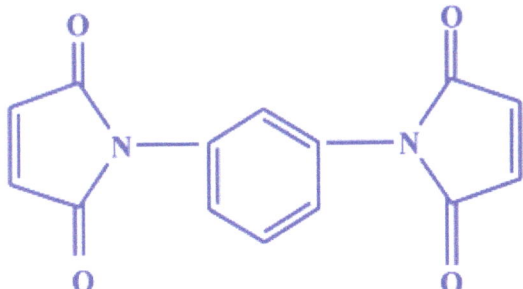

Figure 1. Chemical structure of N,N'-meta phenylene dimaleimide (Maleide F).

2.1. Preparation of Rubber Compounds

The formulations of the mixes with designations are displayed in Table 1. All the compounds were prepared using an internal mixer (Brabender Plastograph, GmbH & Co. KG, Duisburg, Germany) with a chamber volume of 50 cc. A fill-factor of 0.8 was taken for the efficient mixing of the ingredients. To prepare the NR-based compound, the neat NR was masticated at 50 °C under 50 rpm for 2 min. To this, the ZnO, stearic acid, and MF were added, and the mixing was continued under the same rotor speed and temperature for another 2 min. The sulfur and CBS were then added and mixed for additional one minute. After the mixing, the compound was discharged and homogenized using a two-roll mill. Similarly, the CR- and the BIIR-based compounds were prepared after masticating them at 50 °C under 50 rpm for 2 min. To the masticated CR, the ZnO, MgO, stearic acid and MF were added, and the mixing was continued for another 2 min. To the masticated BIIR, only the ZnO and MF were added and mixed for 2 more minutes. Later, the mixes were discharged and homogenized using two-roll mill as in the case of NR. To prepare the blend-based compounds, the individual rubbers were masticated separately for 2 min under the same processing conditions. The pre-masticated rubbers were mixed for 1 min. To this, the ZnO, MgO, stearic acid and MF were added, and the mixing was continued for 2 more minutes. Finally, the sulfur and CBS were added and mixed for an additional minute. After the mixing, the compound was discharged and homogenized using a two-roll mill. It was then molded into sheets with a thickness of 2 mm by applying a constant force of 200 N using a compression molding heat press LaBEcon 300 (Fontijne Presses, Delft, The Netherlands). To avoid the interference of reversion, a molding temperature of 170 °C was selected in this study.

Table 1. Formulation of the mixes.

Mix No.	Mix ID	NR	CR	BIIR	ZnO	MgO	St. Acid	Sulfur	CBS	MF
1	NR-CV	100	-	-	5	-	2	2.5	0.5	-
2	NR-CVMF$_3$	100	-	-	5	-	2	2.5	0.5	3
3	CR-ZnO	-	100	-	5	4	0.5	-	-	-
4	CR-ZnOMF$_3$	-	100	-	5	4	0.5	-	-	3
5	BIIR-ZnO	-	-	100	5	-	-	-	-	-
6	BIIR-ZnOMF$_3$	-	-	100	5	-	-	-	-	3
7	NR/CR-CV	50	50	-	5	2	1.25	1.25	0.25	-
8	NR/CR-CVMF$_3$	50	50	-	5	2	1.25	1.25	0.25	3
9	NR/BIIR-CV	50	-	50	5	-	1	1.25	0.25	-
10	NR/BIIR-CVMF$_3$	50	-	50	5	-	1	1.25	0.25	3

2.2. Characterization

2.2.1. Cure Characteristics

Maximum torque: M_H, minimum torque: M_L, the difference between maximum and minimum torque: ΔM, scorch time: T_{S2}, optimum cure time: T_{90} (the time required for the torque to reach 90% of the maximum torque) of the rubber compounds were determined from the cure curves from a moving die rheometer (MDR-3000, Mon Tech, Buchen, Germany) at 170 °C as per ASTM D 5289. The cure rate index (CRI), a measure of the rate of curing, was calculated using Equation (1).

$$CRI = 100/(T_{90} - T_{S2}) \qquad (1)$$

2.2.2. Cure-Strain Sweep Analysis

Being a viscoelastic material, rubber possess both the elastic (G') and the viscous (G'') moduli. After curing, the elastic component will be the predominant one. Therefore, G' can give an idea of the strength of the cured network. Generally, the higher the G', the higher will be the strength of the cured network. Using a rubber process analyzer (RPA) it is possible to measure both G' and G'' of rubber compounds at a wide range of temperatures, strains and frequencies. To compare the strength of the blends cured with CV and CV/bismaleimide, a special test configuration in the form of 'cure-strain sweep' was created using a Premier RPA (Alfa Technologies, Hudson, OH, USA). In this test, the uncured sample was subjected to cure up to a predetermined time. After this, the sample is cooled down to 40 °C within the cavity of RPA die and we conducted a strain sweep experiment by varying the strain from 0.5% to 100% at a constant frequency of 1 Hz.

2.2.3. Swelling Behavior

Samples with a dimension of 20 mm × 30 mm × 2 mm size with an initial weight (W_i) were swelled in toluene at room temperature until they reached an equilibrium state of swelling. The swelled samples were then taken out and the adhered solvent was wiped off from the surface using a filter paper, and the weights (W_s) were immediately recorded. From the values of W_i and W_s, and the molecular weight of the solvent (M_w), the equilibrium swelling in percentage and the solvent uptake in mol percentage were calculated using Equations (2) and (3), respectively, [9,19]. To understand the speed of swelling, the solvent uptake of the blends cured with CV and CV/bismaleimide were also measured at different time intervals from 0 to 2880 min.

$$Equilibrium\ swelling\ (\%) = \frac{Ws - Wi}{Ws} \times 100 \qquad (2)$$

$$Solvent\ uptake\ (mol\%) = \frac{1}{Mw}\left(\frac{Ws - Wi}{Wi}\right) \times 100 \qquad (3)$$

2.2.4. Mechanical (Tensile) Properties

The stress-strain behavior and the corresponding tensile properties of the vulcanizates of the blends were measured using a universal testing machine (Testometric M350, Testometric Company, Ltd., Rochdale, UK). The testing was performed under ambient conditions at a crosshead speed of 500 mm/min as per ISO 37 using S2 type specimen with a thickness of 2 mm. The results were reported at an average of six tested specimens.

2.2.5. Hardness Testing

Cured samples having smooth surfaces were used to measure the indentation hardness using a Shore-A hardness tester (Bareiss Durometer, Oberdischingen, Germany) as per ASTM D 2240. Indentations were made on different areas of the samples by applying constant pressure for 3 s. Five readings were taken from different areas of the sample, and we reported the average value.

3. Results and Discussion
3.1. Curing Behavior of Neat NR with CV and MF

Represented in Figure 2 are the curing curves of NR-CV and NR-CVMF$_3$ at 170 °C for 1 h. Their cure characteristics are displayed in Table 2. NR-CV attains a maximum torque in 6.13 min and then exhibits a sharp declination in the rheometric torque with time due to reversion. The reversion was calculated and reported in percentage using Equation (4) and the values are depicted in Table 3.

$$\text{Reversion (\%)} = \frac{S'_{max} - S'_{60}}{S'_{max}} \times 100 \tag{4}$$

where S'_{max} is the maximum torque and S'_{60} is the torque at 60 min.

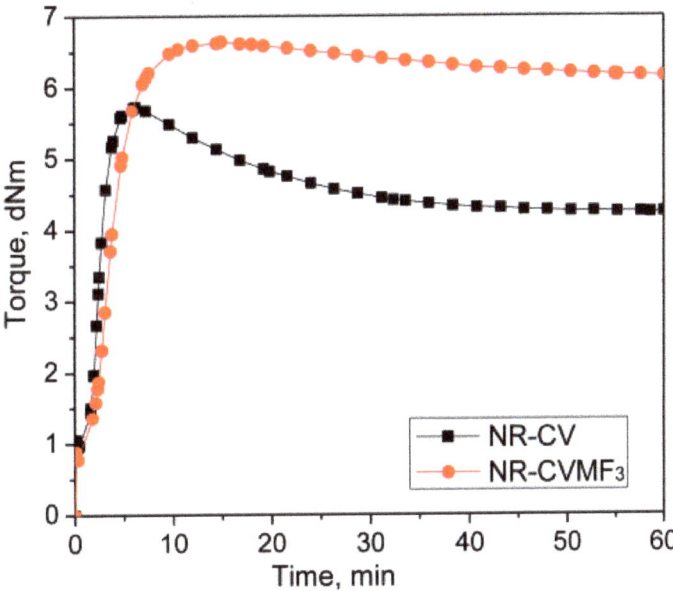

Figure 2. Curing curves of mixes 1 and 2 at 170 °C for 1 h.

Table 2. Cure characteristics of the mixes at 170 °C, 1 h.

Mix ID	M_L (dNm)	M_H (dNm)	ΔM (dNm)	T_{S2} (min)	T_{90} (min)	CRI (min^{-1})
NR-CV	0.96	5.73	4.77	1.95	3.88	51.81
NR-CV-MF$_3$	0.77	6.64	5.87	2.32	6.86	22.02
CR-ZnO	0.74	4.90	4.16	6.26	43.83	2.66
CR-ZnO-MF$_3$	0.64	10.14	9.50	2.18	37.25	2.85
BIIR-ZnO	1.20	2.53	1.33	10.64	13.92	30.48
BIIR-ZnO-MF$_3$	1.24	4.89	3.65	2.71	8.89	16.18
NR/CR-CV	0.64	3.91	3.27	3.08	15.18	8.26
NR/CR-CV-MF$_3$	0.53	6.41	5.88	2.92	22.36	5.14
NR/BIIR-CV	0.87	2.66	1.79	4.97	9.95	20.08
NR/BIIR-CV-MF$_3$	0.82	5.87	5.05	3.10	18.05	6.68

Table 3. Reversion in compounds at 170 °C.

Mix ID	Reversion (%)
NR-CV	25.8
NR-CVMF$_3$	7.5
NR/CR-CV	3.83
NR/CR-CVMF$_3$	No reversion
NR/BIIR-CV	7.9
NR/BIIR-CVMF$_3$	No reversion

The nature of the cure curve of NR-CVMF$_3$ was almost similar to NR-CV during the initial stage of curing. However, NR-CVMF$_3$ exhibits around 16% higher curing torque from the point where the reversion started in NR-CV. As a result, the state of cure in terms of Δ torque in NR-CVMF$_3$ was improved by 19% compared to NR-CV.

From Table 3, it is clear that NR-CV exhibits around 26% reversion at the end of the given curing time. The intensity of reversion in NR-CV could be significantly reduced to 7.5% after incorporating 3 phr MF (NR-CVMF$_3$). It has been reported that the curing process of diene rubbers with the CV system generates polysulfidic crosslinks in the cured network. These polysulfic crosslinks are unstable at elevated temperatures and rearrange to produce a certain amount of conjugated dienes on the rubber backbone, particularly at the point of reversion [14]. Therefore, the possibility of Diels–Alder reaction between the in situ formed conjugated diene at the point of reversion and the maleimide moieties of MF is responsible for the enhanced Δ torque during the curing of NR-CVMF$_3$ [20–22]. The chemical stability of the bismaleimide-based bonds generated in the vulcanized network of NR-CVMF$_3$ can be considered as its enhanced reversion resistance at elevated temperatures.

3.2. Curing Behaviors of Halogenated Elastomers with Metal Oxide and MF

It is well-known that halogenated elastomers such as CR and BIIR can be cured with metal oxides. Represented in Figure 3a,b are the curing behaviors of CR and BIIR as per the formulations given in the mixes 3–6. Their cure characteristics are displayed in Table 2.

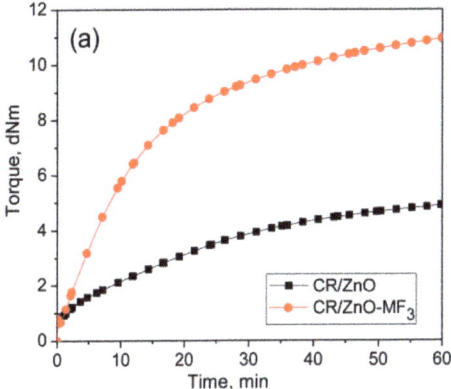 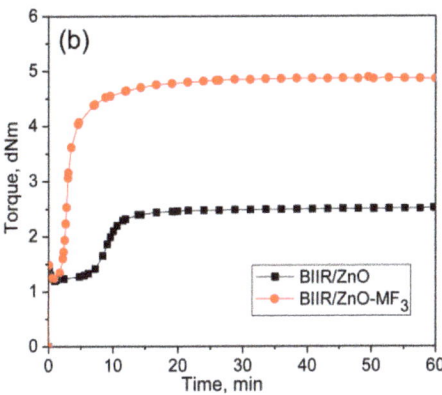

Figure 3. Curing curves of (**a**) Mixes 3 and 4 (**b**) mixes 5 and 6 at 170 °C for 1 h.

CR with metal oxide (ZnO/MgO) exhibit a marching modulus curing behavior with a high induction period (T_{S2} = 6.26 min) and time to optimum cure (T_{90} = 43.8 min). Similar behavior was also observed during the curing of BIIR with ZnO. Here also the induction period was very high (T_{S2} = 10.6 min). However, the curing of BIIR with ZnO exhibits a plateau type curing behavior. As a result, the T_{90} (13.9 min) of BIIR/ZnO was relatively low compared to CR/ZnO/MgO. It is interesting to note that the curing efficiency of ZnO in both the CR and BIIR becomes enhanced after incorporating 3 phr of MF. For instance, the extent of curing in terms of Δ torque in CR-ZnO has been improved by about 144% with the addition of 3 phr MF (CR-ZnO/MF$_3$). Similarly, the Δ torque in BIIR/ZnO was improved by around 174% in the presence of 3 phr MF. Moreover, the induction period, T_{S2} of CR-ZnO has economically been reduced from 6.3 min to 2.2 min, and for BIIR/ZnO, it is reduced from 10.6 min to 2.7 min, respectively, after adding 3 phr MF. Literature reported that the halogenated rubbers such as CR and BIIR can also produce conjugated dienes when it is heating with ZnO [23,24]. Hence, the efficiency in the curing reaction between CR and BIIR with ZnO in the presence of MF can also be ascribed to the Diels–Alder reaction between the in situ formed dienes generated on the polymer with the maleimide moieties of MF [25–27].

3.3. Curing Behaviors of NR/CR (BIIR) Blends with CV and MF

From the curing behaviors of the virgin NR with CV/MF$_3$ and the halogenated elastomers (CR and BIIR) with ZnO/MF$_3$, it has been confirmed that MF can substantially improve the extent of curing by utilizing the in situ formed dienes because of the so-called Diels–Alder reaction. Therefore, an attempt has been made to exploit the in situ formed dienes from these elastomers to enhance the cure compatibility and other physico-mechanical properties of their blends.

To check whether MF can act as a compatibilizing (co-curing) agent, 50/50 blends of NR with CR and BIIR have been prepared as per the formulations corresponding to the mixes 7–10. Depicted in Figure 4a,b are the representative cure curves of NR/CR-CV, NR/CR-CVMF$_3$, NR/BIIR-CV and NR/BIIR-CVMF$_3$ at 170 °C for 1 h. Their cure characteristics are also displayed in Table 2. Both NR/CR-CV and NR/CR-CVMF$_3$ exhibit almost the same speed of curing up to 4.4 min. Later, NR/CR-CV achieved a maximum torque of 3.91 dNm at about 25 min followed by a slight declination in the rheometric torque due to reversion. However, NR/CR-CVMF$_3$ exhibits a marching modulus curing behavior and attained a maximum torque of 6.41 dNm nearly at the end of the given curing time (at 60 min). As a result, the extent of cure in terms of Δ torque in NR/CR-CVMF$_3$ was 80% higher compared to NR/CR-CV. Because of the marching modulus curing behavior, the T_{90} of NR/CR-CVMF$_3$ was higher than NR/CR-CV. Interestingly, no reversion was observed in NR/CR-CVMF3 till the end of the given curing time. The shape of the cure curves NR/BIIR-CV and NR/BIIR-CVMF3 depicted in Figure 4b were similar to NR/CR-CV and NR/CR-CVMF3, respectively. However, the speed of cure in terms of T_{S2} and the extent of cure in terms of ΔM in NR/BIIR-CV were much lower than NR/CR-CV. For instance, the T_{S2} and ΔM values of NR/CR-CV were 3.08 min and 3.27 dNM. On the other hand, the T_{S2} value of NR/BIIR-CV was 4.97 min and its ΔM was around 45% lower compared to NR/CR-CV. Here also, the NR/BIIR-CVMF3 exhibit a higher speed in the early stage of curing followed by a slight marching modulus curing behavior with a 182% higher extent of cure in terms of ΔM compared to NR/BIIR-CV. As in NR/CR-CVMF3, no reversion was also observed in NR/BIIR-CVMF3 till the end of the given 60 min of curing. The possibility of Diels–Alder reaction can be considered high to explain the enhanced state of cure in NR/CR-CVMF3 and NR/BIIR-CVMF3. From the knowledge of the curing behaviors of NR/CVMF3 and CR (BIIR)/ZnOMF3 as discussed earlier, it is reasonable to believe that the Diels Alder reaction might take place in NR/CR-CVMF3 or NR/BIIR-CVMF3 in three ways. One may be the in situ formed diene from the NR phase with the maleimide moieties of MF. The second one may be the in situ formed diene from the CR (or BIIR) phase with the maleimide moieties of MF. The third possibility could be a simultaneous Diels–Alder reaction between the dienes generated in the NR phase and CR (or BIIR) phase with either end of the maleimide moieties of MF as depicted in Figure 5. In this reaction, MF acts as a coupling/compatibilizing agent between NR and CR (or BIIR), which is essential to enhance the compatibility between these rubbers in their blends.

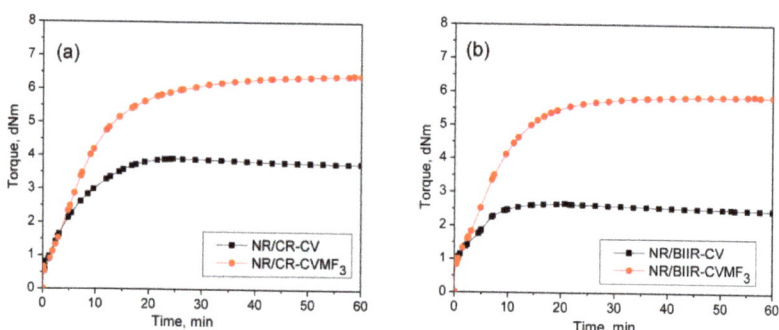

Figure 4. Curing curves of (a) mixes 7 and 8 (b) mixes 9 and 10 at 170 °C for 1 h.

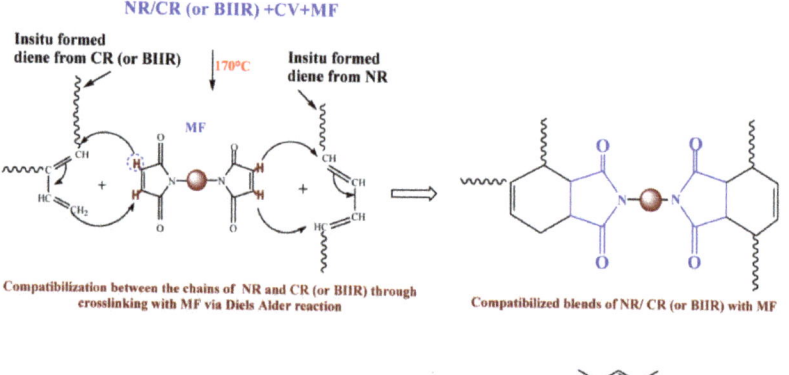

Figure 5. Plausible mechanism proposed for the compatibilisation (co-curing) effect of MF between NR and CR (BIIR) via Diels–Alder reaction.

3.4. Cure-Strain Sweep Analysis of NR/CR (BIIR) with CV and MF

From the curing curves of the blends depicted in Figure 4a,b, it can be noticed that the torque corresponds to the T_{90} (15.2 min) of NR/CR-CV is 3.56 dNm. To reach the same level of torque, NR/CR-CVMF$_3$ took only 8 min. Similarly, the torque at T_{90} (9.95 min) of NR/BIIR-CV is 2.48 dNm. To reach the same torque, NR/BIIR-CVMF$_3$ took only 5 min. Based on the nature of these rheometer cure curves, it has been assumed that the strength of the cured network of the blends NR/CR-CVMF$_3$ and NR/BIIR-CVMF$_3$ might be higher than NR/CR-CV and NR/BIIR-CV, respectively, even if the magnitudes of their curing torque are the same. To check the validity of this assumption, a qualitative rheological test in the form of a cure-strain sweep was conducted as per the testing protocol described in the experimental Section 2.2.2 using the RPA. Depicted in Figure 6a are the RPA cure curves of NR/CR-CV up to its T_{90} (15 min) and NR/CR-CVMF$_3$ up to 8 min at 170 °C. Their shear storage modulus (G′) vs. strain sweep curves are depicted in Figure 6b. It is well known that the rheological parameter G′ indicates the elastic response of viscoelastic material to an applied oscillatory strain. Hence, the term G′ can be considered as the strength of the cured network. Generally, the higher the G′ higher will be the strength of the cured network. It is interesting to note that at a given strain, the G′ of NR/CR-CVMF$_3$ was considerably higher than NR/CR-CV even if the 15 min cured network of NR/CR-CV and the 8 min cured network of NR/CR-CVMF$_3$ exhibited the same RPA ΔM of around 0.13 dNm. For instance, at 10% strain, the NR/CR-CV exhibits a G′ of 324 kPa. At the same strain, the G′ of NR/CR-CVMF$_3$ was 3.7% higher than NR/CR-CV. Represented in Figure 6c,d are the RPA cure curves of NR/BIIR-CV up to its T_{90} (10 min) and NR/BIIR-CVMF$_3$ cured up to 5 min at 170 °C and their G′ vs. strain sweep curves. As seen in the case of NR/CR-CV and NR/CR-CVMF$_3$, similar behaviors were also observed in the G′ vs. strain sweep curves of NR/BIIR-CV and NR/BIIR-CVMF$_3$. For instance, at 10% strain, the G′ of NR/BIIR-CVMF$_3$ was around 25% higher than NR/BIIR-CV even though both the T_{90} cured NR/BIIR-CV and the 5 min cured NR/BIIR-CVMF$_3$ exhibited the same RPA ΔM of around 0.07 dNm. This RPA cure-strain sweep results support the fact that MF can act as a compatibilising (co-curing) agent through the formation of bismaleimide adduct by utilizing the in situ generated dienes on the chains of NR and CIIR (or BIIR) via Diels–Alder reaction as described in Figure 5.

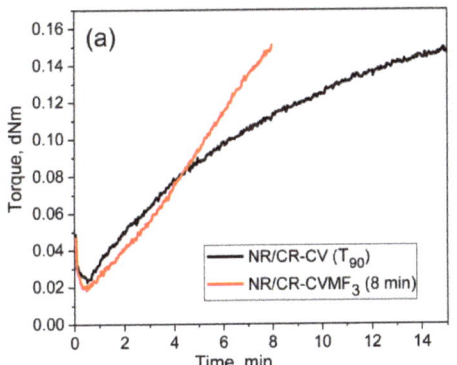 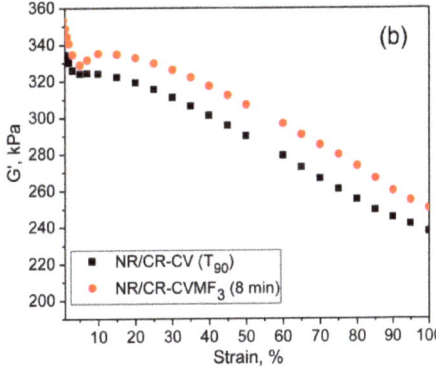

Figure 6. Cont.

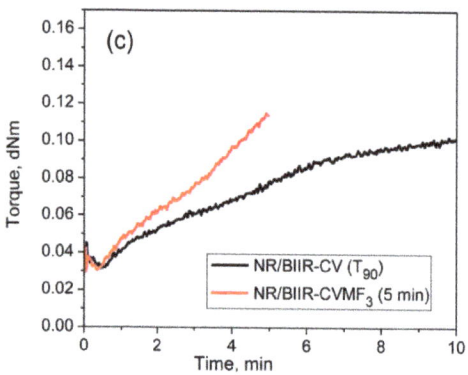
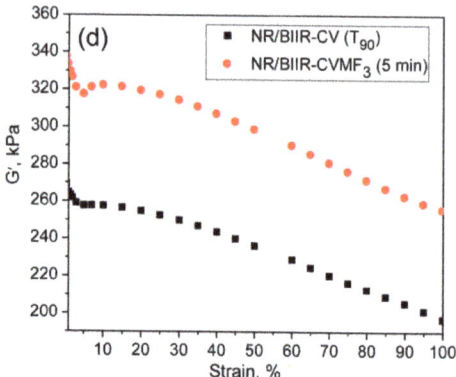

Figure 6. RPA cure curves at 170 °C and the corresponding strain-sweep curves at 40 °C of (**a**,**b**) T_{90} cured NR/CR-CV and 8 min cured NR/CR-CVMF$_3$ and (**c**,**d**) T_{90} cured NR/BIIR-CV and 5 min cured NR/BIIR-CVMF$_3$.

3.5. Swelling Behavior

Depicted in Figure 7 are the solvent uptake in mol% and the swelling index (percentage swelling) of the blends cured at different spans of time at 170 °C.

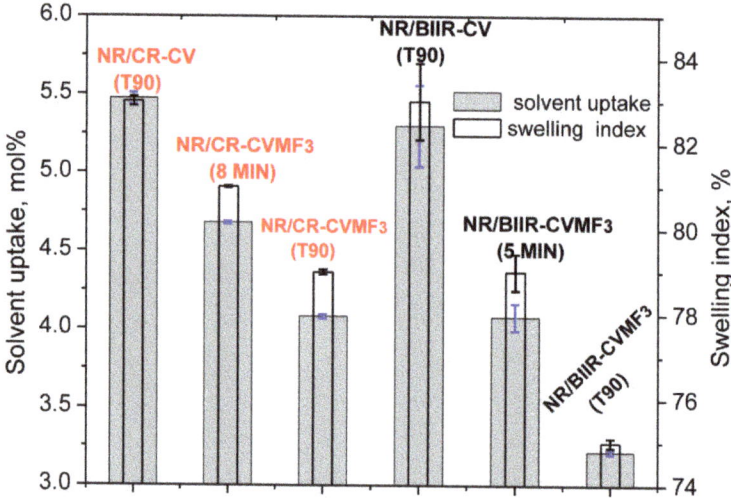

Figure 7. Equilibrium swelling index and the solvent uptake of the blends molded at different curing time.

It can be seen that the solvent uptake and the swelling index of the blend NR/CR-CV cured at its T_{90} was considerably higher than the corresponding NR/CR-CVMF$_3$ cured at its T_{90}. This swelling behavior is quite expected because the ΔM value of the T_{90} cured NR/CR-CV was around 44% lower than the T_{90} cured NR/CR-CVMF$_3$. In general, the lower the ΔM value, the lower will be the crosslink density and, hence, the higher will be the solvent uptake and swelling index. However, it is interesting to note that the solvent uptake and the swelling index of NR/CR-CVMF$_3$ cured up to 8 min was also lower than the T_{90} cured NR/CR-CV even though the ΔM produced in NR/CR-CVMF$_3$ after 8 min of curing and the ΔM of NR/CR-CV cured up to its T_{90} exhibit almost the same value of around 3.0 dNm (Figure 2). Similarly, the solvent uptake of the T_{90} (10 min) cured

NR/BIIR-CV was around 23 mol% higher than the 5 min cured NR/BIIR-CVMF$_3$ even if both these blends cured up to the specified time exhibited the same ΔM of around 1.6 dNm.

From the above-mentioned equilibrium swelling studies of the blends, it has been confirmed that the blends cured with CVMF$_3$ exhibit a higher swelling resistance in terms of solvent uptake compared to the blends cured with only CV. As already explained, one of the reasons for this might be a tightened network structure formed due to the formation of bismaleimide-based adducts between the chains of NR and CR (or BIIR). To check whether the network is really tightened or not right from the beginning of curing owing to the formation of the bismaleimide bonds as shown in Figure 5, we have monitored the solvent uptake of the blends at different intervals of time. Represented in Figure 8a,b is the solvent uptake in g/cm^3 of the T_{90} cured NR/CR-CV, 8 min cured NR/CR-CVMF$_3$, T_{90} cured NR/BIIR-CV and the 5 min cured NR/BIIR-CVMF$_3$. It can be seen that the blend NR/CR-CV exhibits a higher speed of swelling in terms of solvent uptake compared to the NR/CR-CVMF$_3$ even though both these blends cured up to the above-mentioned curing time exhibited the same magnitude of ΔM. For instance, after 5 min of swelling, the solvent uptake of the T_{90} cured NR/CR-CV was around 5% higher than NR/CR-CVMF$_3$. Similarly, after 5 min of swelling, the solvent uptake of the T_{90} cured NR/BIIR-CV was around 25% higher than the 5 min cured NR/BIIR-CVMF$_3$. This low amount of solvent uptake at the early stage of swelling of the blends cured with CVMF$_3$ supports the fact that MF can act as a compatibilizer (co-curing agent) between NR and CR (or BIIR) by utilizing the in situ formed dienes via Diels–Alder reaction, as depicted in Figure 5.

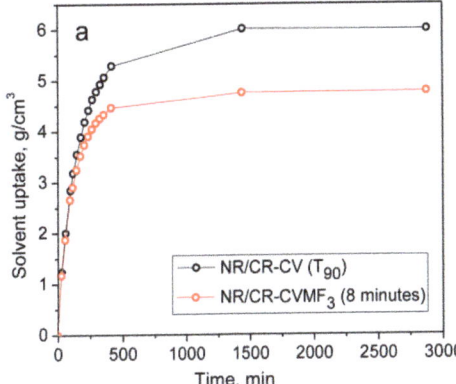
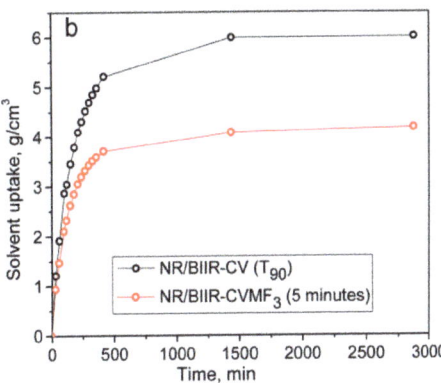

Figure 8. The solvent uptake of (**a**) T_{90} cured NR/CR-CV and 8 min cured NR/CR-CVMF$_3$ (**b**) T_{90} cured NR/BIIR-CV and 5 min cured NR/BIIR-CVMF$_3$.

3.6. Mechanical Properties

Depicted in Figure 9a–d are the mechanical properties such as the tensile strength (TS), elongation at break (EB), modulus at different percentage elongations and the shore-A hardness of NR/CR-CV, NR/CR-CVMF$_3$, NR/BIIR-CV and NR/BIIR-CVMF$_3$ molded at 170 °C as per their T_{90}. NR/CR-CV shows a tensile strength of 2.85 MPa with a breaking elongation of 511%. However, the TS of NR/CR-CVMF$_3$ was 174% higher than NR/CR-CV. Moreover, both the EB and modulus at different percentage elongations of NR/CR-CVMF$_3$ were also considerably higher than NR/CR-CV. Similarly, the TS of NR/BIIR-CVMF$_3$ was 107% higher than the corresponding NR/BIIR-CV. The modulus of NR/BIIR-CVMF$_3$ at different percentage elongations were also significantly higher than NR/BIIR-CV. The hardness of the blends cured with CV and CVMF$_3$ is represented in Figure 9d. The knowledge of the hardness of a rubber material is very essential, particularly when it is used in seals and gaskets. For sealing applications, the rubber compounds should be soft enough for better sealing ability, yet hard enough to sustain the loading force.

Generally, rubbers with a tightly cross-linked network structure exhibit a high resistance to indentation. From Figure 9d, it is clear that the hardness of both the NR/CR-CVMF$_3$ and NR/BIIR-CV MF$_3$ were significantly improved compared to their respective NR/CR-CV and NR/BIIR-CV.

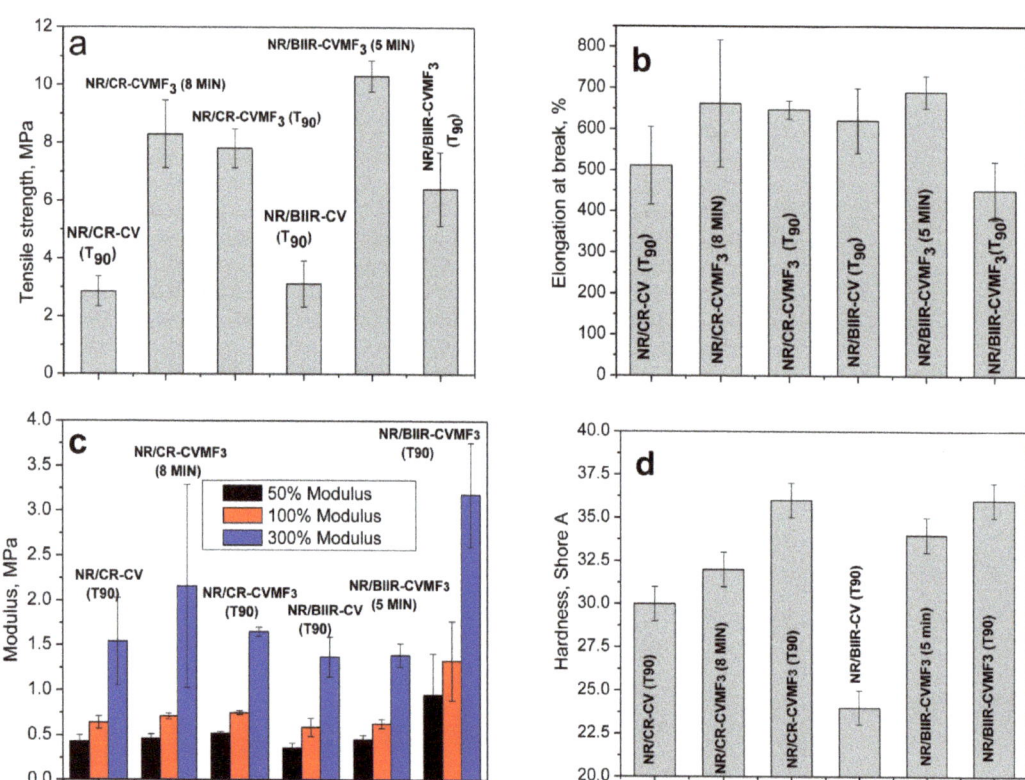

Figure 9. Mechanical properties of the blends moulded at different curing times (**a**) Tensile strength (**b**) Elongation at break (**c**) Modulus at different percentage elongations and (**d**) Shore-A hardness.

The above-mentioned mechanical properties were also evaluated after molding NR/CR-CVMF$_3$ up to 8 min and NR/BIIR-CVMF3 up to 5 min. It is important to note that the ΔM value of the T_{90} cured NR/CR-CV and the 8 min cured NR/CR-CVMF$_3$ are the same. Similarly, the ΔM value of the T_{90} cured NR/BIIR-CV and the 5 min cured NR/BIIR-CVMF$_3$ are also the same. Interestingly, the TS of NR/CR-CVMF$_3$ cured up to 8 min was around 191% higher than the T_{90} cured NR/CR-CV. Similarly, the TS of 5 min cured NR/BIIR-CVMF$_3$ was 230% higher than the T_{90} cured NR/BIIR-CV. Moreover, the EB, modulus at different percentage elongations and shore-A hardness of the 8 min cured NR/CR-CVMF$_3$ and the 5 min cured NR/BIIR-CVMF$_3$ were considerably higher than the T_{90} cured NR/CR-CV and NR/BIIR-CV, respectively. These mechanical property data gives additional support for the enhanced compatibilization between NR and CR (or BIIR) with CVMF via Diels–Alder reaction.

4. Conclusions

The cure characteristics of 50/50 blends of NR/CR and NR/BIIR with a combination of conventional accelerated-sulphur (CV) and 3 phr of a bismaleimide (MF3) gives a strong indication that a co-curing has been taken place between NR/CR and NR/BIIR via

Diels–Alder reaction. One of the primary pieces of evidence for this was the significant enhancement in the rheometer torque during the curing of these blends with $CVMF_3$ compared to the same cured with only the CV system. Through Diles–Alder reaction, the rubber chains in their blends are interconnected via maleimide-based adducts. These crosslinks are believed to be stronger both mechanically and thermally than the sulphur crosslinks in the CV cured blends. Therefore, the networks formed by the $CVMF_3$ cured blends were expected to be stronger than those cured with only the CV system. To check this further, the shear storage modulus (G') of the blends were evaluated by conducting a specially designed cure-strain sweep analysis using a rubber process analyser. The results reveal that the G' values of $CVMF_3$ cured blends were higher than those cured with only the CV system, even though both the $CVMF_3$ and the CV cured blends exhibited the same magnitude of crosslinking torque. This confirms the fact that the network formed in the $CVMF_3$ cured blends is stronger than the CV cured blends. The reversion that was observed in the CV cured blends completely disappeared after curing these blends with $CVMF_3$. This supports that the thermal stability of the $CVMF_3$ cured blends was enhanced. The mechanical properties, particularly the tensile strength, modulus and hardness of the blends cured with $CVMF_3$ exhibited significant improvements compared to the blends cured with only the CV system. Moreover, the swelling resistance of the $CVMF_3$ cured blends improved significantly. All these results support the fact that the compatibility between NR and CR (BIIR) has enhanced in their blends via the proposed Diels–Alder reaction after curing them with $CVMF_3$.

Author Contributions: Conceptualization, S.G.S.; methodology, S.G.S.; software, M.P.; validation, S.G.S. and R.S.; formal analysis, M.P.; investigation, M.P.; resources, R.S.; data curation, M.P.; writing—original draft preparation, S.G.S.; writing—review and editing, S.G.S. and R.S.; visualization, S.G.S.; supervision, R.S.; project administration, R.S.; funding acquisition, R.S. All authors have read and agreed to the published version of the manuscript.

Funding: This work was supported by the Ministry of Education, Youth and Sports of the Czech Republic—DKRVO (RP/CPS/2020/004).

Institutional Review Board Statement: Not applicable.

Informed Consent Statement: Not applicable.

Data Availability Statement: The data presented in this study are available on request from the corresponding author.

Conflicts of Interest: The authors declare no conflict of interest.

References

1. Decker, G.E.; Wise, R.W.; Guerry, D. An oscillating disk rheometer for measuring dynamic properties during vulcanization. *Rubber Chem. Technol.* **1963**, *36*, 451–458. [CrossRef]
2. Dick, J.; Pawlowski, H. Application for the curemeter maximum cure rate in rubber compound development, process control, and cure kinetic studies. *Polym. Test.* **1996**, *15*, 207–243. [CrossRef]
3. Dick, J.; Vare, A.; Harmon, C. Quality assurance of natural rubber using the rubber process analyser. *Polym. Test.* **1999**, *18*, 327–362. [CrossRef]
4. Barick, A.K.; Tripathy, D.K. Effect of organically modified layered silicate nanoclay on the dynamic viscoelastic properties of thermoplastic polyurethane nanocomposites. *Appl. Clay Sci.* **2011**, *52*, 312–321. [CrossRef]
5. Kittur, M.I.; Andriyana, A.; Ang, B.C.; Ch'ng, S.Y.; Mujtaba, M.A. Swelling of rubber in blends of diesel and cottonseed oil biodiesel. *Polym. Test.* **2021**, *96*, 107116. [CrossRef]
6. Crabtree, J.; Kemp, A.R. Weathering of soft vulcanized rubber. *Rubber Chem. Technol.* **1946**, *19*, 712–752. [CrossRef]
7. Anggaravidya, M.; Akhmad, A.; Arti, D.K.; Kalembang, E.; Susanto, H.; Hidayat, A.S.; Limansubroto, C.D. Properties of natural rubber/chloroprene rubber blend for rubber fender application: Effects of blend ratio. *Macromol. Symp.* **2020**, *391*, 1900150. [CrossRef]
8. Quang, N.T.; Hung, D.V.; Linh, N.P.D.; Chuong, B.; Duong, D.L. Detailed study on the mechanical properties and activation energy of natural rubber/chloroprene rubber blends during aging processes. *J. Chem.* **2020**, *2020*, 7064934.
9. Hayeemasae, N.; Salleh, S.Z.; Ismail, H. Utilization of chloroprene rubber waste as blending components with natural rubber: Aspect on metal oxide contents. *J. Mater. Cycles Waste Manag.* **2019**, *21*, 1095–1105. [CrossRef]

10. Salleh, S.Z.; Hanafi, I.; Zulkifli, A. Study on the effect of virgin and recycled chloroprene rubber (vCR and rCR) on the properties of natural rubber/chloroprene rubber (NR/CR) blends. *J. Polym. Eng.* **2013**, *33*, 803–811. [CrossRef]
11. Naba, K.D.; Tripathy, D.K. Miscibility studies in blends of bromobutyl rubber and natural rubber. *J. Elastomers Plast.* **1993**, *25*, 158–179.
12. Kruzelak, J.; Sykora, R.; Hudec, I. Sulfur and peroxide vulcanisation of rubber compounds-overview. *Chem. Pap.* **2016**, *70*, 1533–1555. [CrossRef]
13. Babu, R.R.; Shibulal, G.S.; Chandra, A.K.; Naskar, K. Compounding and vulcanization. In *Advances in Elastomers I. Advanced Structured Materials*; Visakh, P., Thomas, S., Chandra, A., Mathew, A., Eds.; Springer: Berlin/Heidelberg, Germany, 2013; Volume 11, pp. 83–138.
14. Akiba, M.; Hashim, A.S. Vulcanization and crosslinking in elastomers. *Prog. Polym. Sci.* **1997**, *22*, 475–521. [CrossRef]
15. Mallon, P.E.; McGill, W.J.; Shillington, D.P. A DSC study of the crosslinking of polychloroprene with ZnO and MgO. *J. Appl. Polym. Sci.* **1995**, *55*, 705–721. [CrossRef]
16. Vukov, R. Zinc oxide cross-linking chemistry of halobutyl elastomers—A model compound approach. *Rubber Chem. Technol.* **1984**, *57*, 284–290. [CrossRef]
17. Kuntz, I.; Zapp, R.L.; Pancirov, R.J. The chemistry of the zinc oxide cure of halobutyl. *Rubber Chem. Technol.* **1984**, *57*, 813–825. [CrossRef]
18. Pöschl, M.; Sathi, S.G.; Stoček, R.; Kratina, O. Rheometer evidence for the co-curing effect of a bismaleimide in conjunction with the accelerated-sulfur on natural rubber/chloroprene rubber blends. *Polymers* **2021**, *13*, 1510. [CrossRef]
19. Ahmed, K.; Nizami, S.S.; Raza, N.Z.; Shirin, K. Cure characteristics, mechanical and swelling properties of marble sludge filled EPDM modified chloroprene rubber blends. *Adv. Mater. Phys. Chem.* **2012**, *2*, 90–97. [CrossRef]
20. Sathi, S.G.; Jang, J.Y.; Yu, H.C.; Huh, Y.I.; Nah, C. Cure characteristics and physico-mechanical properties of a conventional sulfur-cured natural rubber with a novel anti-reversion agent. *J. Polym. Res.* **2016**, *23*, 237–248.
21. Sathi, S.G.; Stoček, R.; Kratina, O. Reversion free high-temperature vulcanization of cis-polybutadiene rubber with the accelerated-sulfur system. *Express Polym. Lett.* **2020**, *14*, 838–847.
22. Sathi, S.G.; Harea, E.; Machů, A.; Stoček, R. Facilitating high-temperature curing of natural rubber with a conventional accelerated-sulfur system using a synergistic combination of bismaleimides. *Express Polym. Lett.* **2021**, *15*, 16–27. [CrossRef]
23. Desai, H.; Hendrikse, K.G.; Woolard, C.D. Vulcanization of polychloroprene rubber. I. A revised cationic mechanism for ZnO crosslinking. *J. Appl. Polym. Sci.* **2007**, *105*, 865–876. [CrossRef]
24. Berry, K.; Liu, M.; Chakraborty, K.; Pullan, N.; West, A.; Sammon, C.; Topham, P.D. Mechanism for cross-linking polychloroprene with ethylene thiourea and zinc oxide. *Rubber Chem. Technol.* **2015**, *88*, 80–97. [CrossRef]
25. Sathi, S.G.; Jang, J.Y.; Jeong, K.U.; Nah, C. Thermally stable bromobutyl rubber with a high crosslinking density based on a 4,4′-bismaleimidodiphenylmethane curing agent. *J. Appl. Polym. Sci.* **2016**, *133*, 44092. [CrossRef]
26. Sathi, S.G.; Jeon, J.; Won, J.; Nah, C. Enhancing the efficiency of zinc oxide vulcanization in brominated poly (isobutylene-co-isoprene) rubber using structurally different bismaleimides. *J. Polym. Res.* **2018**, *25*, 108–121. [CrossRef]
27. Sathi, S.G.; Jang, J.Y.; Jeong, K.U.; Nah, C. Synergistic effect of 4,4′-bis(maleimido) diphenylmethane and zinc oxide on the vulcanization behavior and thermo-mechanical properties of chlorinated isobutylene–isoprene rubber. *Polym. Adv. Technol.* **2017**, *28*, 742–753.

Article

Influence of Ultraviolet Radiation on Mechanical Properties of a Photoinitiator Compounded High Vinyl Styrene–Butadiene–Styrene Block Copolymer

Sanjoy Datta [1,*], Radek Stocek [1] and Kinsuk Naskar [2]

1 Centre of Polymer Systems, Tomas Bata University in Zlín, tr. Tomase Bati 5678, 760 01 Zlin, Czech Republic; stocek@utb.cz
2 Indian Institute of Technology Kharagpur, Kharagpur 721302, West Bengal, India; knaskar@rtc.iitkgp.ac.in
* Correspondence: sdatta@utb.cz; Tel.: +420-775925027

Abstract: Ultraviolet curing of elastomers is a special curing technique that has gained importance over the conventional chemical crosslinking method, because the former process is faster, and thus, time-saving. Usually, a suitable photoinitiator is required to initiate the process. Ultraviolet radiation of required frequency and intensity excites the photoinitiator which abstracts labile hydrogen atoms from the polymer with the generation of free radicals. These radicals result in crosslinking of elastomers via radical–radical coupling. In the process, some photodegradation may also take place. In the present work, a high vinyl (~50%) styrene–butadiene–styrene (SBS) block copolymer which is a thermoplastic elastomer was used as the base polymer. An attempt was made to see the effect of ultraviolet radiation on the mechanical properties of the block copolymer. The process variables were time of exposure and photoinitiator concentration. Mechanical properties like tensile strength, elongation at break, modulus at different elongations and hardness of the irradiated samples were studied and compared with those of unirradiated ones. In this S-B-S block copolymer, a relatively low exposure time and low photoinitiator concentration were effective in obtaining optimized mechanical properties. Infrared spectroscopy, contact angle and scanning electron microscopy were used to characterize the results obtained from mechanical measurements.

Keywords: ultraviolet radiation; thermoplastic elastomer; high vinyl S-B-S; photoinitiator; mechanical properties

1. Introduction

Light-induced polymerization is one of the most effective methods to generate three-dimensional polymer networks, because of the high initiation rates reached under intense illumination [1–3]. In most UV-curing applications, a solvent-free liquid resin is converted quasi-instantly into a highly crosslinked polymer, selectively in the exposed areas, to produce protective coatings, quick-setting adhesives or high-resolution relief images. The photochemical process has been widely used to crosslink solid polymers with polymerisable functional groups on their backbones [4], e.g., cinnamates [5], epoxides [6,7] and acrylates [8]. A distinct advantage of photoinitiation is to afford precise control of the chemical process. The crosslinking reaction is instantaneous, and starts immediately with the impingement of light of suitable frequency, and it can be stopped by switching off the UV lamp. The rate of reaction of course varies as a function of UV beam intensity.

UV-crosslinking of epoxy [9–11] or acrylate [12,13] functionalized natural rubber in presence of suitable photoinitiators have been studied successfully, but natural rubber alone does not show any reaction under such condition because of the low reactivity of the amylene double bond. In styrene–butadiene–styrene (SBS), it had been recently reported that the vinyl-functionalized mid-block can be readily photocrosslinked by UV irradiation at ambient temperature in the presence of a suitable photoinitiator. The pendent vinyl double bonds are known to be more reactive than the in-chain butene double bonds of the polybutadiene segments [14]. They are thermoplastic elastomers in nature and exhibit mouldability like thermoplastics at elevated temperatures and the functional performance of elastomer at ambient temperatures. Literature survey shows that very limited work related to UV-curing of SBS copolymer is openly published because of the commercial sensitivity. In particular, photocuring of SBS polymers was extensively studied by Decker et al. [15–17].

The authors have successfully presented a comprehensive investigation using statistical design of experiments (DOE), using design expert software pertaining to response surface methodology (RSM) to identify the influential effects of the process variables on the final physical properties of UV-photocured SBS block copolymer [18]. The aim of that work was to mathematically understand the effects of process parameters (time and distance) as a function of photoinitiator (PI) concentration and molecular characteristics (vinyl content) on the physico-mechanical properties of UV-cured SBS block copolymer. In that study, it was found that relatively lower exposure time at lower photoinitiator concentration with a closer distance from the UV source on a higher vinyl content polymer produced the optimum condition for the overall balance of mechanical properties.

Based on the results obtained in the previous work, this study was framed. The main objective of the present work was to study the effect of UV radiation on the mechanical properties of the polymer using 4,4′ dihydroxybenzophenone as the photoinitiator at various concentrations, each of the batches subjected to two different exposure time of 15 s and 30 s and to correlate the results obtained with the previous paper based on the optimisation of photoinitiator concentration and time. Further, the authors were interested to see improvement in the mechanical properties with the use of the new photoinitiator over the previously used benzophenone serving the same purpose. The results obtained were supported through attenuated total refraction (ATR) Fourier transform infrared (FT-IR) spectroscopy, contact angle and microscopic characterisation of the batches used for the study.

2. Materials and Methods

2.1. Materials

Styrene–butadiene–styrene (S-B-S) block copolymer Kraton DKX222 was obtained from Kraton Polymers, Belgium. It contains 18 weight percent bound styrene and 82 weight percent bound butadiene. The microstructure of the polybutadiene midblocks is about 50% 1,4(trans, cis) and 50% 1,2(vinyl) insertion in a random sequence. It has a density of 910 kg/m^3 and weight average molecular weight <Mw> = 71,000 [19]. The structure of the polymer is shown in Figure 1 [14]. It is also seen from Figure 1 that there are dangling groups in the polymer main chains and these groups are due to 1,2(vinyl) insertion during polymerisation. In this figure, PS represents the end block polystyrene units while the midblock polybutadiene units (represented by the curved lines are seen to house the dangling vinyl units shown as smaller protruding straight lines.

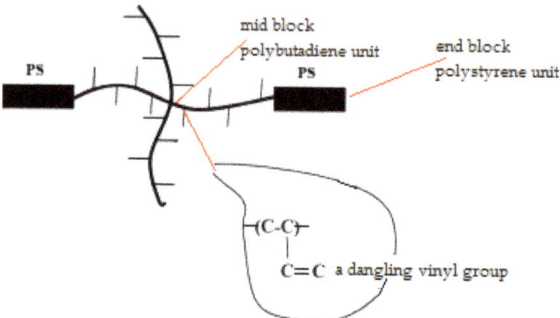

Figure 1. A two-dimensional structure of high vinyl styrene–butadiene–styrene (SBS) block copolymer.

2.2. Preparation of the Batches

Batches of the high vinyl SBS copolymer varying in the PI concentration were prepared in a Haake Rheomix OS 600 (Thermo Fisher Scientific GmbH, Karlsruhe, Germany), with a mixer chamber volume of 85 cm^3. Each batch size was around 55 g and the mixer temperature was kept between 90–100 °C. A constant rotor (cam type) speed of 65 rpm was applied. After 2 min of homogenization of the polymer mass, appropriate amount of the PI was added and the mixing was completed in 6 min. Immediately after each mixing, the composition was removed from the mixer, and while still in hot condition, passed once through a cold two-roll mill to achieve a sheet of about 2 mm thickness. The sheet was cut and pressed (2 mm) in a compression molding machine (George Moore press, UK), at 120 °C, for 5 min and 3.94×10^4 kg/m^2 ram diameter pressure. While molding, TeflonVR sheets were placed between the sheet and the hot plates. The sheet was then cooled to room temperature by circulating cold water through the press plates.

The sheets with a thickness of around 2 mm were subjected to ultraviolet treatment of appropriate doses as shown in Table 1.

Table 1. Batch compositions in phr *.

Components, phr *	\multicolumn{13}{c}{Sample Designation}												
	$k_{0,0}$	$k_{UV,0.2,15}$	$k_{UV,0.4,15}$	$k_{UV,0.6,15}$	$k_{UV,0.8,15}$	$k_{UV,1.0,15}$	$k_{UV,1.5,15}$	$k_{UV,0.2,30}$	$k_{UV,0.4,30}$	$k_{UV,0.6,30}$	$k_{UV,0.8,30}$	$k_{UV,1.0,30}$	$k_{UV,1.5,30}$
SBS block copolymer	100	100	100	100	100	100	100	100	100	100	100	100	100
Photoinitiator	0	0.2	0.4	0.6	0.8	1.0	1.5	0	0.2	0.4	0.6	0.8	1.0

* phr is parts per hundred rubber by mass.

UV radiation was carried out using an Ultraviolet Medium Pressure Quartz Lamp with a wavelength of about 250–350 nm (Advanced Curing System, Bangalore, India). Samples were exposed to the radiation under 1800W mercury lamp, in the presence of air, at a defined time and packing height. The maximum light intensity at the sample position was measured by radiometry (IL-390 light bug) to be 600 mW cm^{-2}. The samples were designated as k_{UV}, k to represent the Kraton polymer used and $_{UV}$ signifying ultraviolet radiation treatment. This was followed by the numbers 0.2, 0.4, 0.06, 0.8, 1.0 and 1.5 corresponding to the PI concentration in phr. Finally, the numbers 15 and 30 showed the time of exposure of the samples to UV radiation.

2.3. Testing Programs

2.3.1. Mechanical Characterization

Tensile tests on the treated and untreated samples were performed according to ASTM D 412 on dumbbell-shaped specimens (Type 2) using a Hounsfield tensile testing machine H10KS (Germany) at a constant crosshead speed of 500 mm/min.

"Shore A" hardness of the samples was measured using a Durometer type A, as per ASTM D 2240.

2.3.2. Spectroscopic Characterization

Fourier Transform Infrared (FT-IR) Spectroscopic Analysis of the unirradiated and irradiated samples were done in attenuated total reflection (ATR)-FTIR spectra in the range of 4000 to 650 cm^{-1} using an infrared spectrophotometer (Nicolet Nexus, Madison, WI, USA). The spectra were obtained at a resolution of 4 cm^{-1} using a zinc selenide crystal. The data obtained from the spectrometer were then fed in an algorithm of baseline creation and subsequent subtraction [20–23] to quantify the disappearance of the vinyl pendant groups showing peak at 909 cm^{-1} [24], which actively participated in the photocrosslinking process. The quantification was done against normalized peak of polybutadiene unit at 965 cm^{-1} [24] of the SBS block copolymer.

2.3.3. Calculation of Surface Energy by Contact Angle Method

The contact angles of different liquids on UV-irradiated samples were obtained using a Ramé Hart contact angle meter. Before the UV treatment, the samples were compression moulded within Mylar (polyester) films to keep them dust-free. Only during the brief time of exposure of the samples to UV, the films were temporarily removed. After the UV treatment, the samples were again covered on both sides with the Mylar films. During the contact angle measurement, the surfaces of the samples were exposed by removing the covers. All investigations were carried out using polymer plates which were cut out from the moulded and UV-crosslinked sheets to obtain dimensions of 10 mm × 10 mm × 2 mm. This produced an almost perfectly flat surface for contact angle measurements.

The sessile drop method employing 4 µL drops of different probe liquids was applied for the contact angle measurements. The liquids used for the contact angle measurements were bi-distilled water, formamide and diiodomethane. Each contact angle value quoted was the mean of at least three measurements with a maximum error of ±1°. All investigations were performed in air at 25 ± 1 °C. the experiments were carried out up to exactly 5 min. Surface energies of the UV-crosslinked samples were calculated equating the measured contact angle (θ) to the free surface energy using the Owens and Wendt equation Equation (1) [25]

$$\cos\theta = -1 + \frac{2(\gamma_s^d \cdot \gamma_l^d)^{1/2}}{\gamma_l} + \frac{2(\gamma_s^p \cdot \gamma_l^p)^{1/2}}{\gamma_l} \quad (1)$$

where γ^d and γ^p are the dispersive and the polar components respectively of the free surface energy of solid and liquid, (s = solid and l = liquid). To find the contact angle, Rame Hart goniometer (Rame Hart Instrument Co, Succasunna, NJ, USA) was used. Bidistilled water, formamide and diiodomethane were selected as the probe liquids. The surface parameters of these probe liquids were taken from the literature for calculating contact angle (θ) [26,27].and are shown in Table 2.

Table 2. Literature data on contact angle probe liquid measurement.

Serial Number	Liquid	γ_d^l (mN·m^{-1})	γ_p^l (mN·m^{-1})	Reference
1	Formamide	39.5	18.7	26 Hefter (06)
2	Diiodomethane	48.5	2.3	27 Tang (2005)
3	Water	21.8	51.0	27 Tang (2005)

2.3.4. Morphological Studies in Raame Hart Camera

To understand the nature of dispersion of UV photoinitiator within the matrix of the variously compounded high vinyl S-B-S block copolymer, visible light was passed through selected samples and the anterio-postirior photographs were captured in a camera attached with the Ramé Hart contact angle equipment. The images were magnified enough to capture the dispersion. However, the camera did not have the provision to register the value of magnification.

2.3.5. Scanning Electron Microscopic Studies

To examine the surface morphology, scanning electron microscopic (SEM) studies were performed on gold-coated samples using a scanning electron microscope JSM 5800 (JEOL, Tokyo, Japan) at 10 kV at a magnification of 5k.

2.3.6. Crosslink Density Calculation

Crosslink densities were measured using the modified Flory Rehner equation by the equilibrium solvent swelling method. In this case, cyclohexane was chosen as the equilibrium solvent due to its solubility parameter of 8.18 (cal/cm^3)$^{1/2}$ which is close to that of butadiene units of S-B-S). Initial weight, swollen weight and de-swollen or dried weight were measured and substituted in Equation (2) which is as follows:

$$\nu = -\frac{1}{v_s} \cdot \frac{\ln(1-v_r) + v_r + \chi(v_r)^2}{(v_r)^{1/3} + 0.5 v_r} (mol \cdot ml^{-1}) \quad (2)$$

where:

$\nu \rightarrow$ = number of moles of effectively elastic chains per unit volume of the polymer [mol/mL] (Overall Crosslink Density),
$V_s \rightarrow$ = molar volume of the solvent (here cyclohexane) used [cm^3/mol],
$\chi \rightarrow$ = polymer-swelling agent interaction parameter (here, 0.3) (Barton 1985) or Flory–Huggin's parameter,
$V_r \rightarrow$ = volume fraction of the polymer in the swollen network, expressed as Vr = 1/(Ar + 1),
$A_r \rightarrow$ = is the ratio of the volume of absorbed solvent (cyclohexane) to that of the polymer after swelling (Flory and Rehner 1943; Naskar 2004).

3. Results and Discussion

3.1. Mechanical

The results obtained for the unirradiated as well as the samples irradiated at 15 s and 30 s at various concentrations of the photoinitiator are presented in Table 3.

Table 3. Mechanical properties of the compounds at varying times and phoinitiator concentration irradiated with UV of a given frequency and intensity.

Components	$k_{0,0}$	$k_{UV,0.2,15}$	$k_{UV,0.4,15}$	$k_{UV,0.6,15}$	$k_{UV,0.8,15}$	$k_{UV,1.0,15}$	$k_{UV,1.5,15}$	$k_{UV,0.2,30}$	$k_{UV,0.4,30}$	$k_{UV,0.6,30}$	$k_{UV,0.8,30}$	$k_{UV,1.0,30}$	$k_{UV,1.5,30}$
S-B-S	100	100	100	100	100	100	100	100	100	100	100	100	100
Photoinitiator	0	0.2	0.4	0.6	0.8	1.0	1.5	0.2	0.4	0.6	0.8	1.0	1.5
Mechanical Properties													
Hardness, Shore A	41	50	50	51	51	51	51	50	51	51	52	52	53
T.S. *, MPa	5.3	7.3	7.0	6.5	6.1	5.6	5.0	7.2	6.3	6.1	5.8	5.4	4.8
M $^\#$ 100, MPa	0.7	0.9	1.0	1.2	1.3	1.3	1.0	0.7	0.8	1.0	1.2	1.3	0.9
M200, MPa	0.9	1.2	1.3	1.4	1.6	1.7	1.3	1.0	1.2	1.4	1.5	1.7	1.3
M300, MPa	1.2	1.6	1.7	1.8	2.2	2.2	1.7	1.3	1.6	1.8	2.2	2.2	1.7
E.B. $^\$$, %	1200	1140	1090	1050	1030	1000	980	1130	1070	1010	1000	940	900
XLD $^\&$, mol·mL^{-1}·10^5		2.37	2.90	3.51	3.95	4.39	4.86	2.41	2.92	3.67	3.89	4.43	4.87

* ultimate tensile strength; # modulus at a specified elongation % of 100, 200, 300; $ elongation at break; & crosslink density.

For the UV-irradiated samples, the tensile strength showed a decreasing trend with an increase in photoinitiator concentration with the 15 s crosslinked samples showing marginal higher values at equivalent photoinitiator concentration over the 30 s crosslinked ones. The maximum tensile strength of 7.3 MPa at photoinitiator concentration of 0.2 phr and irradiation time of 15 s was a clear indication of improvement over the unirradiated control sample which had a tensile strength value of 5.3 MPa. The 30 s exposed sample at the same photoinitiator concentration showed a tensile strength of 7.2 MPa from where it may be inferred that an additional exposure time of 15 s, which consumed more energy did not produce any improvement in tensile strength. The results obtained are better visualized in Figure 2.

Though all the photoinitiator compounded samples got crosslinked with the generation of free radical sites on the polymer in accordance with Scheme 1 [28], yet with an increase in the photoinitiator concentration the probability of crack initiation in the polymer matrix and subsequent crack propagation was perhaps the most plausible explanation for the reduction in tensile strength.

Arguably though the surface consumed up all the photoinitiator as will be discussed in the subsequent part of the results and discussion section, yet there was a large excess of un-reacted photoinitiatior in the bulk, acting as an impurity. This excess amount was further supported through photographs captured in a Ramě–Hart camera attached with the contact angle goniometer which is presented in Figure 3. These were photographs taken in the antirio-postirior direction with visible light passing through the polymer.

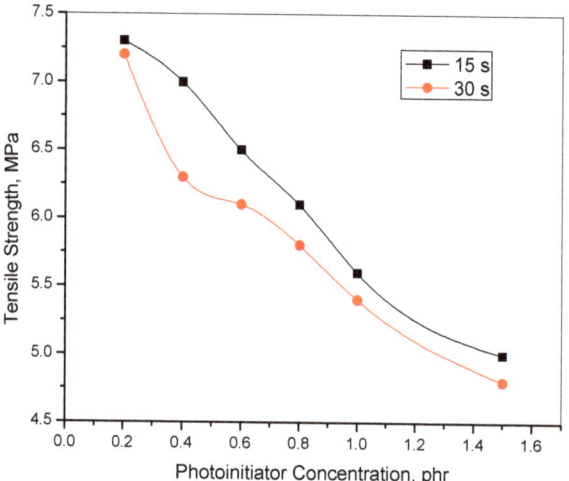

Figure 2. Tensile strength as a function of photoinitiator concentration.

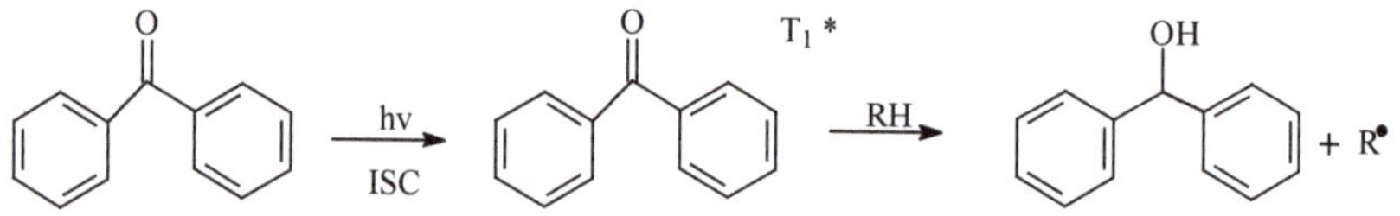

Scheme 1. Plausible reaction scheme of 4,4′dihydroxybenzophenone under UV irradiation.

The photographs captured the presence of aggregates of the photoinitiator embedded within the polymer matrix. They clearly proved that with an increase in the photoinitiator concentration, the average size as well as the population of the photoinitiator both increased. The photographs show a gradual increase in darkening shades with an increase in photoinitiator concentration, which conclusively proved the assumption of residual photoinitiator in the bulk.

The elongation at break also showed a decreasing trend as a function of the photoinitiator concentration as is evident from Figure 4.

It was reasoned out that the elongation at break was the determining parameter for the tensile strength. Additionally, from Table 3, better understood through Figure 5, it is seen that the M100, M200 and M300 for both 15 and 30 s UV-irradiated samples increased almost linearly from 0.2 to 1.0 phr of the photoinitiator and then decreased at the highest concentration of 1.5 phr.

Here also, it was supposed that at very high photoinitiator concentration, the polymer housed many big aggregates of the unreacted photoinitior in the bulk, which served as potential areas of weaknesses to decrease the magnitude of modulus through the phenomenon of multipoint crack initiation and subsequent crack propagation.

Usually for lowly crosslinked thermoplastic elastomers, as is the case in the present study and as will be shown through the crosslink density calculation, the modulus even at reasonably high elongations increases as a function of crosslink density while the tensile strength increases to a maximum and then decreases. This is because modulus is a function of crosslink density only, while tensile strength depends simultaneously on the crosslink density as well as the amount of energy that can be dissipated from the polymer matrix.

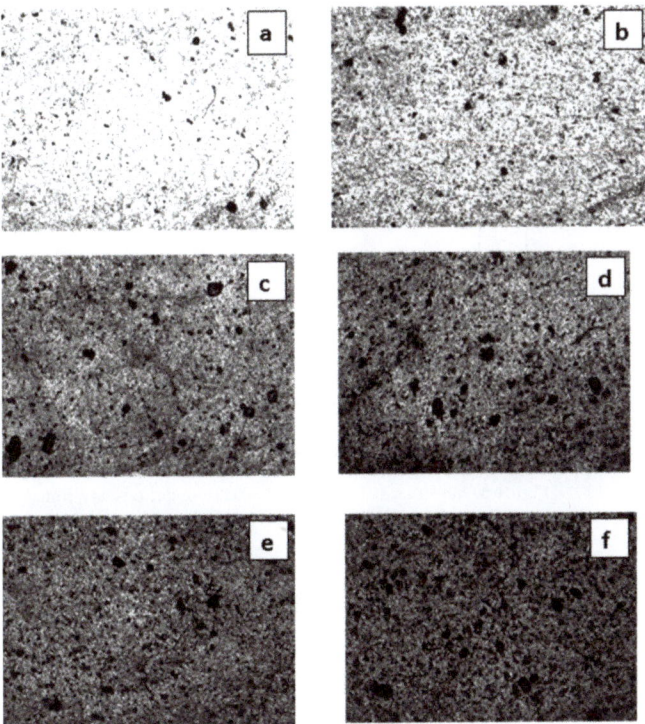

Figure 3. Antirio-posterior photographs of UV-cured samples at (**a**) 0.2, (**b**) 0.4, (**c**) 0.6, (**d**) 0.8 (**e**) 1.0 and (**f**) 1.5 phr of the antioxidant (magnified but not quantified).

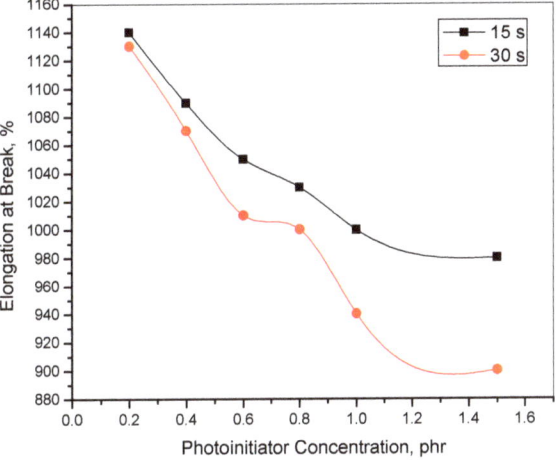

Figure 4. Elongation at break as a function of photoinitiator concentration.

However, in the present study, it was observed that the crosslink density continuously increased as a function of the photoinitiator concentration, though at a little lesser rate from 1.0 to 1.5, as is shown in Figure 6, but the modulus at any of the three defined elongations decreased from 1.0 to 1.5 phr of the photoinitiator concentration.

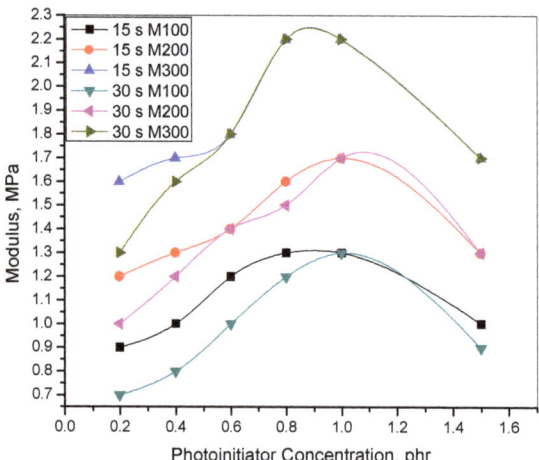

Figure 5. Modulus at 100, 200 and 300% elongations as a function of photoinitiator concentration.

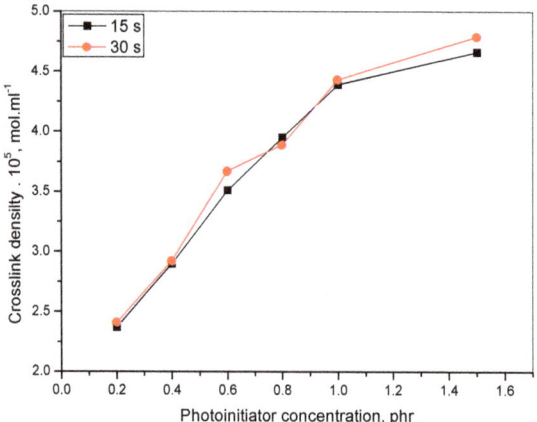

Figure 6. Crosslink density as a function of photoinitiator concentration at 15 and 30 s exposure time.

In general, crosslink levels should be high enough to prevent failure by viscous flow but low enough to avoid brittle failure. However, in this study, the crosslink density was in the low order of 10^{-5}, which could not have enhanced the brittle failure of the polymer. Still, it was observed that the modulus after an initial steady increase finally decreased. The explanation has already been given in terms of crack initiation and propagation.

For elongation at break as well as modulus, the 15 and 30 s crosslinked samples showed comparable results as is evident from Table 3, with the maximum modulus obtained at 1 phr of photoinitiator for both. The modulus values produced marked improvement over the control sample as can be observed from the table.

Hardness measured on the surfaces of the irradiated samples ranging from 50–53 "Shore A" also reflected an increase over the control sample where the hardness was only 41. In general, an increase in hardness is accompanied by an increase in modulus values. However, as was argued earlier, the modulus at the highest photoinitiator concentration decreased as a result of the role played by the residual aggregates of photoinitiator in the bulk.

Along with the postulate of aggregate size playing a role in determining the mechanical properties, another radical explanation may be given to understand the trends observed.

Generally, when irradiated with UV light, benzophenone and substituted benzophenones absorb energy and are excited to singlet state which is not stable. So, they rapidly relax to the more stable triplet state by intersystem crossing (ISC). Studies have found that the excited triplet states are efficient hydrogen abstracting species as shown in Scheme 1. These in turn lead to the formation of polymeric free radicals by absorbing hydrogen from the liable sites [28]. According to Scheme 1, it can be said that during photochemical reaction involving benzophenone type photoinitiator, the total number of macroradical sites depends on the concentration of the photoinitiator.

Thus, the negative shift in the physical properties as a function of photoinitiator concentration as was found in the results of the experiments can be reasoned out by the fact that an increase in photoinitiator concentration has two possible effects. On one hand, it accelerates the crosslinking reaction by the formation of more reactive species. On the other hand, it steepens the cure depth gradient, especially for thick samples, called as "inner shield effect" [17,29]. Hence it was ascertained that as the photoinitiator concentration increased from 0.2 to 1.5 phr, a compromise between effective and insufficient crosslinking in the inner or middle layers of the samples took place and consequently the physical properties were affected in a negative manner. This was true for tensile strength for all the concentrations and correct for modulus above a concentration of 1.0 phr. However, as discussed earlier, this was the second effect, the first one attributed to the aggregate residues of the unreacted photoinitiator in the bulk again due to the inner shielding effect. However, insufficient did not mean that the crosslink density would decrease after a certain concentration of the used photoinitiator.

This phenomenon of insufficient crosslinking in the bulk was conclusively proved by carrying out some interesting sol–gel experiments. Cylohexane is a suitable solvent for the SBS block copolymer and when uncrosslinked, the polymer mass formed a monophasic solution in it. However, it was found that each UV-irradiated sample in the presence of the photoinitiator generated two strings after immersing the samples in cyclohexane for a period of 48 h at ambient temperature. The photograph of one such swollen sample at equilibrium with cyclohexane showing the generated strings are shown in Figure 7.

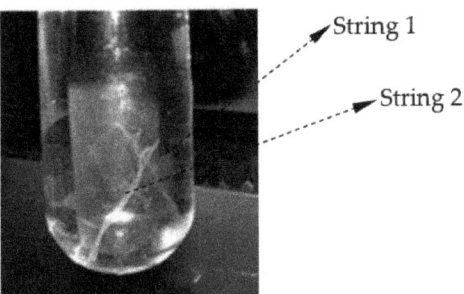

Figure 7. Photograph of UV-crosslinked polymer showing the generation of two strings in cyclohehane after a period of 48 h from the initial time of immersion.

Due to the inner shielding effect, the inside of each of the samples did not get crosslinked and was thus dissolved in the solvent. The two outer surfaces that got crosslinked were rendered insoluble and naturally in their highest entropic condition were manifested as strings. Since the test specimens for the sol–gel experiments were scissor cut from the original UV-irradiated samples, thus they were necessarily cuboidal in shapes, with a length, a breadth (dependent on the cutting) and a thickness of about 2 mm (original moulded thickness before exposure to UV radiation), the strings which were generated after 48 h immersion in cyclohexane were actually very thin UV-crosslinked elements (much thinner than 2 mm), but with the same lengths and breadths of the specimens used before immersion.

Based on this assumption, a very simple but an innovative method was adopted to find the thickness of cure of such a crosslinked system. The relation m = ρ/v, was used for the purpose, where ρ was the density of the compounded and crosslinked sample, m the mass and v the volume all expressed in appropriate units. Knowing the density to be 0.91 g cm^{-3} for the uncompounded and uncrosslinked polymer and assuming that the density did not vary considerably after the addition of a small amount of the photoinitiator in the range of 0.2 to 1.5 phr, it was further assumed, for all practical calculations that the densities of the crosslinked samples were also 0.91 g cm^{-3}.

After the formation of the strings (actually each was a cuboidal volume with a definite length, breadth and thickness) and subsequent drying, it was found that the masses of the two strings for each of the photoinitiator concentrations were almost the same. This proved that both the sides exposed to the same time of exposure to UV radiation got crosslinked to the same extent. The volume of a string was obtained by dividing the mass by the density. Since the length and the breadth of the string were assumed to be the same as that of the sample before it was immersed in the solvent, the area of the string remained the same. Dividing the calculated volume by this area provided the thickness of crosslinking of the string.

This was then a direct means to find the thickness of crosslinking without going for any other instrumental methods. For the samples under investigation, the thickness of crosslinking was about 0.30 µm (each side) at photoinitiator concentration of 0.2 phr and 0.22 µm at 1.5 phr. There was no reportable difference between the thickness of crosslinking due to a variation in the exposure time.

The vinyl double bond in the SBS block copolymer is more reactive than the in-chain butene double bonds. However, a close look in the vinyl chemistry shows that, in addition to the intermolecular crosslinking, an intra-molecular cyclization or cyclopolymerization process may take place in the vinyl bonds located on the same polybutadiene chains [15]. The latter reaction leads to the formation of hard brittle clusters or domains, which in turn act as stress concentration points to reduce the mechanical properties. Hence, the ultimate physical properties, along with what is discussed so far were again a compromise between the inter- and intra-molecular crosslinking for the high vinyl-SBS samples (Scheme 2) [15]. In addition to that, it may also be reasoned out that a significant rise in the temperature took place due to an exothermic crosslinking reaction. The resulting increase in molecular mobility would favour more crosslinking and hence the formation of a tighter network structure [15].

The rise in temperature, though not monitored in the present study can yet be appreciated according to some researches on some other polymers under the broad head of thermal and thermomechanical properties of block copolymer [30,31].

From what has been discussed so far, it may be inferred that the polymer was productive to UV radiation in the range of 250–350 nm in the presence of 4,4' dihydroxybenzophenone as the photoinitiator. Although the mechanical properties were very close, it was s the 15 s irradiated samples that showed marginal better tensile properties over the 30 s irradiated ones.

If the overall mechanical properties were improved in a 2 mm thick sample where inner shielding effect prohibited the bulk of the polymer to get crosslinked, then definitely the properties would have been much better if the experiments were carried out with thin films of dimensions in the order of µm.

In fact, the present study with a 2 mm thick sample, compounded by melt mixing process was advantageously used to understand both, what happened under such a condition, and also to predict what may happen if the solvent casting is chosen to get a thin film of the polymer in which the photoinitiator will be more homogeneously distributed with a uniform crosslinking of virtual no inner shielding effect due to the thin film.

3.2. Infrared Spectroscopic Studies

Furthermore, in order to support the assumption of crosslinking via vinyl double bonds in the UV-irradiated samples, ATR FT-IR studies were done. Figure 8a,b show the unsubtracted spectra of some selected samples in the wavenumber range of 1050 to 680 cm^{-1}.

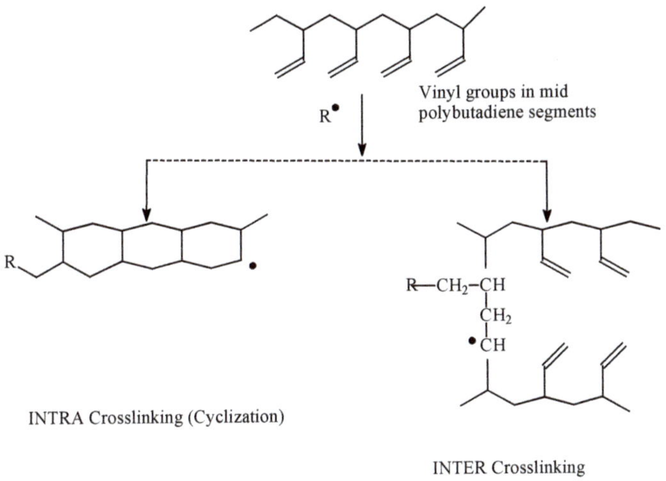

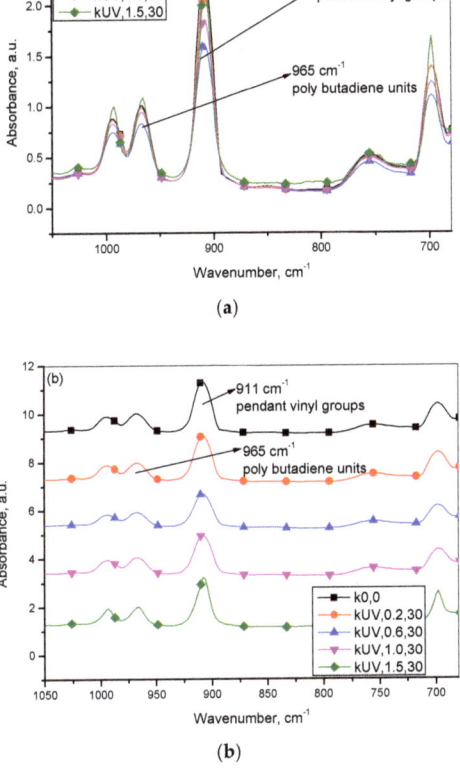

Scheme 2. Plausible reaction scheme of vinyl functionality in the butadiene segments.

Figure 8. Un-subtracted spectra of the control sample and selected 30 s irradiated samples (**a**) superimposed and (**b**) the spectra stacked by Y offsets.

These spectra were of limited use in understanding the active participation of the dangling vinyl groups in photocrosslinking as they were not baseline subtracted. Thus, baseline subtraction was done by using an algorithm of baseline fitting and subsequent subtraction [20–23]. The baseline subtracted spectra of the same selected samples in the wavenumber range of 1050 and 680 cm^{-1} are presented in Figure 9a,b.

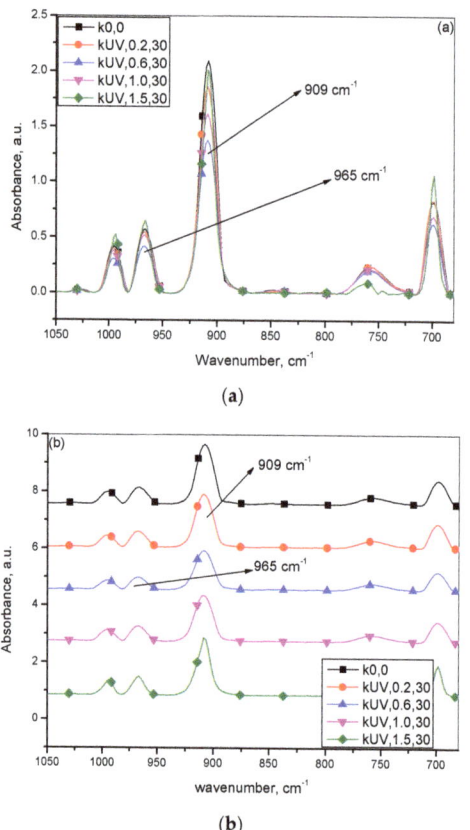

Figure 9. Baseline-fitted and subtracted spectra of the control sample and selected 30 s irradiated samples (**a**) superimposed and (**b**) the spectra stacked by Y offsets.

After baseline subtraction, the characteristic absorbance peak height for the main chain polybutadiene at 965 cm^{-1} was normalized and against this normalized peak height, the absorbance peak height of the vinyl pendant group at 909 cm^{-1} was calculated. The baseline subtracted peak heights and the subsequent calculations are presented in Table 4.

Table 4. Baseline subtracted characteristic peak height ratios of vinyl to polybutadiene for the control sample and some selected for the 30 s UV-irradiated samples.

Sample Designation	Infrared Absorbance Peak Heights		Vinyl/Polybutadiene
	Vinyl, 909 cm^{-1}	Polubutadiene, 965 cm^{-1}	
$k_{0,0}$	2.0944	0.5703	3.6728
$k_{UV,0.2,30}$	1.8572	0.5451	3.4074
$k_{UV,0.6,30}$	1.3804	0.4197	3.2889
$k_{UV,1.0,30}$	1.6156	0.5178	3.1202
$k_{UV,1.5,30}$	2.0113	0.6456	3.1155

It is observed from the table that from the control sample to 1.0 phr of the photoinitiator, the ratio decreased in almost a linear manner. However, the ratio became almost the same for the 1.0 and the 1.5 phr photoinitiator compounded samples [15].

Since the ratio of the peak heights at 1 and 1.5 phr were almost the same, it was inferred that at this highest concentration, some unreacted photoinitiator remained on the surface of the polymer. The above analysis suggested that the vinyl double bonds were the active sites of crosslinking. In the process, some intramolecular cyclisation might have also taken place which is shown in Scheme 2.

Superimposed, baseline subtracted FT-IR spectra of the control sample and the sample at photoinitiator concentration of 0.2 phr irradiated for 30 s, in the wavenumber range of 4500 to 650 cm^{-1} is shown in Figure 10. It depicts that there was no observable appearance of oxidized groups in the UV-treated sample. Thus, it proved that during the process of UV irradiation in air, no significant oxidation occurred.

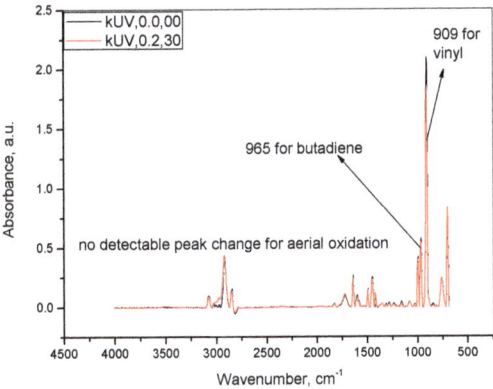

Figure 10. Superimposed, baseline subtracted spectra of the control sample and 0.2 phr incorporated 30 s irradiated sample to study oxidation of the UV-irradiated sample.

These two baseline-subtracted spectra were already shown in Figure 9 amongst other spectra, but in a smaller wavenumber range

3.3. Surface Phenomenon through Contact Angle Studies

The disappearance of the vinyl pendant groups without any aerial oxidation was further supported through surface energy calculation (Table 5), by measuring equilibrium contact angle in selected solvents with the variously compounded and UV-crosslinked samples.

Table 5. Surface energy (mJ·m^{-2}) using water, formamide and diiodomethane as probe liquids.

Sample Designation	Water and Formamide	Formamide and Diiodomethane	Water and Diiodomethane
$k_{UV,0.2,15}$	41.43	41.91	42.06
$k_{UV,0.4,15}$	40.68	40.38	40.29
$k_{UV,0.6,15}$	39.96	39.12	38.89
$k_{UV,0.8,15}$	37.36	37.27	37.26
$k_{UV,1.0,15}$	36.59	35.65	35.45
$k_{UV,1.5,15}$	35.05	34.65	34.55
$k_{UV,0.2,30}$	43.56	41.34	40.68
$k_{UV,0.4,30}$	40.32	40.06	39.99
$k_{UV,0.6,30}$	39.60	38.49	38.20
$k_{UV,0.8,30}$	35.89	36.64	36.82
$k_{UV,1.0,30}$	36.59	35.29	35.00
$k_{UV,1.5,30}$	35.05	34.30	34.11

Since the maximum crosslinking occurred on the surface due to the phenomenon of cure depth gradient, contact angle measurements were done to calculate the surface energy to understand the changing nature of the surface. Figure 11a,b show surface energy as a function of photoinitiator concentration for the 15 s and 30 s crosslinked samples respectively.

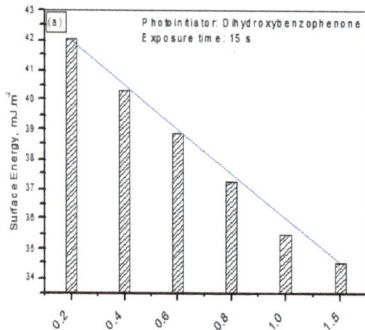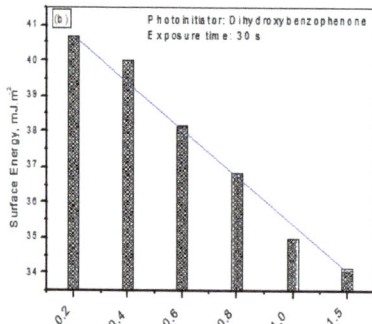

Figure 11. Surface energy as a function of photoinitiator concentration at (**a**) 15 s exposure time and (**b**) 30 s exposure time.

It is seen from both figures that surface energy decreased with an increase in the photoinitiator concentration. Since $\gamma_{s(t)}$ which is the total surface energy of the solid sample is the sum total of γ_s^d (dispersive part of the solid component) and γ_s^p (polar part of the solid component), a decrease in $\gamma_{s(t)}$ may be due to either a decrease in γ_s^d or a decrease in γ_s^p or a decrease in both. Table 6 shows that in the present case, the decrease in the total surface energy with an increase in photoinitiator concentration up to photoinitiator concentration of 1.0 phr was mainly due to a higher rate of decrease of the polar component.

Table 6. Dispersive and polar components of the total surface energy using water and diiodomethane as probe liquids.

Sample Designation	γ_t (mJ·m^{-2})	γ_d (mJ·m^{-2})	γ_p (mJ·m^{-2})
k$_{UV,0.2,15}$	42.06	33.83	8.23
k$_{UV,0.4,15}$	40.29	32.64	7.65
k$_{UV,0.6,15}$	38.89	31.95	6.94
k$_{UV,0.8,15}$	37.26	31.50	5.76
k$_{UV,1.0,15}$	35.45	30.14	5.31
k$_{UV,1.5,15}$	34.55	28.89	5.66
k$_{UV,0.2,30}$	40.68	33.20	7.48
k$_{UV,0.4,30}$	39.99	32.88	7.11
k$_{UV,0.6,30}$	38.20	31.60	6.60
k$_{UV,0.8,30}$	36.82	30.89	5.93
k$_{UV,1.0,30}$	35.00	29.51	5.49
k$_{UV,1.5,30}$	34.11	28.26	5.85

In the polymer under study, polystyrene and the midblock 1,4 polybutadiene were non-polar while midblock 1,2 vinyl insertions contributed mainly to the polarity. This was due to the presence of sp^2 hybridised carbon atom attached to sp^3 hybridised carbon atom in the pendent vinyl groups. The faster rate of decrease of the polar component was then attributed to the disappearance of the pendent vinyl groups from the midblock polybutadiene during the process of crosslinking.

Also, with an increase in the photoinitiator concentration up to 1 phr, the total surface energy as well as the polar component, both decreased. Had there been residual photoinitiator, which was polar in nature due to the presence of carbonyl group and 4, 4′ hydroxy substitutions present on the surface of the polymer after crosslinking had taken place, then the polar component would have increased. This only happened at 1.5 phr.

This study conclusively revealed that with an increase in photoinitiator concentration up to 1 phr, more vinyl groups participated in photocrosslinking, with no residual photoinitiator remaining on the surface. A marginal increase in the polar component at 1.5 phr was attributed to some residual photoinitiator remaining unreacted at this concentration.

This unreacted mass was the reason behind failure through crack propagation. That is why the modulus value increased up to 1 phr and then decreased in all the cases.

Finally, it can be said that the polar component would not have decreased to such an extent or might have marginally increased, if, during the process of crosslinking, aerial oxidation would have occurred through which some carbonyl groups would have been formed. Thus, no such event markedly happened during the process.

3.4. SEM Studies on the Surfaces

Surface analyses were performed by scanning electron microscopy to understand the surface changes due to irradiation. Figure 12a,b show the surface photomicrographs of 15 s and 30 s UV-exposed polymer samples respectively at 1.5 phr of photoinitiator concentration.

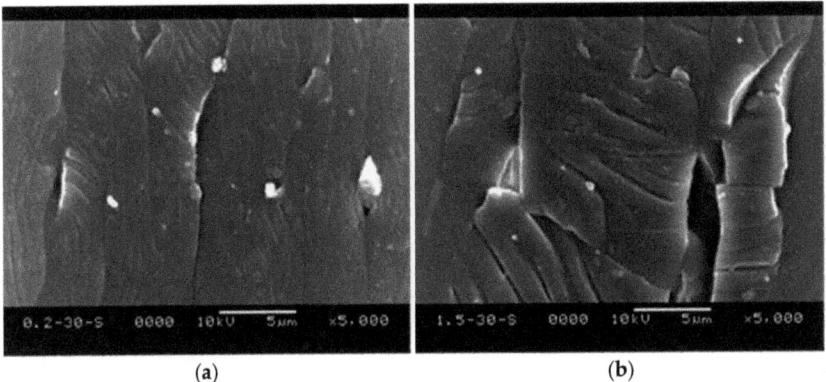

Figure 12. Surface photomicrographs of 15 s (**a**) and 30 s (**b**) UV-exposed polymer samples at 1.5 phr of photoinitiator concentration.

It was observed that the 15 s exposed sample showed the formation of some micro surface cracks which were much more pronounced in the case of the samples exposed for 30 s. From these observations it was inferred that along with the process of photoinitiator induced crosslinking which enhanced the tensile properties, photodegradation of the surface also occurred simultaneously under the condition of irradiation with UV light of given intensity and frequency. This resulted in the breakage of polymer bonds producing fragments [32]. Thus, the samples exposed to higher time, i.e., 30 s showed marginally lower tensile strength and modulus at equivalent photoinitiator concentrations when compared with the samples irradiated for 15 s.

4. Conclusions

The effects of ultraviolet radiation on the mechanical properties of a high vinyl SBS block copolymer were studied. The process variables were time of exposure to ultraviolet radiation and photoinitiator concentration in the polymer matrix at a fixed predetermined distance from the UV lamp.

The polymer showed positive reactivity towards ultraviolet radiation in the frequency range 250–350 nm in the presence of 4, 4′dihydroxybenzophenone as the photoinitiator. Both tensile strength and modulus showed improvement upon treatment with ultraviolet radiation over the control sample without any UV treatment. Even with an improvement over the control sample, the ultimate tensile strength decreased as a function of photoinitiator concentration while the modulus at 100, 200 and 300% increased from 0.2 to 1.0 phr of the photoinitiator concentration and then decreased at a concentration of 1.5 phr.

Inner shielding effect and some intramolecular cyclization were responsible for the reduction in the tensile properties. The overall balance of properties was thus a compromise between effective crosslinking and photoinduced degradation. The best results were obtained at a lower exposure time of 15 s and a photoinitiator concentration of 1 phr.

This study was purely experimental with an incorporated photoinitiator only, deliberately avoiding the use of any photosensitizer. This yielded instances of micrometer-thick crosslinking only. This very small thickness was effectively ascertained using a novel but very simple sol–gel experiment.

Further research aims in using a suitable photosensitizer along with the photoinitiator of interest.

Author Contributions: Conceptualization, S.D. and K.N.; methodology, S.D.; software, S.D.; validation, S.D.; formal analysis, S.D.; investigation, S.D.; resources, R.S.; data curation, S.D.; writing—original draft preparation, S.D., K.N., and R.S.; writing—review and editing, S.D., R.S., and K.N.; visualization, S.D.; supervision, R.S.; project administration, R.S.; funding acquisition, R.S. All authors have read and agreed to the published version of the manuscript.

Funding: This work was supported by the Ministry of Education, Youth and Sports of the Czech Republic—DKRVO (RP/CPS/2020/004).

Institutional Review Board Statement: Not applicable.

Informed Consent Statement: Not applicable.

Data Availability Statement: The data presented in this study are available on request from the corresponding author.

Acknowledgments: The authors acknowledge Board of Research in Nuclear Sciences (BRNS), Department of Atomic Energy (DAE), India, and Xavier Muyldermans from Kraton Polymers They also acknowledge Suman Chakraborty, Mechanical Engineering Department, Indian Institute of Technology Karagpur, for the contact angle measurements.

Conflicts of Interest: The authors declare no conflict of interest. The funders had no role in the design of the study; in the collection, analyses, or interpretation of data; in the writing of the manuscript, or in the decision to publish the results.

References

1. O'Donnell, J. Radiation Curing of Polymeric Materials. In *ACS Symposium Series*; Hoyle, C.E., Kinstle, J.F., Eds.; American Chemical Society: Washington, DC, USA, 1990; no. 417; p. xiv + 567.
2. Decker, C. Photoinitiated crosslinking polymerisation. *Prog. Polym. Sci.* **1996**, *21*, 593–650. [CrossRef]
3. Scranton, A.B.; Bowman, C.N.; Peiffer, R.W. Photopolymerization: Fundamentals and Applications. In *ACS Symposium Series*; American Chemical Society: Washington, DC, USA, 1997; no. 673.
4. Green, G.E.; Stark, B.P.; Zahir, S.A. Photocross-linkable resin systems. *J. Macromol. Sci. Rev. Macromol. Chem.* **1982**, *C21*, 187–273. [CrossRef]
5. Reiser, A.; Egerton, P.L. Mechanism of crosslink formation in solid polyvinylcinnamate and related photopolymers. *Photogr. Sci. Eng.* **1979**, *23*, 144–150.
6. Puskas, J.E.; Kaszas, G.; Kennedy, J.P. New transparent flexible UV-cured films from polyisobutylene-polyisoprene block polymers. *J. Macromol. Sci. Chem.* **1991**, *A28*, 65–80. [CrossRef]
7. Crivello, J.V.; Yang, B. Synthesis and photoinitiated cationic polymerization of epoxidized elastomers. *J. Macromol. Sci. Chem.* **1994**, *A31*, 517–533. [CrossRef]
8. Xuan, H.L.; Decker, C. Photo-cross-linking of acrylated natural-rubber. *J. Polym. Sci. Pol. Chem.* **1993**, *31*, 769–780. [CrossRef]
9. Decker, C.; LeXuan, H.; Viet, T.N.T. Photo-cross-linking of functionalized rubber. 2. Photoinitiated cationic polymerization of epoxidized liquid natural-rubber. *J. Polym. Sci. Pol. Chem.* **1995**, *33*, 2759–2772. [CrossRef]
10. Decker, C.; LeXuan, H.; Viet, T.N.T. Photo-cross-linking of functionalized rubber. 3. Polymerization of multifunctional monomers in epoxidized liquid natural rubber. *J. Polym. Sci. Pol. Chem.* **1996**, *34*, 1771–1781. [CrossRef]

11. Decker, C.; Viet, T.N.T.; LeXuan, H. Photo-cross-linking of functionalized rubber. 6. Cationic polymerization of epoxidized rubber. *Eur. Polym. J.* **1996**, *32*, 1319–1331. [CrossRef]
12. Decker, C.; Viet, T.N.T.; LeXuan, H. Photocuring of functionalized rubbers. 4. Synthesis of rubbers with acrylate groups. *Eur. Polym. J.* **1996**, *32*, 549–557. [CrossRef]
13. Decker, C.; Viet, T.N.T.; LeXuan, H. Photocuring of functionalized rubbers. 5. Radical polymerization of rubbers with acrylate groups. *Eur. Polym. J.* **1996**, *32*, 559–567. [CrossRef]
14. *Brochure of Kraton: Kraton Fact Sheet*; K0406; Kraton Polymers US LCC: Houston, TX, USA, 2006.
15. Decker, C.; Viet, T.N.T. Photocrosslinking of functionalized rubbers IX. Thiol-ene polymerization of styrene-butadiene-block-copolymers. *Polymer* **2000**, *41*, 3905–3912. [CrossRef]
16. Decker, C.; Viet, T.N.T. High-speed photocrosslinking of thermoplastic styrene-butadiene elastomers. *J. Appl. Polym. Sci.* **2000**, *77*, 1902–1912. [CrossRef]
17. Decker, C.; Viet, T.N.T. Photocrosslinking of functionalized rubbers, 8 The thiol-polybutadiene system. *Macromol. Chem. Phys.* **1999**, *200*, 1965–1974. [CrossRef]
18. Datta, S.; Babu, R.R.; Bhardwaj, Y.K.; Sabharwal, S.; Naskar, K. Optimization of Mechanical properties of Photocrosslinked of Styrene Butadiene Styrene Block Copolymers using Statistical Experimental Design. *Tpe Mag.* **2011**, *4*, 232–241.
19. Mayenez, C.; Muyldermans, X. Flexographic Printing Plates from Photocurable Elastomer Compositions. EP 0 696 761 B1 8 April 1998.
20. Datta, S.; Antos, J.; Stocek, R. Smart numerical method for calculation of simple general infrared parameter identifying binary rubber blends. *Polym. Test.* **2017**, *57*, 192–202. [CrossRef]
21. Datta, S.; Antos, J.; Stocek, R. Characterisation of ground tyre rubber by using combination of FT-IR numerical parameter and DTG analysis to determine the composition of ternary rubber blend. *Polym. Test.* **2017**, *59*, 208–217. [CrossRef]
22. Datta, S.; Harea, D.; Harea, E.; Stocek, R. An advanced method for calculation of infrared parameter to quantitatively identify rubber grade in a multi-component rubber blend. *Polym. Test.* **2019**, *73*, 208–217. [CrossRef]
23. Datta, S.; Harea, E.; Stocek, R.; Kratina, O.; Stenicka, M. Configuration of Novel Fractographic Reverse Engineering Approach Based on relationship between Spectroscopy of Ruptured Surface and Fracture Behaviour of Rubber Sample. *Materials* **2020**, *13*, 4445. [CrossRef]
24. Litminov, V.M.; Dey, P.P. *Spectroscopy of Rubbers and Rubbery Materials*; Rapra Technology: Shawbury, UK, 2002.
25. Spelt, J.K.; Neumann, A.W. Solid-surface tension—The equation of state approach and the theory of surface-tension components—theoretical and conceptual considerations. *Langmuir* **1987**, *3*, 588–591. [CrossRef]
26. Hefer, A.W.; Basin, A.; Little, D.N. Bitumen Surface Energy Characterization using a Contact Angle Approach. *J. Mat. Civ. Eng.* **2006**, *18*, 757–767. [CrossRef]
27. Tang, S.; Kwon, O.J.; Choi, H.S. Surface characteristics of stainless steel after an AISI 304L atmospheric pressure plasma treatment. *Surf. Coat. Technol.* **2005**, *195*, 298–306. [CrossRef]
28. Deng, J.P.; Yang, W.T.; Ranby, B. Surface photografting polymerization of vinyl acetate (VAc), maleic anhydride, and their charge transfer complex. I. VAc(1). *J. Appl. Polym. Sci.* **2000**, *77*, 1513–1521. [CrossRef]
29. Decker, C. Kinetic-study of light-induced polymerization by real-time UV and IR spectroscopy. *J. Polym. Sci. Pol. Chem.* **1992**, *30*, 913–928. [CrossRef]
30. Basfar, A.A.; Mosnacek, J.; Shukri, T.M.; Bahattab, M.A.; Noireaux, P.; Courdreuse, A. Mechanical and thermal properties of blends of low-density polyethylene and ethylene vinyl acetate crosslinked by both dicumyl peroxide and ionizing radiation for wire and cable applications. *J. Appl. Polym. Sci.* **2008**, *107*, 642–649. [CrossRef]
31. Lu, H.; Wei, M.H. On the origin of the Vogel-Fulcher-Tammann law in the thermo-responsive shape memory effect of amorphous polymers. *Smart Mater. Struct.* **2013**, *22*, 105021. [CrossRef]
32. Kaczmarek, H. Changes to polymer morphology caused by UV irradiation. 1. Surface damage. *Polymer* **1996**, *37*, 189–194. [CrossRef]

Article

Future-Oriented Experimental Characterization of 3D Printed and Conventional Elastomers Based on Their Swelling Behavior

Klara Loos [1,*], Vivianne Marie Bruère [1,†], Benedikt Demmel [1,2,†], Yvonne Ilmberger [1,2,†], Alexander Lion [1] and Michael Johlitz [1]

1. Institute of Mechanics, Bundeswehr University Munich, 85579 Neubiberg, Germany; vivianne.bruere@unibw.de (V.M.B.); benedikt.demmel@unibw.de (B.D.); yvonne.ilmberger@unibw.de (Y.I.); alexander.lion@unibw.de (A.L.); michael.johlitz@unibw.de (M.J.)
2. Bundeswehr Research Institute for Materials, Fuels and Lubricants, 85435 Erding, Germany
* Correspondence: klara.loos@unibw.de
† These authors contributed equally to this work.

Abstract: The present study investigates different elastomers with regard to their behavior towards liquids such as moisture, fuels, or fuel components. First, four additively manufactured materials are examined in detail with respect to their swelling in the fuel component toluene as well as in water. The chemical nature of the materials is elucidated by means of infrared spectroscopy. The experimentally derived absorption curves of the materials in the liquids are described mathematically using Fick's diffusion law. The mechanical behavior is determined by uniaxial tensile tests, which are evaluated on the basis of stress and strain at break. The results of the study allow for deriving valuable recommendations regarding the printing process and postprocessing. Second, this article investigates the swelling behavior of new as well as thermo-oxidatively aged elastomers in synthetic fuels. For this purpose, an analysis routine is presented using sorption experiments combined with gas chromatography and mass spectrometry and is thus capable of analyzing the swelling behavior multifacetted. The transition of elastomer constituents into the surrounding fuel at different aging and sorption times is determined precisely. The change in mechanical properties is quantified using density measurements, micro Shore A hardness measurements, and the parameters stress and strain at break from uniaxial tensile tests.

Keywords: swelling; absorption; infrared spectroscopy; mass spectrometry; gas chromatography; mechanical behavior; synthetic aviation fuels; 3D printed elastomers

1. Introduction

Nowadays, polymers are used in a wide range of applications such as hoses, sealings, and membranes, where they come into contact with various surrounding liquids such as water [1,2], oils [3], acids [4], organic solvents [5], and fuels [6,7]. Polymers tend to swell, i.e., to absorb liquids, which is usually associated with an increase in volume. This leads not only to a change in physical parameters such as density and hardness but also to a change in mechanical properties.

The present paper investigates the swelling behavior of soft elastomers on the basis of various aspects. Firstly, attention is paid to the swelling behavior of additively manufactured soft polymers. The new additive manufacturing processes not only expand their field of application but also raise fundamental questions about the optimal choice of printing process parameters. The present study on the swelling behavior represents an enormously important step in the analysis of the additive manufacturing process. The printed end product differs greatly depending on whether a wet or dry filament was printed. Only the knowledge of how to set the correct printing parameters enables an optimal printing result.

The application of soft polymers in sealings and hoses, especially fuel-conveying hoses, motivates the further part of the presented study. Here, pristine as well as thermo-oxidatively aged elastomers are subjected to a study on their swelling behavior. The concrete question is ultimately the operational stability of fuel hoses in which new types of synthetic fuels are conveyed. The diffusion of fuel components into the polymer as well as the diffusion of polymer components into the surrounding liquid are investigated in detail in this study. The influence of swelling on the mechanical properties before and after aging is quantified in detail.

2. Swelling Behavior of Additively Manufactured Soft Polymers in Water and in the Fuel Component Toluene

Additive manufacturing (AM) gradually comes into focus for the fabrication of functional parts. With respect to soft polymers, the limitations of the AM technologies regarding both the type and the nature of the employed materials hinder the use of conventional elastomers. Instead, 3D printers may operate with other types of existing materials as well as with newly developed ones. AM parts show significantly different mechanical properties compared to the same parts produced with conventionally produced elastomers, e.g., natural rubber. Therefore, generating knowledge on this subject is a fundamental step toward understanding the mechanical behavior of these materials from 3D printing processes and the feasibility of their applications. This contribution aims to give a brief overview of the additive manufacturing of elastomers and to show the influence of liquid media, such as moisture and fuels, on the printing process as well as on the subsequent application. Regarding their chemical composition, the investigated materials are initially analyzed, while the focus of this work lies on swelling tests with water and toluene followed by uniaxial tensile tests.

2.1. Elastomers in the AM Scenario

Popularly known as 3D Printing, the origins of AM are associated with the prototyping industry. Nevertheless, it has become a more and more interesting alternative for the production of parts with technical applications in addition to conventional manufacturing methods, such as machining and injection molding. The layer-wise process of joining materials to form parts from 3D model data allows a less wasteful, on-site manufacturing along with the exemption of individual tooling and reduction of postprocessing [8–10]. AM is particularly advantageous for the small-scale fabrication of small and complex custom-made parts. As considerable progress is constantly achieved, not only on the technologies themselves but also on the employed materials, a wider range of products and applications is increasingly allowed.

In this context, we can find the elastomers. Several current technologies are able to print parts based on rubber-like materials. It is important to point out, though, that every AM technology demands a specific type and nature of material, which generally does not allow the use of traditional, vulcanized rubber. One of the main reasons is that the vulcanization process cannot easily be transferred to AM. Some technologies, however, allow the use of conventional liquid silicone rubber. Alternatively, thermoplastic elastomers (TPEs) and photopolymers are employed in the printing of soft, elastic components.

One of the downsides associated with those alternative materials is typically the still inferior mechanical behavior [11,12] in terms of operational performance and service life compared to conventional materials. The combination of large deformations with complete recovery at a required demanding stress along with long-term stability is not always fully feasible, for instance. Nevertheless, the continuous development in material science changes the everyday scenario; materials are improved, and new ones are created with the purpose of enhancing the mechanical properties. At the moment, research on such materials is of considerable significance in order to generate both knowledge and a better understanding of their behavior from an engineering point of view. In this way, it is possible to be aware of the properties of current materials for AM of elastomeric parts and help in the optimization of the related printing processes, so that they can also be of use in functional parts instead of being limited to prototyping and demonstration.

Some of the today's AM technologies that are able to process elastic materials are fused filament fabrication (FFF), PolyJet, and liquid additive manufacturing (LAM), which are explored in this section and sketched in Figure 1. FFF—also denoted by the trademarked name fused deposition modeling (FDM) by the company Stratasys®—is one of the most popular, low-cost AM processes. In this technology, the material in filament form is fed through a heated nozzle up to melting and pushed in it by a motor, characterizing the deposition by means of extrusion, after which the material cools down and solidifies into the desired geometry. Since this process demands the use of thermoplastics, TPEs are the choice for the printing of elastomeric parts. PolyJet is a material jetting process, working similar to a standard 2D inkjet printer and equipped with multiple nozzles. These nozzles deposit droplets of photopolymers that are then cured by exposure to UV light. The LAM technology is based upon the extrusion of a liquid or high-viscosity material onto a build plate, where the two-part component materials are mixed in a screw-like manner right before the deposition. Then, heat is provided to carry out the crosslinking by thermal energy.

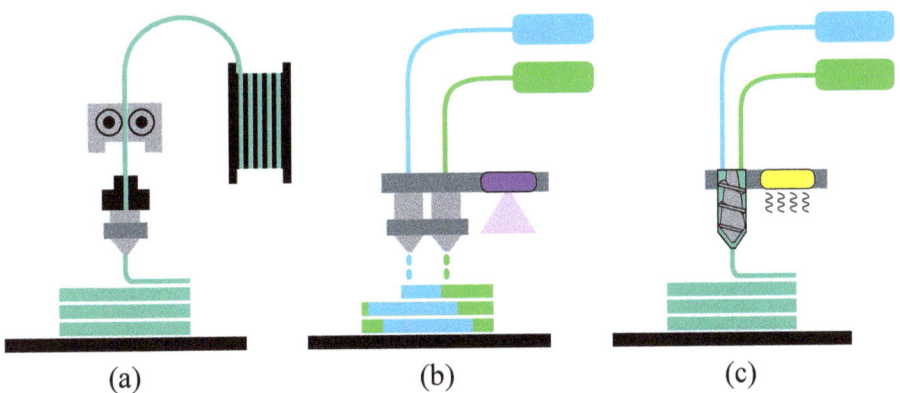

Figure 1. AM processes for (**a**) fused filament fabrication (FFF), (**b**) PolyJet, and (**c**) liquid additive manufacturing (LAM).

Overall, AM is a relatively young manufacturing process in the elastomeric field that has a lot to be explored, although very promising. Both the market and the industry can benefit themselves not only with the fabrication of new parts on demand, reducing warehousing costs, but also with the faster replacement of damaged components, particularly those discontinued by original manufacturers. Despite being more cost-effective for small batch sizes, the evolution of the technologies and materials helps in gradually enabling the use of AM for mass production at a competitive market price in comparison with the traditional fabrication methods. Nevertheless, ensuring the quality of the final printed part is imperative to establish AM in the field of rapid manufacturing.

2.2. Overview of the Investigated 3D Printed Elastomers

The following subsections present chemical and mechanical investigations for 3D printed elastomers of three main types of currently available materials: liquid silicone rubber (LSR), thermoplastic polyurethane (TPU), and photopolymers. There are four printable materials of interest for this research. SILASTIC™LC 3335, supplied by Dow® (Midland, MI, USA) is a two-component viscous silicone with a hardness of 50 Shore A and printed with the LAM process in an innovatiQ's LiQ 320 printer (Feldkirchen, Germany). The second material is Filaflex 70 A, a TPU from company Recreus (Elda, Spain), with a hardness of 70 Shore A. Using this material, parts can be manufactured by a FFF process, which was performed in an Original Prusa i3 MK3S+ 3D printer (Prag, Czech Republic) for this work. The third and fourth materials can be processed with the PolyJet technology. They consist of the same type of photopolymer of resinous material with the name TangoBlackPlus combined with VeroClear (referred here simply as Tango+), supplied by Stratasys® (Rechovot, Israel), but vary in hardness with 70 and 50 Shore A, respectively. For those, a Stratasys® Object500 Connex3 printer (Material Jetting—MJ) was used. Table 1 shows some of the mechanical properties of these polymers according to the supply companies. All samples were printed with a 100% infill. A 90°-line orientation was used for LSR and TPU, and drops were applied for Tango+.

Table 1. Properties of investigated materials provided by the manufacturers.

Property	SILASTIC	Filaflex 70 A	Tango+ 70	Tango+ 50
Hardness [Shore A]	50	70	70	50
Tensile Strength [MPa]	9.5	32	3.5–5.0	1.9–3.0
Elongation at Break [%]	480	900	65–80	95–110
Printing Technology	LAM	FFF	MJ	MJ

A chemical characterization was performed using infrared (IR) spectroscopy. The changes in weight caused by moisture absorption and exposure to toluene were investigated by means of water uptake and solvent absorption tests. At last, tensile tests allowed comparisons of the mechanical properties of the different elastomer samples before and after the water uptake tests. For the solvent absorption tests, however, no tensile properties were able to be evaluated since the intense toluene swelling damaged the samples.

2.3. Chemical Composition Analysis via IR Spectroscopy

To characterize the polymers, Fourier Transform infrared spectrometer (FTIR) spectra were recorded with an attenuated total reflectance (ATR) unit. The measuring device was a Bruker Tensor 27 with Platinum ATR and 32 scans were taken per sample in the range between 400 and 4000 cm^{-1} with a resolution of 4 cm^{-1}. Figure 2 shows the IR spectra of the four above mentioned polymers.

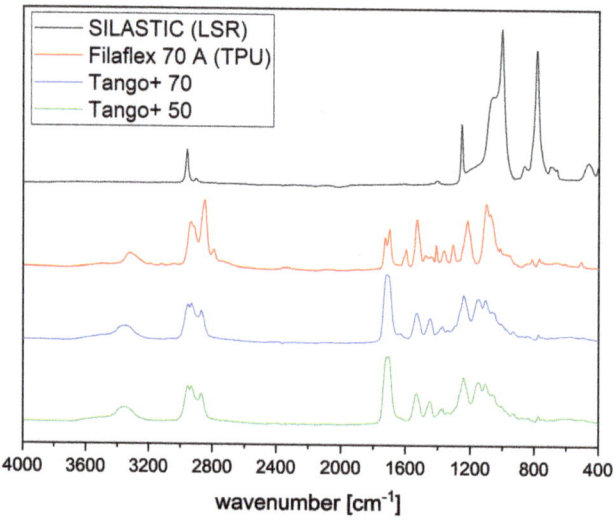

Figure 2. FTIR spectra for SILASTIC, Filaflex 70 A, Tango+ 70, and Tango+ 50.

The absorption bands for the silicone SILASTIC at 1065 and at 1007 cm^{-1} are assigned to the Si-O-Si backbone. The bending band of the Si-CH$_3$ is found at 1260 cm^{-1}, and the coupling stretching of Si-C and rocking band of -CH$_3$ is assigned at 788 cm^{-1}. The absorption band at 2962 cm^{-1} belongs to the stretching vibration of CH$_3$. These determined bands can be assigned to the specific bands of silicones [13,14].

The TPU Filaflex 70 A shows pronounced absorption bands at 2939 and at 2851 cm^{-1}, which are the characteristic bands of the aliphatic C-H asymmetric and symmetric stretching. In the range of 1105 cm^{-1}, aliphatic ether groups are displayed. The spectrum also exhibits bands at 1731 and at 1702 cm^{-1} due to free and hydrogen-bonded urethanic C=O stretching. The band based on the N-H stretching vibration of the urethane amide was observed at 3323 cm^{-1}. The sharp band at 1530 cm^{-1} corresponds to the C≡N stretching and N-H bending. The band at 1220 cm^{-1} is in accordance with the stretching vibration of urethanic -C-(C=O)-O-. These observations indicate that the TPU used in this work is a polyether polyurethane [13].

As the spectra of the Tango+ 70 and the Tango+ 50 are very similar, they are considered together. Just like the TPU, Tango+ also shows the specific bands at 1530 and at 1240 cm^{-1} for urethane content. The band at 1710 cm^{-1} indicates the presence of both acrylate and urethane. The characteristic bands of aliphatic C-H asymmetric and symmetric stretching are in the range from 2956 to 2873 cm^{-1}. This indicates that the Tango+ material used in this study is a polyurethane acrylate [13,15].

2.4. Absorption Experiments with Toluene and Water

The geometries of tensile test specimens according to the Standard DIN 53504:2017-03-S2 [16] for SILASTIC, Filaflex 70 A and Tango+ were used for the absorption experiments carried out according to the Standard DIN ISO 1817:2016-11 [17] with the solvent toluene and water. Prior to the absorption experiments, the test specimens were conditioned under a standard climate (23 °C and 50% humidity), and the initial masses were determined. The mass measurements were made using a SARTORIUS Secura® 225D-1S precision balance with a range of 120 g and a resolution of 0.0001 g. Each specimen was then stored in a screw glass containing 50 mL of toluene or water, respectively. After defined periods of time, the sample was taken out, dried with a lint-free paper, and weighed. After this, it was stored again in the solvent. To minimize the evaporation of the fluid out of the specimen, special care was taken by an immediate weight measurement. Three specimens were used for each material, and the results of the measurements were expressed in terms of the mean average values and standard deviations

2.4.1. Absorption Test with Toluene as Solvent

Figure 3 shows the amount of toluene absorbed in relation to the initial mass in terms of the percentage mass gain as a function of time. To account for Fickian diffusion, the sorption times on the abscissa are reported as the square root of time $t^{0.5}$. In the beginning, an almost linear mass absorption takes place until the diffusion process slows down, and finally, in the equilibrium state, the mass does not change anymore. The results indicate a pronounced swelling of about 135% for SILASTIC and Filaflex 70 A. The saturation was reached after 24 h ($\approx$294 s$^{0.5}$). The procedure applied to SILASTIC and Filaflex 70 A was not suitable for the Tango+ materials. Tango+ 50 and Tango+ 70 were so heavily swollen by toluene that they could not hold their shape and broke after 3.5 h ($\approx$112 s$^{0.5}$) and 4.5 h ($\approx$127 s$^{0.5}$), respectively. This is due to their drop-like construction. Toluene accumulates at the boundary surface of the droplets. This softens the bonding between the drops. Since a little pressure already let the bonds break and, thus, the integrity of the samples was jeopardized, the measurement was aborted.

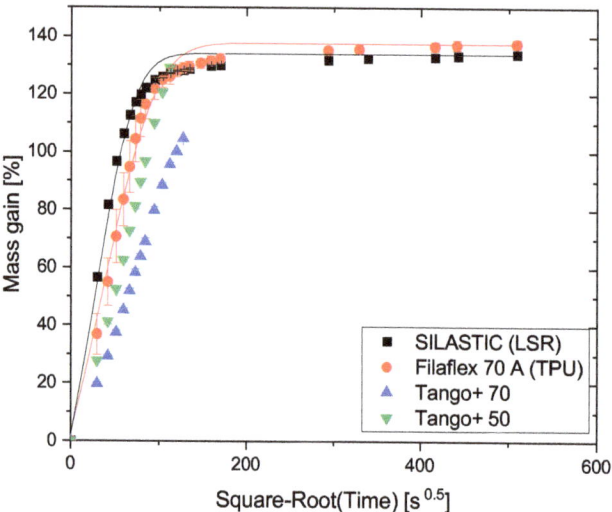

Figure 3. Absorption curves of the four AM polymers in toluene.

The solid lines in Figure 3 show the theoretical sorption curves normalized to the initial mass calculated by iteratively fitting Equation (1) to the available measurement data. This formula, solved with the method of least squares, describes an ideal model of absorption in accordance with Fick's Law [18], where M_t and M_∞ represent the mass of the absorbed solvent at the time t and in the equilibrium state, respectively, D is the diffusion coefficient, h is the sample thickness, and n is the summation index.

$$\frac{M_t}{M_\infty} = 1 - \frac{8}{\pi^2} \sum_{n=0}^{\infty} \left(\frac{1}{(2n+1)^2} \exp\left(\frac{-(2n+1)^2 \pi^2 D t}{h^2} \right) \right) \quad (1)$$

Figure 3 shows that toluene has different effects on the three polymer types used, due to their different chemical structure. It can also be seen that SILASTIC and Filaflex 70 A comply with Fick's Law. As the complete testing was not possible for the Tango+ materials, no statements can be made in this regard. However, attention must also be paid to the printing parameters as they also influence the swelling behavior. Changing the infill pattern, the degree of infill, or the number of perimeters leads to different absorption properties.

2.4.2. Absorption Test with Water

The hygroscopic nature of the four chosen elastomeric materials was analyzed in a Water uptake test. The results are plotted along with the fitted curves based on Fick's Law from Equation (1) in Figure 4. It can be observed that the photopolymers and the TPU are particularly hygroscopic, while the LSR samples exhibit a much lower water uptake less than 10% of the others. It can also be seen that the TPU follows Fick's Law. No statement can be made about the behavior of the SILASTIC material, as the individual measured values scatter strongly in the initial range. In comparison with TPU, the two more hydrophilic photopolymers Tango+ show a different absorption behavior: the polymer chains form more hydrogen bonds with the water molecules and free unbound water molecules in the voids between the chains show Fickian diffusion behavior [19].

A considerable part of the absorption for all materials occurs during the first hours of the experiment; in the first 24 h, more than half of the saturation is already achieved. This evidences the importance of the environment in functional applications. While water does not play a major role in the conventional rubber-like material LSR, the same cannot be affirmed for the photopolymers and the TPU. In fact, it is already known that TPEs in general are hygroscopic, which should be taken into consideration, notably before and during the FFF printing process. Filaments with increased moisture levels lead to extrusion failures and the generation of voids on the streaming, caused by steam formation on the heated nozzle [20]. Moreover, depending on the printing environment and the exposure time, the filament spools in use may constantly absorb air moisture. Figure 5 visualizes the consequences of printing with a moist filament, which gives a surface that is more opaque and with less uniform strand deposition than for geometries created with a dry filament. The excess of material on the left-hand side of the printing with a dry filament (green) is due to the accumulation of random start and end points of the deposition. Therefore, the material should be properly stored not only before/after printing but also while the object is being constructed to ensure good printing quality.

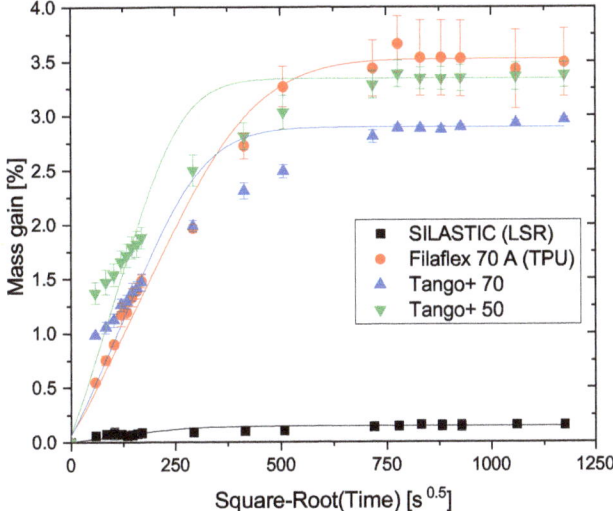

Figure 4. Curves for water absorption evolution.

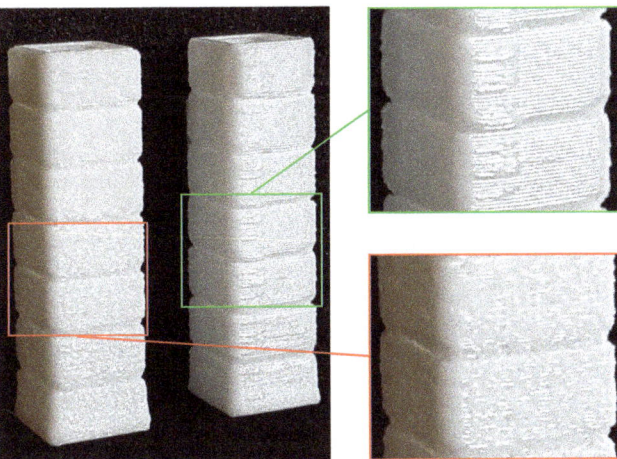

Figure 5. Influence of material condition prior to FFF printing on final geometry for a moist (red) and a dried (green) Filaflex 70 A filament.

2.5. Mechanical Behavior Analysis via Uniaxial Tensile Tests

The effect of moisture on the mechanical behavior of the printed samples was evaluated by tensile tests on a Zwick Roell 1445 universal testing machine with a force sensor of 500 N and an optical extension sensor ProLine lightXtens 2-1000 at ambient temperature, a preload of 0.1 MPa and a strain rate of 200 mm min^{-1}, according to the Standard DIN 53504:2017-03 [16]. The stress–strain curves of the samples before and after the water absorption test in the saturation state were measured. The average results of three samples for each material and condition are displayed in Figure 6. The values for the ultimate stress and the elongation at break can be found in Table 2. Due to the pronounced swelling and the resulting low strength or breakage of the specimens, the tensile test after toluene absorption was omitted.

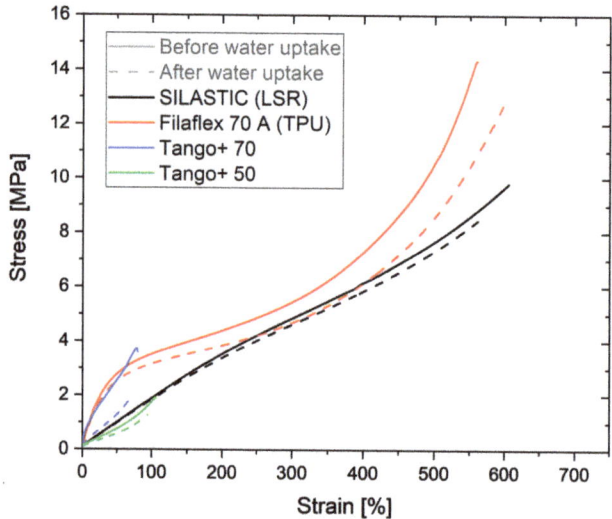

Figure 6. Stress–strain curves until break for the four different AM polymers before and after water absorption.

Table 2. Ultimate stress and strain at break before and after water absorption.

Condition	Property	SILASTIC	Filaflex 70 A	Tango+ 70	Tango+ 50
Before water uptake	Stress [MPa]	11.69 ± 1.46	15.82 ± 1.62	3.73 ± 0.09	1.94 ± 0.02
	Strain [%]	674.47 ± 59.10	575.02 ± 16.57	78.23 ± 0.50	105.93 ± 0.71
After water uptake	Stress [MPa]	10.28 ± 1.23	14.57 ± 1.41	1.84 ± 0.02	1.35 ± 0.04
	Strain [%]	636.61 ± 51.35	630.99 ± 26.93	69.05 ± 0.28	97.27 ± 3.70

The curves show a decrease in stiffness and ultimate stress for the photopolymers Tango+, which is more pronounced for Tango+ 70 than for Tango+ 50. Indeed, there is a decrease of 50% on the tensile strength for Tango+ 70, although the elongation at break has a change of only 12%. For Tango+ 50, the tensile stress is lower by 31%, while the elongation at break decreases by 8%. It could be inferred that the higher the hardness for Tango+, the greater the effect of water saturation on the ultimate strength. Further investigations could corroborate that hypothesis.

Filaflex 70 A also shows that moisture influences its tensile properties. While the stress decreases by only 8%, the elongation at break increases by 10%. This is due to the hydrophilic character of the polymer chains. The water surrounds the chains like a lubricant and allows them to slide off each other more easily when pulled. This results in a higher elongation at break.

The LSR elastomer does not show any indication of alterations in the behavior before water uptake and after water saturation. The deviations in the stress–strain curves can be attributed to the variability of the individual samples in general. In fact, the small influence on the tensile results could already be expected due to the low saturation content for the LSR.

2.6. Intermediate Conclusions and Remarks

The present work explores the effects of absorption of toluene and water on three types of available 3D printed elastomers: liquid silicone rubber, thermoplastic elastomer, and photopolymer. Moreover, chemical analyses via IR Spectroscopy and mechanical analyses via uniaxial tensile tests complement the investigations. The results reveal the intense swelling response of all materials when exposed to toluene, while more diverse outcomes are displayed for water absorption. Furthermore, not only the chemical composition of the materials but also the nature of the printing technology have a direct impact on both the absorption behavior and the mechanical performance.

It should be considered that the AM processes are subject to variations not only on the tensile results but also on mechanical testing in general, according to the printing conditions as well as the postprocessing. The LSR and the photopolymers, for example, can go through a postcuring stage, increasing their stiffness; FFF printers allow the user to change the path and the direction of the material deposition. AM can lead to greater process deviations than the conventional and established manufacturing techniques. For that reason, 3D printing optimization is one of the current challenges in order to deliver functional parts in similar and reliable performance.

In addition, new materials emerge as a way to improve the quality of the printed parts. Nevertheless, since AM is a flexible process, the printing of the geometry can always be adapted to exhibit specific properties. As an example, we find the percentage of infill influencing on the compression of the part. For elastomeric materials with higher hardness, it is possible to increase the compressibility by decreasing the infill percentage. Figure 7 exemplifies that two geometrically identical cylinders printed with the same material (SILASTIC, 50 Shore A) in different infill percentages of 25% and 50% exhibit distinct performances when compressed with a force of 50 N.

The investigations presented here are a significant step toward the study of the feasibility of 3D printed elastomeric parts for technical applications. Knowledge on water uptake and toluene absorption, for instance, is beneficial when deciding the printing material for a defined geometry in a specific application. It can be noted that special attention must be taken when using the studied photopolymers and TPU in water applications, as moisture absorption cannot be neglected. Silicone could be better suited for moist environments, on the other hand. The toluene absorption tests show the relevance on applications involving contact with fuels, in which the silicone and the TPU revealed an intense swelling behavior, while the photopolymers were not able to withstand the solvent. With such a piece of information, the process of selecting the material and manufacturing options in AM becomes accessible in future works.

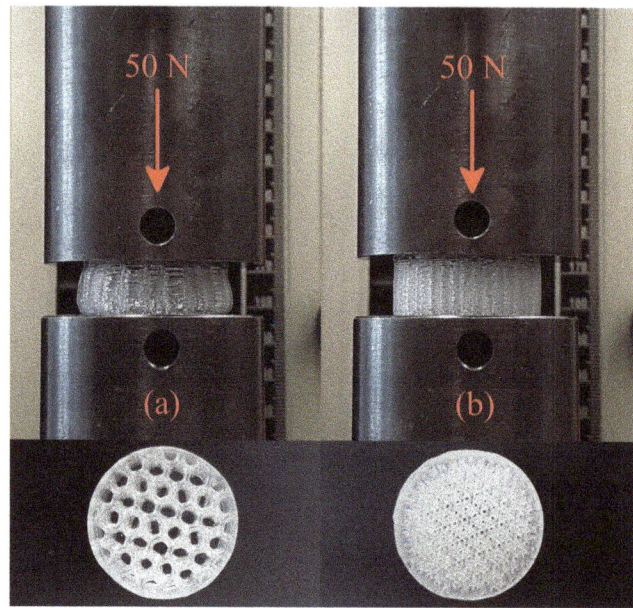

Figure 7. Compression of SILASTIC cylinders printed with infill percentages of (**a**) 25% and (**b**) 50%.

3. Swelling Behavior of Pristine and Thermo-Oxidatively Aged Elastomers in Synthetic Fuel

Synthetic fuels are attracting considerable interest as alternative energy sources in the aviation sector [21,22]. The main advantages are their high volumetric energy density and renewable feedstocks. One of the crucial topics is a possibly different or adverse interaction between the fuels and construction materials in the aircraft, in comparison to conventional kerosene [23]. Especially soft materials, such as elastomers in sealings, tank hoses, and linings, respond distinctly to fuel contact by swelling, extraction of plasticizers [24], and a change in mechanical properties. So far, most studies focus on the interaction of fuels with pristine elastomers. Graham et al. [6] investigated the influence of selected aromatics blended in a synthetic jet fuel on the volume swell by correlation with partition coefficients determined with gas chromatography/mass spectrometry (GC/MS). Blivernitz et al. [7] developed a method for the simultaneous and time-resolved quantification of sorption and extraction processes of individual model fuel components and elastomer additives via GC/MS assisted sorption experiments. Acrylonitrile-butadiene-rubber (NBR) is a material typically used for applications with nonpolar fuels and is part of many aircraft seals. With long aircraft service times, especially in the military sector, the elastomer properties change due to aging [25]. This contribution presents an advanced analytical method to study diffusion processes of NBR in contact with liquid media after thermo-oxidative aging. The focus lies on an in-depth understanding of the diffusion processes of single substances and substance classes in model and real fuels by gas chromatography/mass spectrometry combined with mechanical testing.

3.1. Kerosene and Synthetic Aviation Fuels

Up to now, conventional kerosene, such as Jet A-1, made from non-renewable crude oil, is the main energy source in the aviation sector [22]. To tackle the environmental crisis by reducing the CO_2 emissions and saving resources, innovative solutions are required. Although other technologies such as hydrogen or fuel-cell-powered [26] and battery-electric aircrafts [27] are considered and in development, synthetic aviation fuels are the most important alternative energy source at the moment [28]. The advantages of liquid fuels are their high gravimetric and volumetric energy density and the potential to use existing aircraft and infrastructure. Furthermore, the competing technologies need further improvements to become competitive, but finally a mix of technologies must be considered and tailored to the respective applications. Liquid aviation fuels in a chemical view are composed of different hydrocarbon classes: linear alkanes, branched isoalkanes, cycloalkanes, and aromatics [29]. The fuel properties are highly dependent on the chemical composition and the distribution of the hydrocarbons. While conventional kerosene contains aromatic substances, certain types of the synthetic fuels do not. Because the aromatic hydrocarbons increase the swelling for a lot of elastomers, there are mainly three strategies to meet the specified range of 8–25 vol% aromatics in fuels: First, blend aromatic-free fuels with petroleum-derived kerosene [30,31]; second, add aromatic additives; or, third, design processes to produce so-called drop-in fuels, which are synthetic fuels with aromatics and capable of replacing kerosene equally. All relevant parameters for conventional kerosene are specified in the Standard ASTM D1655 [32] and Defence Standard 91-091 [33]. To stay abreast of the current developments regarding synthetic fuels from various renewable feedstock, the Standard ASTM D7566 [34] and Annex D of Defence Standard 91-091 [33] were established. The specification process for new fuels is strictly regulated [35], because of the high safety standards in the aviation sector.

3.2. Testing the Interactions of Elastomers with Synthetic Fuels

However, some fuels are already certified and tested thoroughly. Apart from fulfilling the fuel specifications, the ongoing challenge is material compatibility, since liquid fuels are in contact with soft elastomeric materials such as sealings, tank hoses, and linings. The next decade is likely to witness a considerable rise in the demand for synthetic fuels attributable to the ambitious goals of different advocacy groups. The aim of the International Air Transport Association (IATA) is to reduce global aviation-caused CO_2 emissions by 50% relative to the level of 2005 until 2050 [36]. One intermediate goal in Germany is to achieve a 2% synthetic fuel stake by 2025 [37]. On account of its multidisciplinary nature, there is an immense potential to investigate this topic experimentally. For instance, fuel properties, the chemical and physical aging of elastomers with fuel contact [38], and swelling and diffusion phenomena [3] may be investigated. The swelling of the elastomer can also be simulated using constitutive models [39]. On a bigger scale, fuel-burning tests [40] or ultimately flight tests are conducted. Regarding material compatibility, mechanical studies are essential to investigate the tightness of seals [41], leakage due to shrinkage, when refueling or mechanical properties in dependence of the swelling status [42]. Whereas previous work is often limited to the interactions of pristine elastomer with fuels or fuel-like substances, the current study aims to estimate long-term effects by conducting experiments with pre-aged elastomers.

3.3. Experimental Part

Carbon black filled and stabilized NBR with a ready-to-use formulation containing 18 wt% acrylonitrile is used here, cf. Table 3. It contains further carbon black N550 (60 phr), the plasticizer DEHP (20 phr), the antioxidant 6PPD (2 phr), sulfur (2 phr) and vulcanizing agents, and the composition are the same as in a previous study [7]. Dumbbell-shaped specimens (S2) with a thickness of 1.2 mm are aged thermo-oxidatively, in a Binder ED56 oven with natural convection at 120 °C for three and seven days, without mechanical load. Applied Research Associates (ARA) produces the synthetic jet drop-in fuel ReadiJet™ used here. Chemically, it is a complex mixture of hydrocarbons similar to the conventional kerosene Jet A-1 with an aromatics content of 21.2 vol% measured according to Standard ASTM D1319-20a [43] and a density of 0.823 g cm^{-3} [29,30] and is used as the immersion fluid.

Table 3. Composition of the investigated NBR18 elastomer.

Component	Content/phr
Perbunan 1846	100
Di(2-ethylhexyl) phthalate (DEHP)	20
N-(1,3-dimethylbutyl)-N'-phenyl-p-phenylenediamine (6-PPD)	2
Carbon black (type: N550)	60
Zinc oxide	5
Stearic acid	1
Sulfur	2
N-Cyclohexyl-2-benzothiazole sulfenamide (CBS)	1.5
TMTM-80 [1]	0.5

[1] 80% tetramethylthiuram monosulfide, 20% elastomer binder and dispersing agents.

3.3.1. Testing Procedure

Each sample was tested as a threefold measurement. Samples are pristine or aged for 3 d as well as for 7 d at 120 °C. For clear identification purposes, the specimens were marked clearly with cut-off edges. At first the samples were weighed, and their thickness was measured. The samples were then immersed in ReadiJet and taken out after distinct sorption times of 30 min (=42 $s^{0.5}$), 2, 6, 24, 48, and 72 h and one week (=778 $s^{0.5}$) (see Figure 8 (1)). To account for Fickian diffusion, the sorption times on the abscissa are reported as the square root of time $t^{0.5}$. In other elastomer–fuel combinations, longer or shorter storage times in the fuel are possibly needed, depending on their compatibility. After the samples were taken out, they were dipped quickly in low boiling benzine 40/60 to remove adhering fuel, dried with a lint-free paper cloth, and then weighed again (2) to determine the mass change. Subsequently for GC/MS analysis, small pieces of ≈8 mg were punched out (3) and stored (4) in 2 mL GC vials that were filled with 1 mL acetone previously to extract the absorbed substances. To transfer the sample as quickly as possible into the vial, a funnel fixed to a stand was used. In the following step, the density using the Archimedes principle (5), micro Shore A hardness (6), according to Standard ISO 7619-1:2010 [44] and tensile properties such as stress at break and elongation at break according to Standard DIN 53504:2017-03 [16] were determined (7). In this case, the solid–liquid extraction of the absorbed substances in acetone was complete in two days. The acetone extract is then analyzed with GC/MS (8) described in the next subsection. To reduce the evaporation of the fuel, it is advisable to conduct the experiments immediately. For short times, the specimens may be covered in plastic bags to bridge transport times. The weighing was performed at the SARTORIUS precision balance mentioned in Section 2.4. Volume and density were determined using the Archimedes principle with the kit VF 4601 from SARTORIUS. The hardness was measured with a digi test II, by bareiss® equipped with a micro Shore A hardness test head. Tensile tests were conducted with the universal testing machine used in Section 2.5. The samples were not preconditioned under cyclic load to preserve the secondary network, formed by aging-induced oxidative crosslinking. The strain rate is 0.167 s^{-1} = 200 mm min^{-1}, respectively.

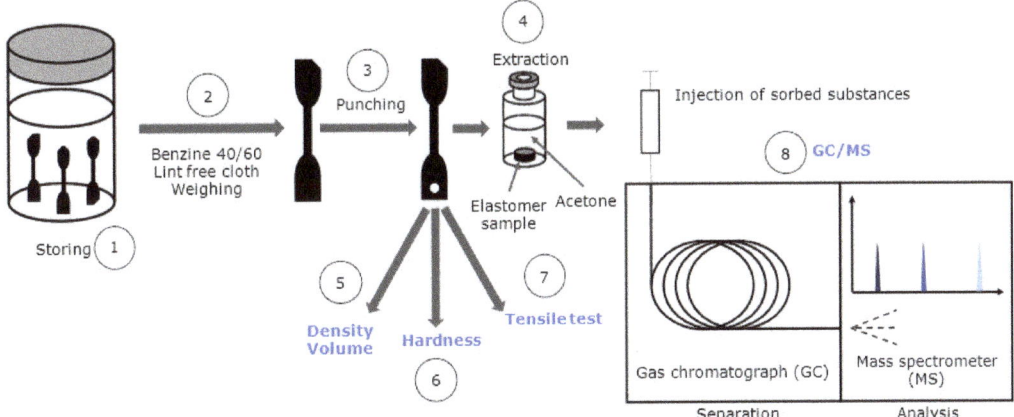

Figure 8. Versatile testing procedure of pristine and aged elastomer samples in contact with synthetic fuel.

3.3.2. Gas Chromatography/Mass Spectrometry (GC/MS)

An Agilent 7890A gas chromatograph coupled with an Agilent 5975 MSD mass spectrometer was used to perform the GC/MS analysis. The column was a 30 m DB-5MS (0.25 mm inner diameter, 0.25 µm film thickness). The GC oven was heated with 50 K min^{-1} from 50 to 320 °C and held for 5 min at 320 °C. By reason of the chemical diversity of fuels, a defined surrogate fuel was prepared and used to calibrate and quantify the aromatics content of unknown samples. First, a mixture (AroMix) with a distribution of aromatic hydrocarbons, comparable to JP-8 jet fuel, was prepared by mixing 25:53:22 vol% of the aromatic liquids 100, 150, and 200 provided by Exxon [45,46]. An aromatic-free coal-to-liquid (CtL) fuel from Sasol was then blended with 10, 30, 50, 70, and 90 vol% of AroMix to yield the surrogate fuels, which were used as calibration standards for determining the aromatics content. Definite mass-to-charge ratios m/z of mass spectra signals were characteristic of aliphatic or aromatic hydrocarbons. Their signal intensities are referred to as $I_{m/z,\text{aliph}}$ and $I_{m/z,\text{aro}}$. After the GC/MS analysis of the standards, the respective characteristic intensities were added up in the retention time range of the fuel. Then, the sum of aliphatic signals $I_{m/z,\text{aliph}}$ was set in relation to the sum of aromatic and aliphatic hydrocarbons $I_{m/z,\text{sum}}$ according to Equation (2). The ratio $I_{m/z,\text{aliph}}/I_{m/z,\text{sum}}$ was plotted versus the aromatics content v of the standard solutions in vol% to yield the calibration graph [46] (see Figure 9):

$$\frac{I_{m/z,\text{aliph}}}{I_{m/z,\text{sum}}} = \frac{\sum_{i=1}^{9} I_{m/z,\text{aliph},i}}{\sum_{i=1}^{9} I_{m/z,\text{aliph},i} + \sum_{i=1}^{7} I_{m/z,\text{aro},i}}$$

$$m/z, \text{aliph} = 41, 55, 57, 69, 71, 83, 85, 97, 99 \qquad n = 9$$
$$m/z, \text{aro} = 91, 105, 115, 119, 120, 128, 142 \qquad n = 7 \qquad (2)$$

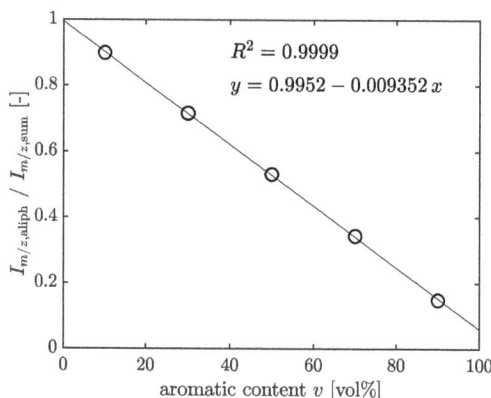

Figure 9. Calibration graph.

It is important that the concentration of the standards is in the same range as the samples. Furthermore, it is beneficial to analyze independent samples with known aromatics content additionally to verify the calibration. The additives DEHP and 6PPD were quantified with external standards of the pure substances in acetone in appropriate concentrations. The known concentrations were plotted versus the peak areas of the additives in the chromatogram to obtain the calibration graph and function [47].

3.3.3. Results

In the following section, the results are shown in dependency on the aging history and sorption time. Pristine samples are represented by light grey (○), aged samples for 3 days at 120 °C by dark grey (◁), and 7 days at 120 °C by black color (□). In Figure 10, the results from density measurements and the calculated volume change (A), micro Shore A hardness (B), elongation at break (C), and the stress at break (D) according to the procedure in Figure 8 are shown.

Before the samples are immersed ($t^{0.5} = 0\,s^{0.5}$), the impact of the aging is evident. The material becomes harder, denser, and thus less resilient with longer aging time. The density increases 5.6% from 1.16 to 1.225 g cm^{-3} after aging for 7 days at 120 °C compared to the pristine sample. Causes for that are volume shrinking due to the loss of additives, and a mass gain, because of oxygen binding to the elastomer. The mass gain is detectable by weighing when elastomers without volatile additives are thermo-oxidatively aged at elevated temperatures. With proceeding sorption time ($t > 0\,s^{0.5}$), the density decreases, whereas the volume increases due to fuel uptake until the swelling equilibrium is reached. The volume increase is less pronounced for the aged samples, because the polymer chains of NBR form a secondary network, when they react with oxygen at elevated temperatures due to crosslinking [48]. The increased crosslink density restricts the intake of fuel into the elastomer. In the swelling equilibrium ($t^{0.5} = 778\,s^{0.5}$), the volume increase is 23.5% smaller for the aged sample (7 d 120 °C) compared to the pristine sample. In addition, the micro Shore A hardness, the elongation at break, and the stress at break decrease with progressing sorption time due to the uptake of fuel and the extraction of the additives. For the aged samples, the change in the properties is smaller because less fuel is absorbed by the elastomer.

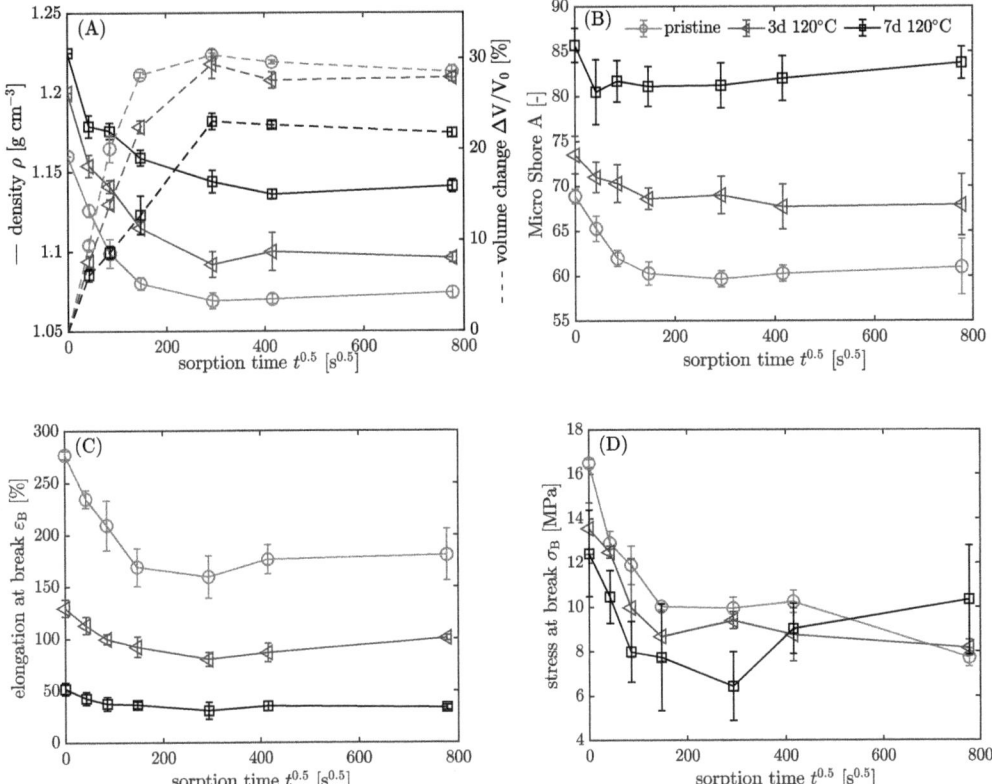

Figure 10. (**A**) Density and calculated volume change, (**B**) micro Shore A hardness, (**C**) elongation at break, and (**D**) stress at break according to the procedure in Figure 8 (step 5–7) in dependency on sorption time $t^{0.5}$. Samples: light-grey circles ○: pristine; grey triangles ◁: 3 d 120 °C; black squares □: 7 d 120 °C.

The contents of the additives DEHP and 6PPD and the fuel ReadiJet, which are present in the elastomer in dependency on the sorption time, are reported in Figure 11A. DEHP and 6PPD are quantified with the external calibration, and the absolute contents are related to the filled elastomer mass m_0 without any further soluble additives. The overall mass change is determined by weighing the sample before and after the immersion in the fuel. It is evident that the changes in the sample mass during the sorption process is caused by the uptake of fuel into the elastomer and simultaneous desorption of the additives DEHP and 6PPD into the fuel which surrounds the NBR sample. Since the overall mass change is a convolution of both processes, it is corrected to only yield the mass uptake of ReadiJet by subtracting the known initial contents of DEHP and 6PPD from the initial elastomer mass: $m_0 = m_{\text{el+additives}} \cdot (1 - w_{\text{DEHP,0}} - w_{\text{6PPD,0}})$. Details of this method are shown by Blivernitz et al. [46,47].

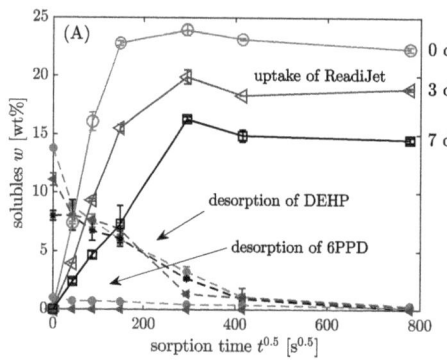 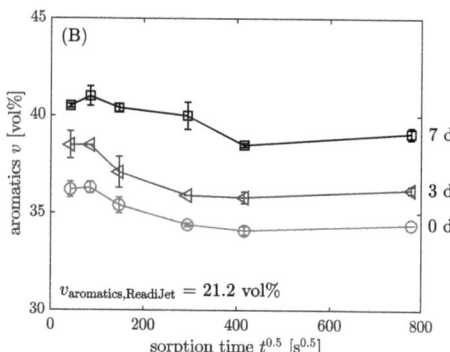

Figure 11. (**A**) Uptake of ReadiJet and desorption of DEHP and 6PPD, (**B**) aromatics content according to the procedure in Figure 8 (step 8) in dependency on sorption time $t^{0.5}$. Samples: light-grey circles ○: pristine; grey triangles ◁: 3 d 120 °C; black squares □: 7 d 120 °C.

As described previously, the aromatics content of the absorbed liquid phase in the elastomer was quantified, as one can see in Figure 11B. The graph shows that the relative aromatics content in the equilibrium is higher than in the surrounding fuel. Whereas the neat ReadiJet contains 21.2 vol% aromatics, the liquid phase in the pristine sample contains 34.4 vol% aromatics at equilibrium. This is due to the higher affinity of aromatics towards NBR, which originates in interactions between the nitrile group of NBR and aromatic electrons [49]. At the beginning of the sorption (36.6 vol%), there is even a higher enrichment of the aromatics. When the elastomer is pre-swollen by aromatics, the other substances also enter the elastomer more easily so that there is a relative decrease in the aromatics content towards the equilibrium. The aromatics content also changes with proceeding sorption time and is different for pristine (34.4 vol%) and aged samples (39.1 vol%). With increasing aging time, the crosslinking and polarity of NBR is increased, making it harder for the fuel to enter the elastomer. Therefore, when the swelling is inhibited, the aromatic hydrocarbons are more enriched due to their high affinity toward NBR.

3.4. Intermediate Conclusions and Remarks

The properties of NBR deteriorate when the elastomer ages, which is expressed through its increased hardness and density as well as its decreased elongation and stress at break. The interaction of the elastomer with the synthetic fuel dynamically leads to a further change in its properties in dependency on the absorbed fuel. For aged elastomers, the fuel uptake is reduced. Gas chromatography coupled with mass spectrometry (GC/MS) adds valuable chemical information to the mechanical testing results. It allows one to separate the overall mass change into the uptake of fuel and the desorption of additives. Furthermore, it is possible to analyze the composition of the absorbed liquid in the elastomer with respect to its chemical nature. In the near future, a significant rise in the demand of synthetic fuels is expected, which makes this topic interesting to investigate experimentally.

4. Conclusions and Outlook

The swelling behavior of soft polymers is of great interest in the wide applications of elastomeric materials. The presented studies investigates the water, the fuel, and the fuel component toluene uptake using versatile testing methods. The first study highlights the pronounced swelling capability of several state-of-the-art additively manufactured materials depending on the solvent. As expected, the absorption of toluene is considerably more pronounced and reaches double the original weight (>100% mass gain). The absorption of water, on the other hand, reaches <5% but is of just as much interest, since water is omnipresent due to humidity and directly influences the printing result. For the additive manufacturing process, a concrete comparison between dry-printed material and wet-printed material is presented. Furthermore, the aspect of thermo-oxidative aging is investigated while the swelling behavior of pristine and aged elastomers in synthetic fuel ReadiJet is quantified. The characterization includes new experimental strategies such as a combination of sorption experiments with gas chromatography and mass spectrometry to precisely analyze the diffusion process from the surrounding liquid in and additives out of the elastomer. The quantified change in density and volume, the change in Shore hardness and the change in mechanical properties (elongation and stress at break) via tensile test highlight their need for experimental investigation.

Author Contributions: Supervision, A.L. and M.J.; writing—original draft, V.M.B., B.D. and Y.I.; writing—review and editing, K.L., A.L. and M.J. All authors have read and agreed to the published version of the manuscript.

Funding: We acknowledge financial support for open access publication by Bundeswehr University Munich. Parts of this research work were funded by dtec.bw—Digitalization and Technology Research Center of the Bundeswehr, which we gratefully acknowledge [project FLAB-3Dprint].

Institutional Review Board Statement: Not applicable.

Informed Consent Statement: Not applicable.

Data Availability Statement: Not applicable.

Acknowledgments: We acknowledge scientific cooperation with Bundeswehr Research Institute for Materials, Fuels and Lubricants, Erding, Germany. Special thanks to Sebastian Eibl, Tobias Förster, and Jens Holtmannspötter.

Conflicts of Interest: The authors declare no conflict of interest.

Abbreviations

The following abbreviations are used in this manuscript:

6PPD	Antioxidant N-(1,3-dimethylbutyl)-N'-phenyl-p-phenylenediamine
AM	Additive Manufacturing
ARA	Applied Research Associates
ATR	Attenuated total Reflectance
CtL	Coal-to-Liquid
DEHP	Plasticizer Di(2-ethylhexyl)phthalate
FDM	Fused Deposition Modeling
FFF	Fused Filament Fabrication
FTIR	Fourier Transform Infrared Spectrometer
GC/MS	Gas Chromatography coupled with Mass Spectrometry
IR	Infrared
IATA	International Air Transport Agency
LAM	Liquid Additive Manufacturing
LSR	Liquid Silicone Rubber
MJ	Material Jetting
NBR	Acrylonitrile-butadiene-rubber
TPE	Thermoplastic Elastomer
TPU	Thermoplastic Polyurethane

References

1. Lacuve, M.; Colin, X.; Perrin, L.; Flandin, L.; Notingher, P.; Tourcher, C.; Hassine, M.B.; Tanzeghti, H. Investigation and modelling of the water transport properties in unfilled EPDM elastomers *Polym. Degrad. Stab.* **2019**, *168*, 108949. [CrossRef]
2. Gong, B.; Tu, Y.; Zhou, Y.; Li, R.; Zhang, F.; Xu, Z.; Liang, D. Moisture absorption characteristics of silicone rubber and its effect on dielectric properties. In Proceedings of the Annual Report Conference on Electrical Insulation and Dielectric Phenomena, Chenzhen, China, 20–23 October 2013.
3. Förster, T.; Blivernitz, A. Migration of mineral oil in elastomers. *J. Rubber Res.* **2021**, *24*, 257–269. [CrossRef]

4. Sun, C.; Negro, E.; Nale, A.; Pagot, G.; Vezzù, K.; Zawodzinski, T.A.; Meda, L.; Gambaro, C.; Di Noto, V. An efficient barrier toward vanadium crossover in redox flow batteries: The bilayer [Nafion/(WO$_3$)$_x$] hybrid inorganic-organic membrane *Electrochim. Acta* **2021**, *378*, 138133. [CrossRef]
5. Zhang, H.; Sun, C. Cost-effective iron-based aqueous redox flow batteries for large-scale energy storage application: A review. *J. Power Sources* **2021**, *493*, 229445. [CrossRef]
6. Graham, J.L.; Striebich, R.C.; Myers, K.J.; Minus, D.K.; Harrison, W.E. Swelling of nitrile rubber by selected aromatics blended in a synthetic jet fuel. *Energy Fuels* **2006**, *20*, 759–765. [CrossRef]
7. Blivernitz, A.; Förster, S.; Eibl, S. Simultaneous and time resolved investigation of diffusion processes of individual model fuel components in acrylonitrile-butadiene-rubber in the light of swelling phenomena. *Polym. Test.* **2018**, *70*, 47–56. [CrossRef]
8. Bikas, H.; Stavropoulos, P.; Chryssolouris, G. Additive manufacturing methods and modelling approaches: A critical review. *Int. J. Adv. Manuf. Technol.* **2016**, *83*, 389–405. [CrossRef]
9. Ford, S.; Despeisse, M. Additive manufacturing and sustainability: an exploratory study of the advantages and challenges. *J. Clean. Prod.* **2016**, *137*, 1573–1587. [CrossRef]
10. Janusziewicz, R.; Tumbleston, J.R.; Quintanilla, A.L.; Mecham, S.J.; DeSimone, J.M. Layerless fabrication with continuous liquid interface production. *Proc. Natl. Acad. Sci. USA* **2016**, *113*, 11703–11708. [CrossRef] [PubMed]
11. Ligon, S.C.; Liska, R.; Stampfl, J.; Gurr, M.; Muelhaupt, R. Polymers for 3D printing and customized additive manufacturing. *Chem. Rev.* **2017**, *117*, 10212–10290. [CrossRef] [PubMed]
12. Lukić, M.; Clarke, J.; Tuck, C.; Whittow, G.; Wells, G. Printability of elastomer latex for additive manufacturing or 3D printing. *J. Appl. Polym. Sci.* **2016**, *133*, 42931. [CrossRef]
13. Hummel, D.O.; Scholl, F.K. *Atlas der Polymer- und Kunststoffanalyse*, 2nd ed.; Hanser: München, Germany, 1988.
14. Salih, S.I.; Oleiwi, J.K.; Ali, H.M. Study the Mechanical Properties of Polymeric Blends (SR/PMMA) Using for Maxillofacial Prosthesis Application. *IOP Conf. Ser. Mater. Sci. Eng.* **2018**, *454*, 012086. [CrossRef]
15. Wang, L.; Ju, Y.; Xie, H.; Ma, G.; Mao, L.; He, K. The mechanical and photoelastic properties of 3D printable stress-visualized materials. *Sci. Rep.* **2017**, *7*, 10918. [CrossRef] [PubMed]
16. DIN Deutsches Institut für Normung e.V. *DIN 53504:2017-03 Testing of Rubber—Determination of Tensile Strength at Break, Tensile Stress at Yield, Elongation at Break and Stress Values in a Tensile Test*; Beuth Verlag GmbH: Berlin, Germany, 2017.
17. DIN Deutsches Institut für Normung e.V. *DIN ISO 1817:2016-11 Rubber, Vulcanized or Thermoplastic—Determination of the Effect of Liquids (ISO 1817:2015)*; Beuth Verlag GmbH: Berlin, Germany, 2016.
18. Crank, J. *The Mathematics of Diffusion*, 2nd ed.; Oxford University Press: Oxford, UK, 1979.
19. Ardebili, H.; Zhang, J.; Pecht, M.G. (Eds.) Characterization of encapsulant properties. In *Encapsulation Technologies for Electronic Applications*, 2nd ed.; William Andrew, Elsevier: Oxford, UK, 2019; pp. 221–258.
20. Herzberger, J.; Sirrine, J.M.; Williams, C.B.; Long, T.E. Polymer design for 3D printing elastomers: Recent advances in structure, properties, and printing. *Prog. Polym. Sci.* **2019**, *97*, 101144. [CrossRef]
21. Wang, W.; Tao, L. Bio-jet fuel conversion technologies. *Renew. Sustain. Energy Rev.* **2016**, *53*, 801–822. [CrossRef]
22. Riedel, M.B.U. Alternative fuels in aviation. *CEAS Aeronaut. J.* **2015**, *6*, 83–93.
23. Muzzell, P.; Stavinoha, L.; Chapin, R. Synthetic Fischer-Tropsch (FT). In *JP-8/JP-5 Aviation turbine Fuel Elastomer Compatibility*; Final Report; Defense Technical Information Center: Fort Belvoir, VA, USA, 2005.
24. Gormley, R.J.; Link, D.D.; Baltrus, J.P.; Zandhuis, P.H. Interactions of Jet Fuels with Nitrile O-Rings: Petroleum-Derived versus Synthetic Fuels. *Energy Fuels* **2009**, *23*, 857–861. [CrossRef]
25. Zhao, J.; Ynag, R.; Iervolino, R.; Barbera, S. Changes of Chemical Structure and Mechanical Property Levels during Thermo-oxidative Aging of NBR. *Rubber Chem. Technol.* **2013**, *86*, 591–603. [CrossRef]
26. Baroutaji, A.; Wilberforce, T.; Ramadan, M.; Ghani, A. Comprehensive investigation on hydrogen and fuel cell technology in the aviation and aerospace sectors. *Renew. Sustain. Energy Rev.* **2019**, *106*, 31–40. [CrossRef]
27. Barzkar, A.; Ghassemi, M. Electric Power Systems in More and All Electric Aircraft: A Review. *IEEE Access* **2020**, *8*, 169314–169332. [CrossRef]
28. Bicer, Y.; Dincer, I. A comparative life cycle assessment of alternative aviation fuels. *Int. J. Sustain. Aviat.* **2016**, *2*, 181–202. [CrossRef]
29. Scheuermann, S.S.; Forster, S.; Eibl, S. In-Depth Interpretation of Mid-Infrared Spectra of Various Synthetic Fuels for the Chemometric Prediction of Aviation Fuel Blend Properties. *Energy Fuels* **2017**, *31*, 2934–2943. [CrossRef]
30. Zschocke, A.; Scheuermann, S.; Ortner, J. *High Biofuel Blends in Aviation (HBBA)*; Interim Report, ENER C; Deutsche Lufthansa AG: Köln, Germany, 2012.
31. Pechstein, J.; Zschocke, A. Blending of Synthetic Kerosene and Conventional Kerosene. In *Biokerosene*; Springer: Berlin/Heidelberg, Germany, 2018; pp. 665–686.
32. ASTM. *ASTM D1655-20: Standard Specification for Aviation Turbine Fuels*; Book of Standards Volume: 05.01; ASTM: West Conshohocken, PA, USA, 2020.
33. Ministry of Defence. *Defence Standard 91-091 Turbine Fuel, Aviation Kerosine Type, Jet A-1 NATO Code: F-35 Joint Service Designation: AVTUR*; Ministry of Defence: Glasgow, UK, 2019.
34. ASTM. *ASTM D7566-16b: Standard Specification for Aviation Turbine Fuel Containing Synthesized Hydrocarbons*; Book of Standards Volume: 05.04; ASTM: West Conshohocken, PA, USA, 2016.

35. Wilson, G.R.; Edwards, T.; Corporan, E.; Freerks, R.L. Certification of Alternative Aviation Fuels and Blend Components. *Energy Fuels* **2013**, *27*, 962–966. [CrossRef]
36. Bisignani, G. Annual Report. International Air Transport Association. 2011. Available online: https://www.iata.org/contentassets/c81222d96c9a4e0bb4ff6ced0126f0bb/annual-report-2011.pdf (accessed on 8 November 2021)
37. Die Ziele von Aireg Bis 2025. Available online: https://aireg.de/themen/aktuelle-projekte/die-ziele-von-aireg-bis-2025/ (accessed on 8 November 2021).
38. Buckley, G.S.; Roland, C.M. Influence of liquid media on lifetime predictions of nitrile rubber. *J. Appl. Polym. Sci.* **2014**, *131*, 40296. [CrossRef]
39. Musil, B.; Demmel, B.; Lion, A.; Johlitz, M. A contribution to the chemomechanics of elastomers surrounded by liquid media: continuum mechanical approach for parameter identification using the example of sorption experiments. *J. Rubber Res.* **2021**, *24*, 271–279. [CrossRef]
40. Corporan, E.; Edwards, T.; Shafer, L.; Dewitt, M.J.; Klingshirn, C.; Zabarnick, S.; West, Z.; Striebich, R.; Graham, J.; Klein, J. Chemical, Thermal Stability, Seal Swell, and Emissions Studies of Alternative Jet Fuels. *Energy Fuels* **2011**, *25*, 955–966. [CrossRef]
41. Kömmling, A.; Jaunich, M.; Pourmand, P.; Wolff, D.; Hedenqvist, M. Analysis of O-Ring Seal Failure under Static Conditions and Determination of End-of-Lifetime Criterion. *Polymers* **2019**, *11*, 1251. [CrossRef] [PubMed]
42. Liu, Y.; Wilson, C.W.; Blakey, S.; Dolmansley, T. Elastomer Compatibility Test of Alternative Fuels Using Stress-Relaxation Test and FTIR Spectroscopy. In Proceedings of the ASME 2011 Turbo Expo: Turbine Technical Conference and Exposition, GT2011, Vancouver, BC, Canada, 6–10 June 2011; pp. 1–11.
43. ASTM. *ASTM D1319-20a: Standard Test Method for Hydrocarbon Types in Liquid Petroleum Products by Fluorescent Indicator Adsorption*; Book of Standards Volume: 05.01; STM: West Conshohocken, PA, USA, 2020
44. DIN Deutsches Institut für Normung e.V. *Rubber, Vulcanized or Thermoplastic—Determination of Indentation Hardness—Part 1: Durometer Method (Shore Hardness) (ISO 7619-1:2010)*; Beuth Verlag GmbH: Berlin, Germany, 2012.
45. DeWitt, M.J.; Corporan, E.; Graham, J.; Minus, D. Effects of aromatic type and concentration in Fischer-Tropsch fuel on emissions production and material compatibility. *Energy Fuels* **2008**, *22*, 2411–2418. [CrossRef]
46. Blivernitz, A. Untersuchung der Verträglichkeit von Elastomeren mit Synthetischen Flugturbinenkraftstoffen Anhand Ablaufender Diffusionsprozesse. Ph.D. Thesis, Universität der Bundeswehr München, Neubiberg, Germany, 2019.
47. Loos, K.; Blivernitz, A.; Musil, B.; Rehbein, T.; Johlitz, M.; Lion, A. Challenges in Material Modelling and Testing of Elastomers. In Proceedings of the International Rubber Conference, London, UK, 3–5 Septembr 2019.
48. Zhao, J.; Yang, R.; Iervolino, R.; Barbera, S. Investigation of crosslinking in the thermooxidative aging of nitrile-butadiene rubber. *J. Appl. Polym. Sci.* **2015**, *132*, 41319. [CrossRef]
49. Lachat, V.; Varshney, V.; Dhinojwala, A.; Yeganeh, M.S. Molecular Origin of Solvent Resistance of Polyacrylonitrile. *Macromolecules* **2009**, *42*, 7103–7107. [CrossRef]

Article

Advanced Characterisation of Soft Polymers under Cyclic Loading in Context of Engine Mounts

Tomáš Gejguš [†] [⬀], Jonas Schröder [†] [⬀], Klara Loos *[⬀], Alexander Lion and Michael Johlitz

Institute of Mechanics, Bundeswehr University Munich, 85577 Neubiberg, Germany; tomas.gejgus@unibw.de (T.G.); jonas.schroeder@unibw.de (J.S.); alexander.lion@unibw.de (A.L.); michael.johlitz@unibw.de (M.J.)
* Correspondence: klara.loos@unibw.de
† These authors contributed equally to this work.

Abstract: The experimental investigation of viscoelastic behavior of cyclically loaded elastomeric components with respect to the time and the frequency domain is critical for industrial applications. Moreover, the validation of this behavior through numerical simulations as part of the concept of virtual prototypes is equally important. Experiments, combined measurements and test setups for samples as well as for rubber-metal components are presented and evaluated with regard to their industrial application. For application in electric vehicles with relevant excitation frequencies substantially higher than by conventional drive trains, high-frequency dynamic stiffness measurements are performed up to 3000 Hz on a newly developed test bench for elastomeric samples and components. The new test bench is compared with the standard dynamic measurement method for characterization of soft polymers. A significant difference between the measured dynamic stiffness values, caused by internal resonance of the bushing, is presented. This effect has a direct impact on the acoustic behavior of the vehicle and goes undetected by conventional measurement methods due to their lower frequency range. Furthermore, for application in vehicles with internal combustion engine, where the mechanical excitation amplitudes are significantly larger than by vehicles with electric engines, a new concept for the identification of viscoelastic material parameters that is suitable for the representation of large periodic deformations under consideration of energy dissipation is described. This dissipated energy causes self-heating of the polymer and leads to the precocious aging and failure of the elastomeric component. The validation of this concept is carried out thermally and mechanically on specimen and component level. Using the approaches developed in this work, the behavior of cyclically loaded elastomeric engine mounts in different applications can be simulated to reduce the time spent and save on the costs necessary for the production of prototypes.

Keywords: engine mount; elastomer characterisation; experimental testing; resonance frequency; dynamic stiffness; parameter identification; electrodynamic shaker; test bench; cogging torque; synchronous machine

Citation: Gejguš, T.; Schröder, J.; Loos, K.; Lion, A.; Johlitz, M. Advanced Characterisation of Soft Polymers under Cyclic Loading in Context of Engine Mounts. *Polymers* **2022**, *14*, 429. https://doi.org/10.3390/polym14030429

Academic Editors: Radek Stoček, Gert Heinrich and Reinhold Kipscholl

Received: 30 November 2021
Accepted: 17 January 2022
Published: 21 January 2022

Publisher's Note: MDPI stays neutral with regard to jurisdictional claims in published maps and institutional affiliations.

Copyright: © 2022 by the authors. Licensee MDPI, Basel, Switzerland. This article is an open access article distributed under the terms and conditions of the Creative Commons Attribution (CC BY) license (https://creativecommons.org/licenses/by/4.0/).

1. Introduction

The development of technically advanced vehicles or machines requires consideration of their surrounding components and of several different effects. Since electric vehicles have become established products, the focus of manufacturers is shifting to improving driving comfort. Elastomeric mounts in electric vehicles carrying motors experience loads with small amplitudes that can be applied at high frequencies. The bushings in conventional vehicles with internal combustion engines are, however, exposed to vibrations with large amplitudes in the lower frequency range. Nowadays, to decrease development costs, it is necessary to develop a concept of virtual prototypes. While the required amount of measurements decreases significantly by using virtual prototypes, some experiments are still necessary on the sample level for the identification of parameters for material models and on a component level for the validation of virtual prototypes. The material behavior of elastomeric samples or components loaded by cyclic loading is dependant on multiple factors, for instance, the static preload, frequency of the loading, dynamic amplitude, ageing condition, ambient temperature, self-heating, loading history and production instabilities. For each specific problem, with regard to the applicability and economical attractivity, different factors should be considered.

Investigations considering nonlinear viscoelasticity in the time and frequency domain can be found in Hausmann and Gergely [1], De Cazenove et al. [2], Nguyen [3] and Suphadon et al. [4], among others. Detailed descriptions of the dynamic mechanical thermal analysis (DMTA) can be found in Ehrenstein et al. [5] or Wollscheid [6]. A significant part of the noise-vibration-harshness (NVH) behavior of electric motor, called cogging, has been studied by Neapolitan and Nam [7], Hanselman [8] and Jagasics and Vajda [9]. Cogging occurs at much higher frequencies than the vibrations caused by unbalance of the rotor of synchronous machines. Lion and Johlitz [10] have shown that internal resonance of elastomeric motor mounts and cogging occur in roughly the same frequency range. Dynamic measurements in audible frequency range are presented in publications of Haeussler et al. [11], Koblar and Boltezar [12], Kari [13,14] and Ramorino et al. [15].

The characterisation of viscoelastic material behavior is the subject of ongoing research, in which Tárrago et al. [16] investigated the properties of elastomeric bearings in radial and axial direction under small amplitudes. Regarding the self-heating behavior, Behnke et al. [17] used uniaxial multi-stage tests for characterisation, whereas Kyei-Manu et al. [18] and Suphadon et al. [19] applied the cyclic stretching process with different predeformations. Mars and Fatemi [20,21] combined axial and radial cyclic loads while identifying material parameters. Carleo et al. [22] stated that, over the range of strains and strain rates that rubber-based automotive components are expected to operate in, the filled vulcanized elastomers exhibit strong nonlinearity and large hysteresis. Further possibilities for identifying parameters are, apart from the FEM, used by Gil-Negrete et al. [23], and the evaluation of measured temperature fields has been done by Balandraud et al. [24], Marco et al. [25] and Glanowski et al. [26].

First, the standardized DMTA is presented. The main advantages, possibilities and requirements of this useful measurement method are described. Additionally, measurements and the mastercurve according to time temperature shift (TTS) principle by Williams et al. [27] are shown.

For applications regarding electric vehicles, the complex NVH behavior necessitates advanced testing procedures in the frequency domain. Next, the new high-frequency test bench and its design are presented. The results from the test bench for dynamic stiffness (TBDS) are compared with the standard mastercurve. Their deviations as well as the need for such measurements for the aforementioned application are discussed along with the effects undetected by the DMTA.

Furthermore, the material characterisation approach in the time domain, with a slightly different application in vehicles with an internal combustion engine, focused on self-heating caused by dissipated energy during cyclic loading, is presented. The experimental setup is motivated, described and the measurement results are presented. The viscoelastic parameters of one characteristic hysteresis are identified through combined measurements. Finally, the validation of the identified parameter set of a component is performed and the results of the validation are discussed.

2. Dynamic Mechanical Thermal Analysis (DMTA)

Dynamic mechanical thermal analysis is a method to investigate dynamic material characteristics by subjecting specimens to periodic, typically sinusoidal, oscillations of small amplitude, usually under preload.

Generally, there are two ways to perform DMTA measurement:
- applying a predefined deformation and measuring the stress response
- applying a predefined stress and measuring the deformation response

The variety of different loading types and setups is based on the different purposes. Each different loading type is characterized by its respective advantages and difficulties. Some of the loading types require dimensionally stable samples, and others aim to investigate weak effects, thin films or anisotropic materials. Tension and compression DMTA is illustrated in Figure 1 with the design of the Gabo Eplexor 500, an experimental setup for both compression and tension testing differs only in the required clamps and sample geometry. The design consists of a static unit at the top; a dynamic unit at the bottom; and the measurement axis, including individual clamps, in the middle. The static unit is responsible for the quasi-static preload of the sample ($\pm$1500 N or $\pm$35 mm), and its measurement enables the study of the amplitude dependence of the dynamic characteristics. The clamps are individually designed for different loading types and optimized for the precise hold of the sample. Excessive clamping can lead to measurement distortion. The dynamic unit is able to load the sample with a periodic continually controlled load ($\pm$500 N or $\pm$1 µm up to 3 mm in range 0.01 Hz–80 Hz) and measure the dynamic response precisely. The configurable plate springs compensate for the preload and hold the coil of the electrodynamic exciter in position, so that the electrodynamic exciter is not pulled out before the dynamic load (primary voltage) is applied. The dynamic elongation is measured with an eddy current elongation sensor. Another part of the Gabo Eplexor 500 is the temperature chamber, which, in combination with liquid nitrogen, allows for measurements to be taken in a temperature range from -150 to $500\ ^\circ$C.

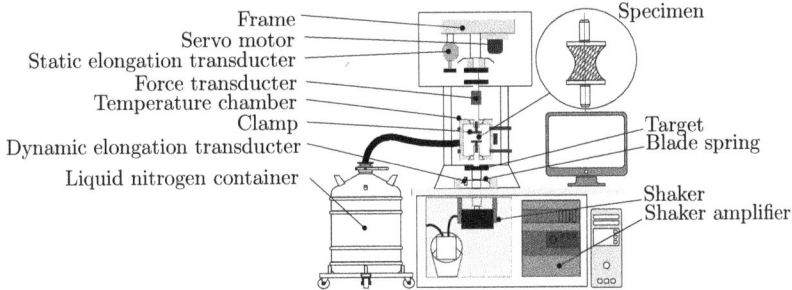

Figure 1. Design principle of Gabo Eplexor 500 N of DMTA testing machine.

2.1. Measurement Procedure

The following considerations should be taken into account when setting up the DMTA:

(a) DMTA experiment setup

The DMTA testing machine is a complex device with many precise components. It is necessary to check for leaks or failures of safety components. The condition of sensors and level of liquid nitrogen are also critical factors. All these factors can lead to failure of the measurement.

(b) Specimen preparation, form and dimensions

If working with dynamic moduli, it is important that the geometry of the specimen meets the requirements of technical standards and is measured precisely.

(c) Measurement parameters

To start a measurement procedure, several measurement parameters have to be set. In terms of mechanical parameters, static preload, dynamic load, frequency range and division of measurement points can be defined. Within thermal parameters, the temperature range, temperature step and soak time need to be chosen. It is also necessary to set the tolerance intervals of all of the aforementioned parameters. There is no universal set of parameters to operate with when starting a new series. Obtaining a suitable set of measurement parameters is a process that consists of making several partial measurements, where every measurement point has to be logged under isothermal conditions. It is important to assure that the loaded sample is still in the so-called linear viscoelastic region (LVR), which is the main assumption that is made in DMTA measurements. In the LVR, the dynamic response (complex modulus) is not amplitude dependent. The LVR is also frequency- and temperature-dependent—lower temperatures or higher frequencies require smaller dynamic amplitudes to stay in LVR. The signal-to-noise ratio also needs to be taken into account in determination of an amplitude of dynamic load that fulfills the required conditions for the whole measurement. In general, the dynamic excitation amplitude must be small, in order for the displacement of the dynamically loaded sample to remain close to its preloaded equilibrium. Additionally, the more elastic the material is, the smaller the LVR is, and vice versa. Only if aforementioned conditions are met, it is possible to shift single frequency sweeps according to time temperature shift principle (TTS) by Williams et al. [27] and obtain a mastercurve, which represents an extrapolated dynamic response for a significantly wider frequency range than the DMTA testing machine is able to measure.

2.2. Dynamic Material Characteristics

After applying the load stress, the strain response can be obtained. The ratio of the dynamic stress σ to dynamic strain ε is called complex tension modulus E^* or complex compression modulus K^*.

The real part E' (K') is the storage modulus and represents the stiffness of the viscoelastic material. It can be interpreted as energy that is stored during loading of the specimen.

The imaginary part E'' (K''), called loss modulus, represents the energy converted or dissipated as heat during mechanical loading process. This energy cannot be recovered.

In the frequency domain, the time delay between loading (dynamic strain ε) and response (dynamic stress σ) is characterized as the phase angle δ. For a purely elastic material, the phase angle δ is equal to zero.

The loss factor $\tan\delta$ is defined as the ratio of the loss modulus to the storage modulus. This characteristic represents internal friction or mechanical damping; therefore, it corresponds to the amount of dissipated energy. In general, an elastic material has low loss factor.

Equations (1)–(4) describe the aforementioned relationships. Further details about dynamic material characteristics can be found in ISO-6721-1:2019 [28], ASTM-D4092:07 [29] and also in the publications by Ehrenstein et al. [5] or Meyers and Chawla [30].

$$|E^*| = \frac{\sigma_A}{\varepsilon_A} = \sqrt{E'^2 + E''^2} \tag{1}$$

$$E' = \frac{\sigma_A}{\varepsilon_A} \cos\delta \tag{2}$$

$$E'' = \frac{\sigma_A}{\varepsilon_A} \sin\delta \tag{3}$$

$$\tan\delta = \frac{E''}{E'} \tag{4}$$

The glass transition temperature T_g is one of the most important material characteristics for polymers and is defined as middle temperature of the region, in which the change from glassy or energy-elastic, to rubbery or entropy-elastic state occurs. The glass state can be interpreted as a stiff state of a polymer at low temperatures caused by immobility of its molecules. In the context of elastomers, the glass transition temperature indicates the lowest temperature at which the elastomer can be deployed for industrial application. The glass transition is also frequency-dependent.

The value of T_g can be determined with several methods and according to several technical norms. One commonly used method is to identify the temperature at which the maximum of loss modulus E''_{max} or maximum of loss factor $(\tan\delta)_{max}$ occurs, as shown in Figure 2, as shown in Ehrenstein et al. [5]. Rieger [31] demonstrated that the T_g obtained by the maximum of loss modulus E''_{max} coincides more with the value obtained with DSC (differential scanning calorimetry) than the value obtained with the maximum of loss factor $(\tan \delta)_{max}$. The beginning of the glass transition is considered as the fall of the E'.

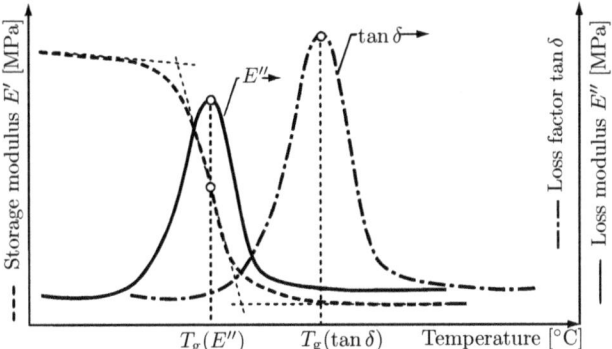

Figure 2. Glass transition temperature evaluation methods [5].

The DMTA measurements are suitable for identification of state changes, temperature dependencies, damping dependencies, blend constituents, thermal limits of the material, anisotropy, changes due to recycling, state of ageing or conditioning, curing, and thermal degradation. For example, through a radical change in the storage modulus E', the energy- or entropy-elastic region can be identified. Different orientations of specimens of anisotropic material will each exhibit a different stiffness. The correlation between water content and change of mechanical properties can be also observed. The degree of curing of duromers can also be obtained based on the changes in moduli E', E'', loss factor $\tan \delta$, and rise of the glass transition temperature T_g. The thermal limits of the material can be identified by radical softening or embrittlement of the specimen. There are also many more applications for complex measurements with the DMTA, as described by Ehrenstein et al. [5].

2.3. Dynamic Mechanical Thermal Analysis (DMTA)—Experiment

Figure 3 displays geometry of the specimen used in this study, along with its dimensions. The material is a carbon-black filled NR-BR (natural rubber–butadiene rubber) blend with hardness of 49 Shore A, which is typically used in rubber bushings. Since an hourglass specimen with inhomogeneous cross-section is used, the measured frequency-dependent characteristic is not the complex modulus E^* but the dynamic stiffness C^*.

The measurement was conducted with the Gabo Eplexor 500. For this specific measurement, the 150 N force sensor and the dynamic elongation sensor with measurement range of 1.5 mm were chosen. The load parameters, such as the static elongation amplitude $\varepsilon_{st_A} = 0.5\%$ and the dynamic elongation amplitude $\varepsilon_{dyn_A} = 0.2\%$, were set up carefully under consideration of the LVR and the initial length of the specimen $L_0 = 25$ mm. The temperature range between -40 °C and 20 °C was used for the temperature sweep of individual isothermal frequency sweeps, which were measured from 1–80 Hz with logarithmic distribution of measurement points 12 Points/decade.

Figure 4 shows the magnitude of the dynamic stiffness $|C^*|$ of the hourglass elastomeric specimen, which was obtained by a temperature-frequency sweep with the DMTA. The corresponding mastercurve was extrapolated by TTS according to Williams et al. [27]. The first vertical line f_1 indicates the frequency range 0–500 Hz. The second line f_2 indicates the frequency range 0–3000 Hz, which will be used for measurements in Section 3.

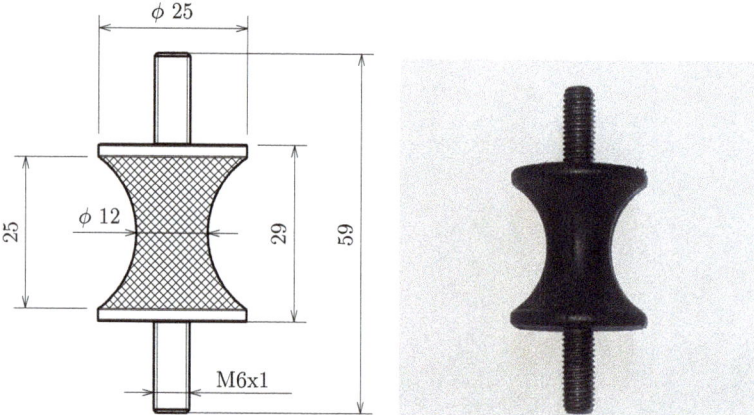

Figure 3. Hourglass specimen used in this study (NR-BR blend).

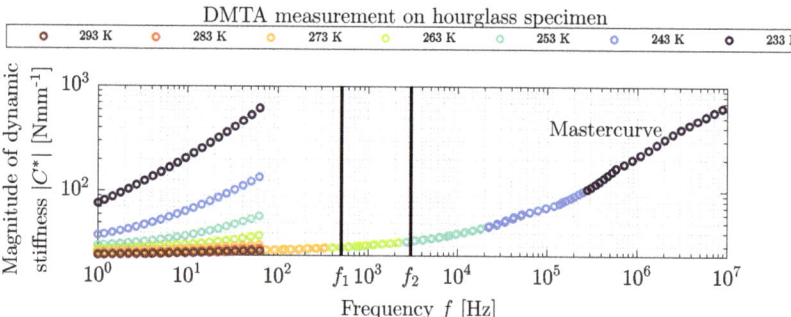

Figure 4. DMTA measurement in frequency range of 1–80 Hz and mastercurve for 1–10^7 Hz of carbon-black filled elastomeric hourglass specimen.

3. High-Frequency Dynamic Testing (Frequency Domain)

There is a new sector of testing machines, which are able to test samples and components at high frequency, loaded by sinusoidal load, at room temperature and non-destructively. This sector of test machines is motivated by NVH behaviour of electric vehicles because the design and working principle of an electric motor is different from an internal combustion engine.

The rotational frequency of a standard internal combustion engine can reach up to 100 Hz. As a result, the relevant frequency region typically used in industry measurements is from 0 to 500 Hz. Permanent magnet synchronous motors (PMSM) of an electric motor can, however, operate with a rotational frequency of up to 300 Hz. Due to the working principle of the PMSM, further superimposed effects can be found in its NVH behavior. In any spectrum of electronic commutated or brushless motors, the pulse width modulation (PWM) effect occurs with the frequency of the motor controller. The cogging torque occurs in every permanent magnet machine and is defined as the torque ripple, which is caused by the uneven attraction of the magnets on the rotor to the teeth and slots on the stator. In the position where the pole aligns with the teeth, the highest attraction or the maximum of magnetic flux density occurs—shown schematically in Figure 5. Neapolitan and Nam [7] state that the cogging torque depends on the rotational frequency, on the number of poles and slots and on the load of the motor. There are several strategies to compensate for this undesired effect. The magnets can be segmented and arranged in several ways or the coils can be switched smoothly, so that they act against the cogging torque, according to Hanselman [8]. The cogging frequency is linked to the least common multiple of the number of poles and slots, as observed by Jagasics and Vajda [9]. Since the vibrations generated by a PMSM occur often at high frequencies, it is necessary to include higher frequencies in the testing standards. These high-frequency effects cause internal resonances of the rubber bushings during the operation of permanent magnet machines, as demonstrated by Lion and Johlitz [10].

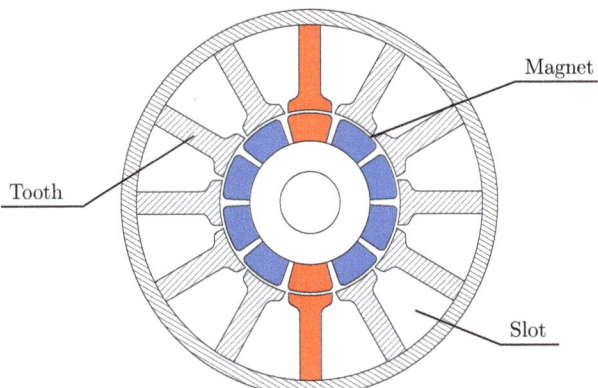

Figure 5. Permanent magnet synchronous motor (PMSM)—cross section (10 Poles, 12 Coils).

3.1. Test Bench for Dynamic Stiffness (TBDS)—Design

There are several scientific publications which deal with high-frequency dynamic testing of elastomers, using different designs of the test apparatus, as can be seen in Ramorino et al. [15] and Koblar and Boltezar [12]. Haeussler et al. [11] performed "free–free" rubber isolator measurements to be able to describe the vibration in virtual points and compute the frequency dependent dynamic stiffness. Testing machines to measure the dynamic stiffness of elastomers within high frequencies are also commercially available. In addition to the approach of Haeussler et al. [11], there are two possibilities for measuring dynamic characteristics of rubber bushing in a wide frequency range, as outlined in Figure 6. The measurement method can be performed, as shown in Figure 6a, directly, utilizing an accelerometer and force sensor, or indirectly, using two accelerometers as in Figure 6b. The disadvantage of the direct method is its more complex design and generally lower resonance frequency of the force sensors.

In the current study, the indirect method was chosen and manufactured on a 500 kg stone table with thread inserts isolated by rubber air springs, as shown in Figure 7a. It is important that the whole apparatus remains isolated from the environment. The seismic mass was hung "free-free" and parallel to the table of dynamic exciter. The solid body resonance of the seismic mass attached with the elastic support cords was outside the measurement range, so that sinusoidal dynamic load could be applied. In this case, the electrodynamic exciter B&K Type 4808 with maximum table acceleration of 71 g powered by B&K Type 2712 was used. This exciter delivers up to 187 N peak sine force in range from 5 Hz to 10 kHz. The accelerometers from PCB with sensitivity of 10.29 mVg^{-1} and B&K with sensitivity of 98.5 pCg^{-1} were utilized. It is preferred that the cables are fixed and have no loops. The measurements were performed on the modular multichannel measurement system PAK MKII by Müller BBM VAS (see Figure 7b) using the corresponding software PAK 6.0, and the excitation velocity amplitude was set to v $= 0.01$ ms^{-1} with tolerance of 5% during the entire measurement from 50–3000 Hz with a frequency step of 1 Hz. The set velocity amplitude was controlled by real-time closed loop offered by PAK 6.0. The sampling rate was set up with respect to the Nyquist criterion. Three analog inputs and one analog output were used to conduct the measurement. The direct computation of dynamic stiffness of the bushing was carried out using the Arithmetic toolbox in PAK 6.0.

The measurement setup in this study is named "Test bench for dynamic stiffness" (TBDS). The TBDS is able to test different specimens and components made from soft or hard polymers since the mass of adaption can be easily compensated. For hard polymers, the dynamic amplitude has to be reduced regarding the power of the dynamic exciter in order to reach 3000 Hz. Concurrently, it is important to consider the measurement noise of the used accelerometers.

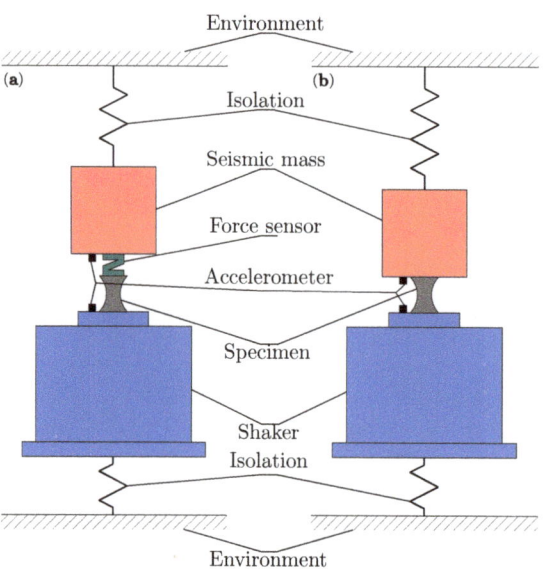

Figure 6. Design concept of test bench for measurement of dynamic stiffness: (**a**) direct method, (**b**) indirect method.

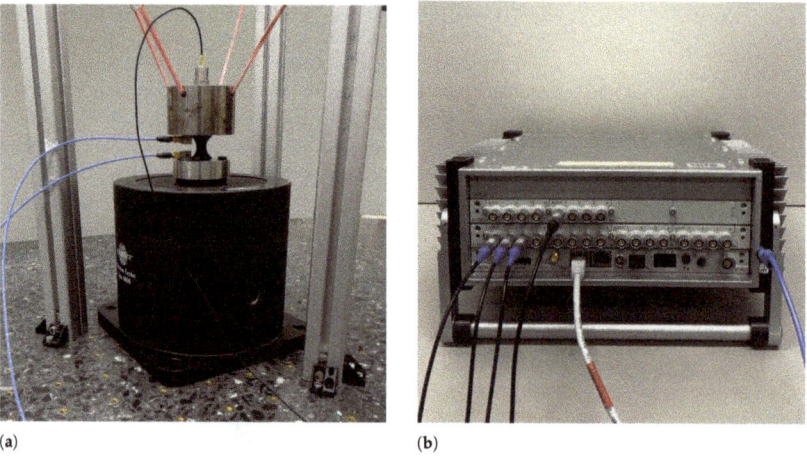

Figure 7. Experiment: (**a**) Test bench for dynamic stiffness (TBDS), (**b**) multichannel measurement system. PAK MKII by Müller BBM VAS.

3.2. Test Bench for Dynamic Stiffness (TBDS)—Experiment

Figure 8 shows a successful isothermal frequency sweep measurement from the TBDS over a large frequency range of 50–3000 Hz of the same hourglass specimen as used in the

DMTA temperature frequency sweep in Section 3. The magnitude of dynamic stiffness $|C^*|$ (Figure 8a), phase angle δ (Figure 8b) and loss factor $\tan\delta$ (Figure 8c) are shown as a factor of frequency. The thick blue vertical lines in Figure 8b at close proximity of 10^3 Hz are caused by measurement noise and plotting method ($-180°$ up to $180°$) of phase angle δ. The first vertical line f_1 marks 500 Hz, which is the region where industrial measurements usually take place. Clearly, the dynamic stiffness can be considered as constant from 50 to 500 Hz. The second vertical line f_3 marks the maximum of the magnitude of dynamic stiffness $|C^*|$ at approx. 1650 Hz, which indicates an internal resonance of the test specimen. This is confirmed by the phase angle δ, which is approximately $90°$ at this resonance frequency.

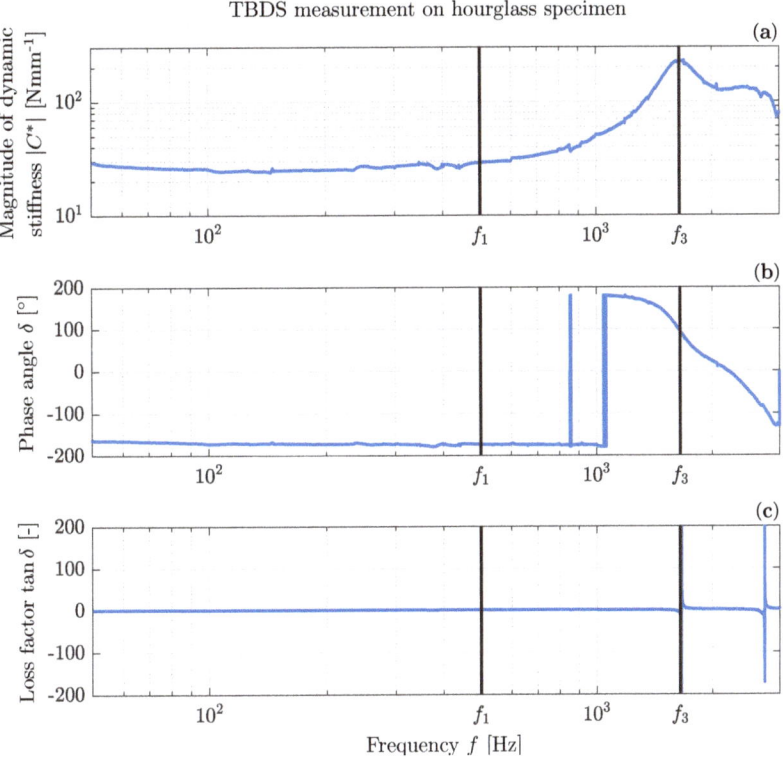

Figure 8. Measurement on hourglass specimen from TBDS.

3.3. Test Bench for Dynamic Stiffness (TBDS)—Validation (DMTA)

In Figure 9, the magnitude of the dynamic stiffness $|C^*|$ measured on our test bench is plotted in comparison with the mastercurve shifted from temperature frequency sweep performed on DMTA with the same specimen, as shown in Figure 4. In the low-frequency range up to 500 Hz marked with line f_1, which is usually used for industrial measurements of rubber bushings, both measurements correlate well with each other. At higher frequencies, a prominent increase of the magnitude of dynamic stiffness $|C^*|$ was measured by TBDS. The first internal resonance of the rubber bushing occurs near 1650 Hz. The standard temperature frequency sweep (DMTA measurement) proceeded up to 100 Hz and then shifted according to TTS (a mastercurve can never yield any information about internal resonance that occurs outside the frequency range of the frequency sweep) . With the mastercurve from the DMTA measurements, the significant increase of dynamic stiffness remains undetected. The DMTA is a versatile useful standard analysis; nevertheless, there is a need, caused by NVH behavior of PMSM, to measure dynamic stiffness at high frequencies for applications in electromobility. Table 1 shows clearly arranged comparison of pros and cons of the TBDS and the DMTA.

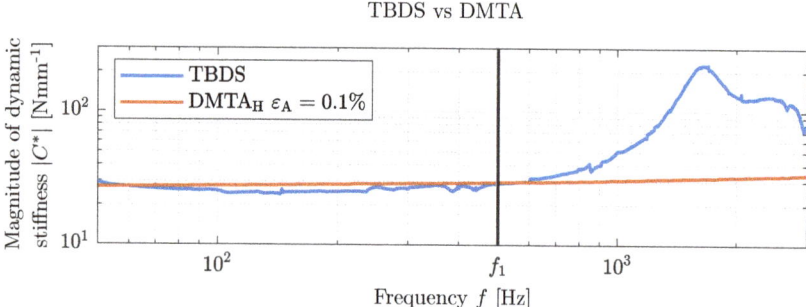

Figure 9. Comparison—TBDS vs. DMTA (hourglass specimen).

Table 1. Pros and cons of TBDS and DMTA.

	TBDS	DMTA
Pros	wide frequency range detection of internal resonance controlled dynamic load open parameters and low costs	controlled preload temperature control controlled dynamic load user-friendly and established
Cons	constant preload no temperature control user comfort	small frequency range component testing difficulties high acquisition costs

4. Dissipative Self-Heating of Engine Mounts (Time Domain)

In contrast to electric machines, bearings used for internal combustion engines (Figure 10) are subjected to tendentially lower frequencies combined with relatively large amplitudes. Cyclic mechanical loads at sufficiently high amplitudes and frequencies lead to significant energy dissipation in viscoelastic materials. In the case of insufficient heat removal, elastomer components may therefore heat up strongly as shown, for example, by Johlitz et al. [32] or Dippel et al. [33]. The occurrence of critical temperatures, which can lead to loss of functionality or to component failure, has to be identified at an early stage in the development process. For reasons of time and cost, simulation is now replacing a large part of the experimental work. However, in addition to suitable material models, this also requires suitable methods and equipment for material characterisation, as the publication by Hartmann [34] describes. This section, therefore, focuses on the viscoelastic material characterisation, which, in the context of dissipative self-heating, has a significant influence on the valency, progress and stability of the resulting temperature field, which is examined in more detail within the study published by Schröder et al. [35]. The present study presents a method for experimental identification of viscoelastic parameters, which can be transferred to complex geometries and multi-axial deformation states and velocities. An overall estimation is made concerning the industrial applicability, which includes not only the duration and quality of the tests but also the total experimental effort for the identification of the material parameters.

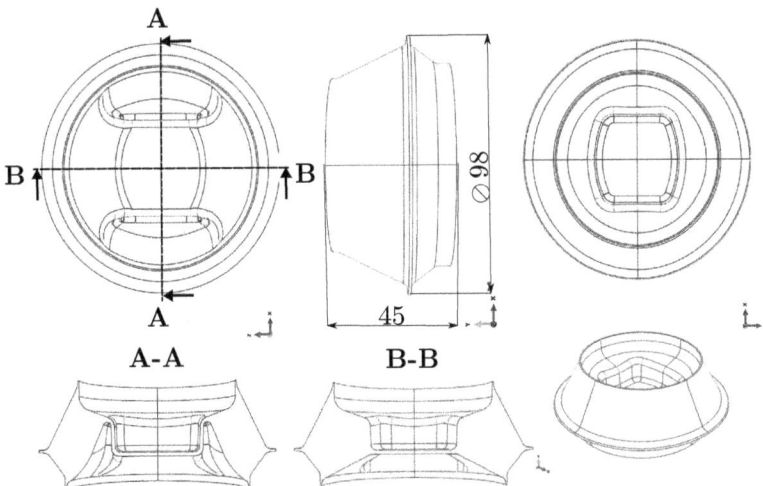

Figure 10. Engine mount for vehicles with internal combustion engines.

4.1. Experimental Setup and Time Domain Procedures

Classical methods such as the relaxation test or the dynamic mechanical analysis do not represent the required relaxation spectrum in the time or frequency range if an amplitude spectrum within a frequency band has to be described. This issue is also addressed by Lion [36] and further investigated experimentally by Dippel et al. [33], among others. Furthermore, a large number of technical elastomer materials show pronounced non-linear viscoelastic material behavior, which is generally not simulated for reasons of efficiency but which must be taken into account during parameter identification. This was investigated, for instance, in the work of Koprowski-Theiß [37] and Scheffer [38], as well as Schröder et al. [39] and many other authors. In the following, the experimental method and the experimental setup are described. The ElectroForce 3200 Series III DMA testing machine from BOSE is used for mechanical testing within the ETI-2 temperature chamber. The following tests are conducted using carbon black-filled natural rubber (NR) samples cross-linked by low-sulphur vulcanisation (Efficient Vulcanisation, EV). Further parameters of the material are the SHORE hardness of 68 ShA, an Elongation at Break of 366%, and a Rebound Resilience of 31%. Periodic uniaxial tensile tests were carried out on simple rectangular specimens 10 mm × 1.8 mm × 50 mm until a stationary temperature field was reached. The process was characterized by a swelling displacement process and varied in terms of frequencies 1, 2, 5 and 10 Hz and strain amplitudes 5, 10 and 15%. The field quantities force and displacement were recorded by sensors. In parallel, the evolution of the surface temperature of the sample was captured by an infrared camera of the type VarioCam and the ambient temperature by a thermal sensor. These measured data provided the basis for the parameter identification, and they were used for the later validation of the parameter set.

4.2. Parameter Identification Using Modified Ellipse Function

The measured data are edited and formatted. The surface temperature evaluated at the location 'Hotspot' where the maximum temperature occurs and the mean value of the oscillating temperature of the last cycles is used as the steady-state equilibrium temperature. This is shown in Figure 11 and summarized in the following as a function of the input process variables. The mechanical measurement data are converted into stretch λ and stress σ by using the geometry of the specimen. In addition to the stationary temperature of the process, a representative cycle is selected and the relevant characteristic properties such as extremal values $(\lambda_{min}/\sigma_{min})$ and $(\lambda_{max}/\sigma_{max})$ and the area A of the associated hysteresis are calculated. Subsequently, this is determined as in Figure 12 and is replaced by an ellipse function $e(\sigma_m, \lambda_m, a, b, \psi, f, t)$ depending on the parameters for transformation σ_m and λ_m, rotation ψ, geometry a and b, frequency f and time t.

$$\begin{bmatrix} \lambda \\ \sigma \end{bmatrix} = \begin{bmatrix} \lambda_m \\ \sigma_m \end{bmatrix} + \frac{1}{2} a \, \cos(2\,\pi\,f\,t) \begin{bmatrix} \cos(\psi) \\ \sin(\psi) \end{bmatrix} + 2\,b\,\sin(2\,\pi\,f\,t) \begin{bmatrix} -\sin(\psi) \\ \cos(\psi) \end{bmatrix} \quad (5)$$

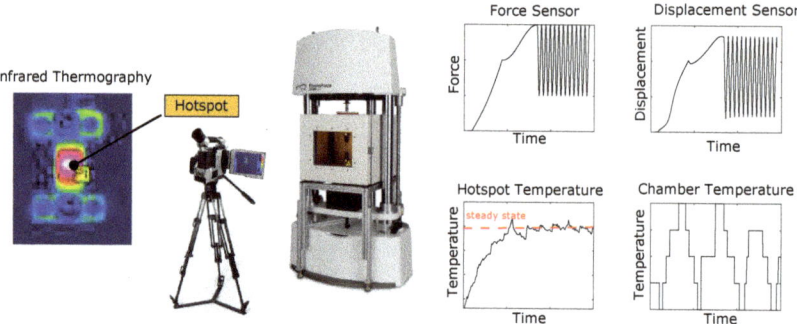

Figure 11. Experimental setup and data acquisition.

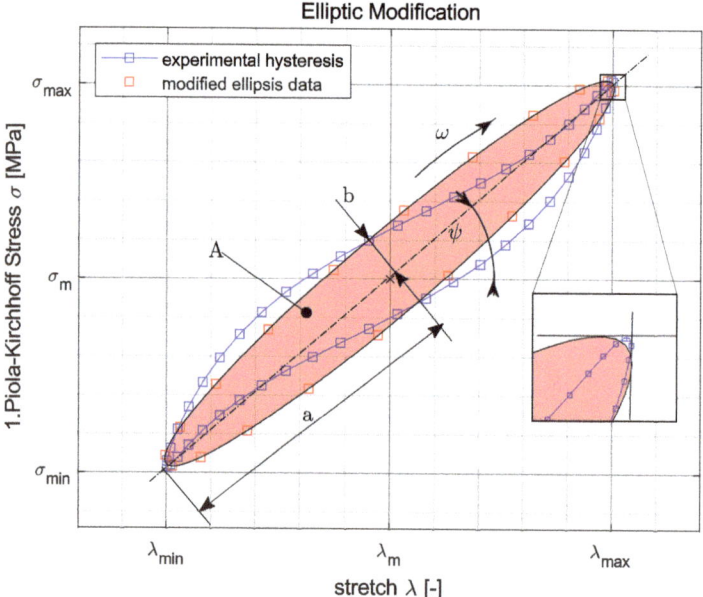

Figure 12. Edited measurements in a hysteresis representation and an elliptical approximation curve.

The target function $z(\mathbf{p}(\eta_i\,\mu_i))$ is defined on the basis of the abstracted data of the measurement $\mathbf{T}^{abs}(\mathbf{p})$ and the process response of the material model $\mathbf{T}^{mod}(\mathbf{p})$, with the viscoelastic model parameters $\mathbf{p}$, viscosities η_i and stiffnesses μ_j being used as variables. To solve the inverse problem, an error function is first introduced, which represents a quality criterion between simulation and experiment.

$$z(\mathbf{p}(\eta_i, \mu_i)) = \left\| \sum_j \mathbf{T}_j^{mod}(\mathbf{p}) - \mathbf{T}_j^{abs} \right\| \to \text{minimal} \quad (6)$$

To optimize the target function, a genetic algorithm based on the mutation-selection principle and the recombination mechanism as well as a gradient-based method are chosen and combined through iteration loops, as Koprowski-Theiß [37] presents in her work. In this manner, advantages of the evolution strategy and the start and extreme value finding can be used to find the best available set of parameters. A numerical simulation is used to replicate the experiment and to validate it at sample level using the determined set of parameters. The extremely satisfactory agreement of the results is shown in Figure 13.

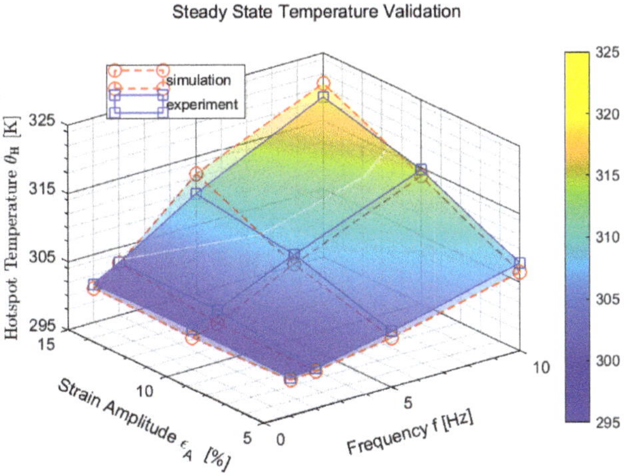

Figure 13. Experimental and simulative steady-state temperatures under not released swell strain conditions.

4.3. Parameter Validation at Component Level

The formulation of the initial and boundary conditions shown in Figure 14 is mandatory for the model within a fully coupled thermomechanical analysis. The analysis is transient with respect to the heat conduction equation, whereas inertia effects are neglected. The heat transfer equation is integrated using a backward-difference scheme. The incrementation is done automatically under the restriction of used increment sizes between $5 \cdot 10^{-5}$ [s] and $5 \cdot 10^{-2}$ [s].

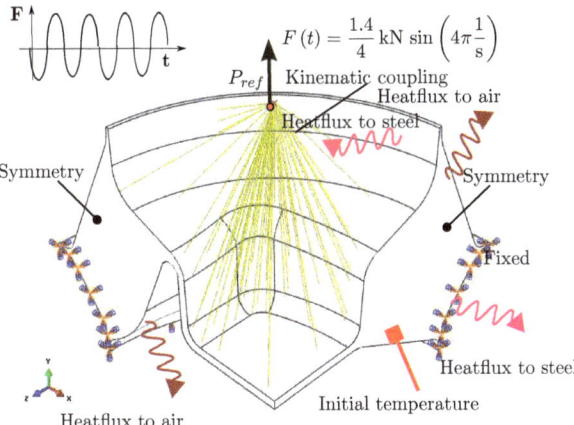

Figure 14. Computational model showing boundary and initial conditions.

A qualitatively valuable discretisation conserves computational resources while ensuring the accuracy of the results. For this purpose, convergence and mesh independence must be proven. To fulfill these requirements, a mesh convergence analysis is carried out. A representative load case is evaluated where the number of elements is varied. The temperature and Mises stress as well as the total computation time for each mesh (model 1–model 10) are determined at four characteristic nodes (Node 1–Node 4). The different discretisations and evaluated locations are shown in Figure 15a.

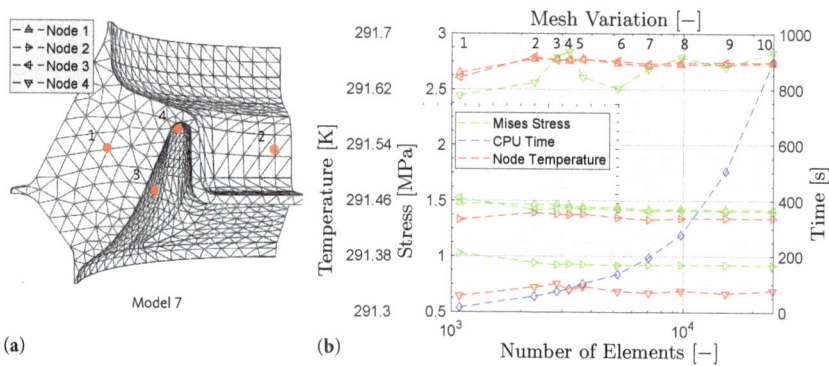

Figure 15. (a) Definition of reference nodes and Model mesh. (b) Convergence study.

With the use of the convergence, the element size or number can be determined where the results do not significantly change any longer. At this point, mesh variation 7, the model is independent of the mesh and leads neither to increased computation times nor to discretisation erros as shown in Figure 15b. The geometry is discretized by 7093 C3D10MHT (three-dimensional 10-node modified temperature-displacement hybrid tetrahedron element with linear pressure and with hourglass control) elements.

The identified parameter set is now used to compute the mechanical and thermal behavior of an engine mount as it is subject to complex load and deformation conditions. These are multi-axial as well as location- and time-dependent. The calculation results are presented in Figures 16b and 17b. For modelling, implementation and optimization of the computation time refer to Schröder et al. [39,40].

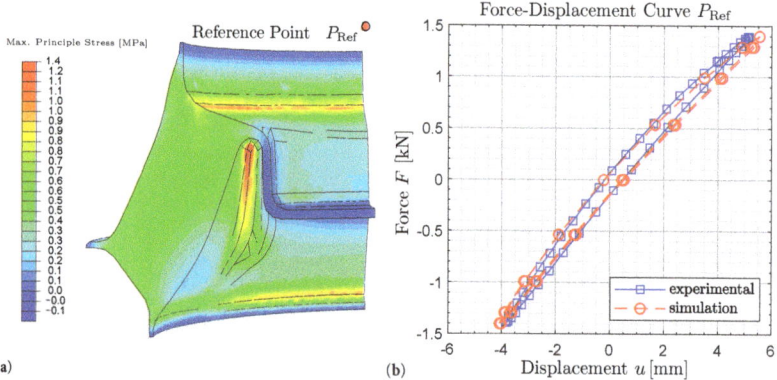

Figure 16. (a) ABAQUS analysis stress visualisation engine mount. (b) Force-displacement curve at the reference point.

In Figure 16b, a nearly identical hysteresis is observed. This means that in addition to the mechanical behavior, the dissipative behavior, which is characterized by the hysteresis area, is also captured. Accordingly, the local loads can be deduced at this point. The agreement of the stationary temperature values between experiment and simulation can be recognized in Figure 17b. It is also observed that the calculation duration is significantly reduced by the suitable selection of the heat capacity. For detailed descriptions and proof of validity, refer to Schröder et al. [40]. The reach of the stationary state is thereby fastened. Furthermore, the local temperature profile in the stationary state can be inferred, as shown in Figure 17b.

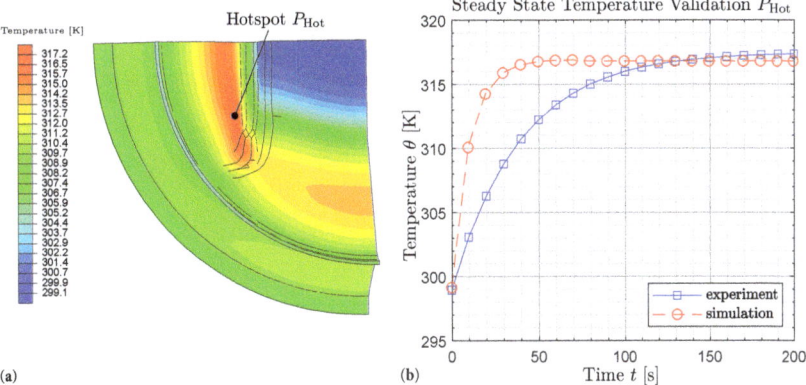

Figure 17. (a) ABAQUS analysis temperature visualisation engine mount. (b) Self-heating under cyclic loading.

5. Summary and Outlook

It is important to verify established procedures since the possibilities and problems are changing. The described DMTA analysis is a useful multifunctional measurement method; however, for several applications, it is necessary to consider effects that remain undetected by this or any other standard method.

For applications in electric vehicles, the vibrations caused by PMSM are of small amplitude with wide frequency range. The characteristic NVH behavior of PMSM is caused by the superposition of multiple mechanical and electromagnetic effects. It was shown on measurements from our beta version of the test bench for measurement of dynamic stiffness and DMTA that the standard measurement procedure has its limitations. The comparison of the two mentioned measurements was discussed, and the reason for the significant difference was stated. The TBDS measurements can be performed on specimens to obtain parameters for material model of frequency dependent moduli and also for validation of simulations of the virtual prototype, where dynamic load is applied on the elastomer bushing. With the help of such models, it is possible to simulate dynamic response of the rubber isolator; compare it with spectrum of the PMSM; and, if it is needed, change the design or position of the isolator to avoid undesired NVH behavior before the start of the production. It is also possible to test aged specimens to obtain information about the change of the spectrum of rubber isolator caused by ageing. In the future, the design will be improved to be able to set up the preload precisely and some components will be replaced by higher quality components. Eventually, the model according to Lion and Johlitz [10] will be fitted and validated on measurements from TBDS.

The second application in this study is adapted on NVH behavior of internal combustion engines. The cyclic, low frequency load, high amplitudes and long operation times are affecting the elastomer bushing. It was shown on measurements that the energy dissipated by loading of elastomer specimen causes its self-heating. At this point, the standard DMTA reaches its limits as well. In this study, the combined mechanical and thermal measurement for the quantification of dissipative heating of elastomer specimen is described and the steady state temperatures are obtained. The results of previously mentioned measurements are used for parameter identification of the hysteresis area for the computation of the characteristic ellipse. The identified parameter set is used for simulation on the component level and validated with respect to combined measurement of engine mount. The effect of dissipative self-heating was characterized by a minimal number of experiments at the specimen level, and the developed model is suitable for transfer to the complex geometry and load cases of the component (engine mount).

With the help of these approaches, the behavior of cyclic loaded elastomer bushing in different applications can be simulated to save the time and costs necessary for the production of prototypes or losses caused by possible complaints from customers.

Author Contributions: Supervision, A.L. and M.J.; Investigation, Writing—original draft, T.G. and J.S.; Writing—review and editing, K.L., A.L. and M.J. All authors have read and agreed to the published version of the manuscript.

Funding: We acknowledge financial support for open access by Universität der Bundeswehr München.

Institutional Review Board Statement: Not applicable.

Informed Consent Statement: Not applicable.

Data Availability Statement: Not applicable.

Conflicts of Interest: The authors declare no conflict of interest.

References

1. Hausmann, G.; Gergely, P. Approximate methods for thermoviscoelastic characterization and analysis of elastomeric lead-lag dampers. In Proceedings of the AAAF, 18th European Rotorcraft Forum, Avignon, France, 15–17 September 1992.
2. De Cazenove, J.; Rade, D.; De Lima, A.; Araújo, C. A numerical and experimental investigation on self-heating effects in viscoelastic dampers. *Mech. Syst. Signal Process.* **2011**, *27*, 433–445. [CrossRef]
3. Nguyen, N.T. Experiments and Inverse Analysis for Determining Non-Linear Viscoelastic Properties of Polymeric Capsules and Biological Cells. Ph.D. Thesis, The University of Michigan, Ann Arbor, MI, USA, 2014.
4. Suphadon, N.; Thomas, A.; Busfield, J. The viscoelastic behavior of rubber under a complex loading. II. The effect large strains and the incorporation of carbon black. *J. Appl. Polym. Sci.* **2010**, *117*, 1290–1297. [CrossRef]
5. Ehrenstein, G.; Riedel, G.; Trawiel, P. *Thermal Analysis of Plastics: Theory and Practice*; Carl Hanser Verlag GmbH & Company KG: Munich, Germany, 2012.
6. Wollscheid, D. Predeformation- and Frequency-Ddependent Material Behavior of Filler-Reinforced Rubber: Experiments, Constitutive Modelling and Parameter Identification. Ph.D. Thesis, Universität der Bundersrwehr München, Neubiberg, Germany, 2014.
7. Neapolitan, R.E.; Nam, K.H. *AC Motor Control and Electrical Vehicle Applications*; CRC Press: Boca Raton, FL, USA, 2018.
8. Hanselman, D.C. *Brushless Permanent Magnet Motor Design*; The Writers' Collective: Orono, ME, USA, 2003.
9. Jagasics, S.; Vajda, I. Cogging torque reduction by magnet pole pairing technique. *Acta Polytech. Hung.* **2016**, *13*, 107–120. [CrossRef]
10. Lion, A.; Johlitz, M. A mechanical model to describe the vibroacoustic behavior of elastomeric engine mounts for electric vehicles. *Mech. Syst. Signal Process.* **2020**, *144*, 106874. [CrossRef]
11. Haeussler, M.; Klaassen, S.; Rixen, D. Experimental twelve degree of freedom rubber isolator models for use in substructuring assemblies. *J. Sound Vib.* **2020**, *474*, 115253. [CrossRef]
12. Koblar, D.; Boltezar, M. Evaluation of the Frequency-Dependent Young's Modulus and Damping Factor of Rubber from Experiment and Their Implementation in a Finite-Element Analysis. *Exp. Tech.* **2013**. [CrossRef]
13. Kari, L. Dynamic transfer stiffness measurements of vibration isolators in the audible frequency range. *Noise Control. Eng. J. Noise Control Eng.* **2001**, *49*, 88–102. [CrossRef]
14. Kari, L. On the dynamic stiffness of preloaded vibration isolators in the audible frequency range: Modeling and experiments. *J. Acoust. Soc. Am.* **2003**, *113*, 1909–1921. [CrossRef]
15. Ramorino, G.; Vetturi, D.; Cambiaghi, D.; Pegoretti, A.; Ricco, T. Developments in Dynamic Testing of Rubber Compounds: Assessment of Non-Linear Effects. *Polym. Test.* **2003**, *22*, 681–687. [CrossRef]
16. Tárrago, M.G.; Kari, L.; Vinolas, J.; Gil-Negrete, N. Frequency and amplitude dependence of the axial and radial stiffness of carbon-black filled rubber bushings. *Polym. Test.* **2007**, *26*, 629–638. [CrossRef]
17. Behnke, R.; Kaliske, M.; Klüppel, M. Thermo-mechanical analysis of cyclically loaded particle-reinforced elastomer components: experiment and finite element simulation. *Rubber Chem. Technol.* **2016**, *89*, 154–176. [CrossRef]
18. Kyei-Manu, W.A.; Tunnicliffe, L.B.; Plagge, J.; Herd, C.R.; Akutagawa, K.; Pugno, N.M.; Busfield, J.J. Thermomechanical Characterization of Carbon Black Reinforced Rubbers During Rapid Adiabatic Straining. *Front. Mater.* **2021**, *8*, 421. [CrossRef]

19. Suphadon, N.; Thomas, A.; Busfield, J. Viscoelastic behavior of rubber under a complex loading. *J. Appl. Polym. Sci.* **2009**, *113*, 693–699. [CrossRef]
20. Mars, W.; Fatemi, A. A novel specimen for investigating the mechanical behavior of elastomers under multiaxial loading conditions. *Exp. Mech.* **2004**, *44*, 136–146. [CrossRef]
21. Mars, W.; Fatemi, A. Factors that affect the fatigue life of rubber: A literature survey. *Rubber Chem. Technol.* **2004**, *77*, 391–412. [CrossRef]
22. Carleo, F.; Barbieri, E.; Whear, R.; Busfield, J.J. Limitations of viscoelastic constitutive models for carbon-black reinforced rubber in medium dynamic strains and medium strain rates. *Polymers* **2018**, *10*, 988. [CrossRef]
23. Gil-Negrete, N.; Vinolas, J.; Kari, L. A simplified methodology to predict the dynamic stiffness of carbon-black filled rubber isolators using a finite element code. *J. Sound Vib.* **2006**, *296*, 757–776. [CrossRef]
24. Balandraud, X.; Toussaint, E.; Le Cam, J.; Grédiac, M.; Behnke, R.; Kaliske, M. Application of full-field measurements and numerical simulations to analyze the thermo-mechanical response of a three-branch rubber specimen. *Const. Model. Rubber VII* **2011**, 45–50. doi:10.1007/978-1-4614-4235-6_37 [CrossRef]
25. Marco, Y.; Le Saux, V.; Jégou, L.; Launay, A.; Serrano, L.; Raoult, I.; Calloch, S. Dissipation analysis in SFRP structural samples: Thermomechanical analysis and comparison to numerical simulations. *Int. J. Fatigue* **2014**, *67*, 142–150. [CrossRef]
26. Glanowski, T.; Le Saux, V.; Doudard, C.; Marco, Y.; Champy, C.; Charrier, P. Proposition of an uncoupled approach for the identification of cyclic heat sources from temperature fields in the presence of large strains. *Contin. Mech. Thermodyn.* **2017**, *29*, 1163–1179. [CrossRef]
27. Williams, M.L.; Landel, R.F.; Ferry, J.D. The Temperature Dependence of Relaxation Mechanisms in Amorphous Polymers and Other Glass-forming Liquids. *J. Am. Chem. Soc.* **1955**, *77*, 3701–3707. [CrossRef]
28. ISO-6721-1:2019. *Plastics—Determination of Dynamic Mechanical Properties—Part 1: General Principles*; International Organization for Standardization: Geneva, Switzerland, 2019.
29. ASTM-D4092:07(2013). *Standard Terminology: Plastics: Dynamic Mechanical Properties*; ASTM International: West Conshohocken, PA, USA, 2013.
30. Meyers, M.A.; Chawla, K.K. *Mechanical Behavior of Materials*; Cambridge University Press: Cambridge, UK, 2009.
31. Rieger, J. The glass transition temperature Tg of polymers-Comparison of the values from differential thermal analysis (DTA, DSC) and dynamic mechanical measurements (torsion pendulum). *Polym. Test.* **2001**, *20*, 199–204. [CrossRef]
32. Johlitz, M.; Dippel, B.; Lion, A. Dissipative heating of elastomers: A new modelling approach based on finite and coupled thermomechanics. *Contin. Mech. Thermodyn.* **2016**, *28*, 1111–1125. [CrossRef]
33. Dippel, B.; Johlitz, M.; Lion, A. Thermo-mechanical couplings in elastomers–experiments and modelling. *ZAMM-J. Appl. Math. Mech./Z. Angew. Math. Mech.* **2015**, *95*, 1117–1128. [CrossRef]
34. Hartmann, S. *From Experiments to Predicting the Component Behavior in Solid Mechanics*; Technical Report, Fakultät für Mathematik/Informatik und Maschinenbau; Technische Universität Clausthal: Clausthal-Zellerfeld, Germany, 2019.
35. Schröder, J.; Lion, A.; Johlitz, M. Numerical studies on the self-heating phenomenon of elastomers based on finite thermoviscoelasticity. *J. Rubber Res.* **2021**, *24*, 237–248. [CrossRef]
36. Lion, A. *Thermomechanik von Elastomeren*; Habilitation, Institut für Mechanik, Universität Kassel: Kassel, Germany, 2000.
37. Koprowski-Theiß, N. Kompressible, Viskoelastische Werkstoffe: Experimente, Modellierung und FE-Umsetzung. Ph.D. Thesis, Naturwissenschaftlich-Technischen Fakultät III, Chemie, Pharmazie, Bio- und Werkstoffwissenschaften der Universität des Saarlandes, Saarbrücken, Germany, 2011.
38. Scheffer, T. Charakterisierung des Nichtlinear-Viskoelastischen Materialverhaltens Gefüllter Elastomere. Ph.D. Thesis, Naturwissenschaftlich-Technischen Fakultät III, Chemie, Pharmazie, Bio- und Werkstoffwissenschaften der Universität des Saarlandes, Saarbrücken, Germany, 2016.
39. Schröder, J.; Lion, A.; Johlitz, M. On the Derivation and Application of a Finite Strain Thermo-viscoelastic Material Model for Rubber Components. In *State of the Art and Future Trends in Material Modeling*; Springer: Berlin/Heidelberg, Germany, 2019; pp. 325–348.
40. Schröder, J.; Lion, A.; Johlitz, M. Thermoviscoelastic modelling of elastomer components in industrial applications. In *Constitutive Models for Rubber XI, Proceedings of the 11th European Conference on Constitutive Models for Rubber (ECCMR 2019), Nantes, France, 25–27 June 2019*; CRC Press: Boca Raton, FL, USA, 2019; p. 294.

MDPI
St. Alban-Anlage 66
4052 Basel
Switzerland
Tel. +41 61 683 77 34
Fax +41 61 302 89 18
www.mdpi.com

Polymers Editorial Office
E-mail: polymers@mdpi.com
www.mdpi.com/journal/polymers

www.ingramcontent.com/pod-product-compliance
Lightning Source LLC
LaVergne TN
LVHW070739100526
838202LV00013B/1269